PROCEEDINGS OF THE INTERNATIONAL ASSOCIATION OF INSTITUTES OF NAVIGATION WORLD CONGRESS 2024

The 2024 IAIN World Congress, held from October 28th–30th, at the Beijing International Convention Center, brought together over 350 participants and 31 delegates from all over the world to discuss cutting-edge advancements in navigation and maritime technology.

Organized by the International Association of Institutes of Navigation (IAIN) and hosted by the China Institute of Navigation (CIN), the Congress delved into the latest technological trends, research, and industry developments in satellite navigation, autonomous shipping, maritime safety, and the integration of 5G, IoT, AI, and big data. This book provides an overview of key discussions, research findings, and milestones in the field, showcasing the collaborative efforts shaping the future of global maritime safety.

This book is intended for professionals in navigation and maritime technology, academic researchers, maritime safety authorities, technology developers, industry enterprises, students, and enthusiasts interested in navigation systems, autonomous shipping, and maritime safety.

PROCEEDINGS OF THE INTERNATIONAL ASSOCIATION OF INSTITUTES OF NAVIGATION
WORLD CONGRESS (IAIN 2024), BEIJING, CHINA, 28–30 OCTOBER 2024

Proceedings of the International Association of Institutes of Navigation World Congress 2024

Expanding Navigation Application and Empowering the Future of Humanity

Edited by

Shijun Chen, Siwen Dong, Weihua Guo, Yan Wang,
Feifei Zhang, Yang Liu, Yuxiang Wang and Erdan Ma
China Institute of Navigation, China

CRC Press
Taylor & Francis Group
Boca Raton London New York Leiden

CRC Press is an imprint of the
Taylor & Francis Group, an **informa** business
A BALKEMA BOOK

First published 2026
by CRC Press/Balkema
4 Park Square, Milton Park, Abingdon, Oxon, OX14 4RN

and by CRC Press/Balkema
2385 NW Executive Center Drive, Suite 320, Boca Raton FL 33431

CRC Press/Balkema is an imprint of the Taylor & Francis Group, an informa business

British Library Cataloguing-in-Publication Data
A catalogue record for this book is available from the British Library

Library of Congress Cataloging-in-Publication Data
A catalog record has been requested for this book

ISBN: 978-1-041-03597-8 (hbk)
ISBN: 978-1-041-03598-5 (pbk)
ISBN: 978-1-003-62442-4 (ebk)

DOI: 10.1201/9781003624424

Typeset in Times New Roman
by MPS Limited, Chennai, India

Proceedings of the International Association of Institutes of Navigation World Congress 2024 – Chen et al. (Eds)

Table of Contents

Proceedings of the International Association of Institutes of Navigation World Congress 2024 – Chen et al. (Eds)

Preface

The IAIN World Congress 2024, held from October 28th to 30th, 2024, at the Beijing International Convention Center, marked a significant milestone in the field of navigation and maritime technology. Organized by the International Association of Institutes of Navigation (IAIN) and hosted by the China Institute of Navigation (CIN), this prestigious event brought together leading experts, researchers, and industry professionals from around the globe to discuss the latest advancements and future trends in navigation technology.

Conference Overview
The IAIN World Congress 2024 served as a platform for international collaboration and knowledge exchange in the field of navigation. With over 350 domestic participants and 31 international delegates from countries such as Egypt, South Korea, the Netherlands, the United States, Norway, Japan, Thailand, Italy, the United Kingdom, and Vietnam, the conference fostered a rich environment for cross-border dialogue and cooperation. The event featured keynote speeches, technical sessions, and panel discussions, all aimed at addressing the challenges and opportunities in the rapidly evolving navigation industry.

Background
The conference was held against the backdrop of significant advancements in global navigation systems, including the widespread adoption of satellite navigation technologies such as GPS, GLONASS, Galileo, and China's BeiDou system. The integration of these systems with emerging technologies like 5G, IoT, AI, and big data has revolutionized the field, enabling new applications in autonomous vehicles, smart transportation, precision agriculture, and more. The IAIN World Congress 2024 provided a timely opportunity to explore these developments and their implications for the future of navigation.

Call for Papers
The conference received a total of 43 papers, including 38 from domestic authors and 5 from international contributors. These papers covered a wide range of topics, from the latest developments in satellite navigation and autonomous shipping to the challenges of maritime education in the face of emerging technologies. The call for papers attracted submissions from leading academic institutions, research organizations, and industry experts, reflecting the global interest in advancing navigation technology.The research directions are maritime traffic safety,positioning and navigation technology,marine environment perception.

Paper Exchange and Collaboration
The conference provided a valuable opportunity for researchers and practitioners to exchange ideas and share their latest findings. The paper sessions were well-attended, with lively discussions on topics ranging from the technical challenges of high-precision navigation to the regulatory and educational implications of autonomous shipping. The exchange of ideas and experiences during these sessions will undoubtedly contribute to the continued advancement of navigation technology.

Awards and Recognition
The conference also recognized outstanding contributions to the field through several awards. The Sadek Award for Best Paper was awarded to a team from Dalian Maritime University for their research on GNSS-based positioning methods in restricted maritime environments. Additionally, the John Harrison Award for Lifetime Achievement was presented to Prof. Dana Goward (ION), and the Necho Award for Special Achievement was awarded to Mr. ZHAI Mo. Furthermore, the Honorary Membership Award was conferred

upon Mr. John Pottle and Mr. Simon Gaskin (RIN). These awards were presented at the closing ceremony of the conference.

The IAIN World Congress 2024 was a resounding success, bringing together the global navigation community to discuss the latest advancements and future directions in the field. The papers presented at the conference reflect the cutting-edge research being conducted around the world and provide valuable insights into the challenges and opportunities facing the navigation industry. We are confident that the knowledge shared and the connections made during this event will continue to drive innovation and collaboration in the years to come.

We extend our heartfelt thanks to all the participants, organizers, and sponsors who made this conference possible. We look forward to the next IAIN World Congress, which will be held in Poland in 2027, and to the continued advancement of navigation technology on a global scale.

Baochen Zhang
IAIN World Congress 2024 Organizing Committee
November 2024

*Proceedings of the International Association of Institutes of Navigation World Congress 2024 –
Chen et al. (Eds)*

Acknowledgements

The successful organization and execution of the **IAIN World Congress 2024** would not have been possible without the collective efforts and contributions of numerous individuals and organizations. We extend our heartfelt gratitude to all those who played a pivotal role in making this event a resounding success.

First and foremost, we would like to express our deepest appreciation to the International Association of Institutes of Navigation (IAIN) for entrusting the China Institute of Navigation (CIN) with the responsibility of hosting this event. Their guidance and support throughout the planning and execution phases were invaluable.

We are profoundly grateful to the Ministry of Transport of the People's Republic of China, the China Association for Science and Technology, China International Science and Technology Exchange Center, China Satellite Navigation and Positioning Association, Chinese Society of Aeronautics and Astronautics, China Air Transport Association, Chinese Society of Astronautics, China Highway and Transportation Society, Chinese Society of Agricultural Engineering,Shanghai Maritime University, CCCC Tianjin Dredging Co., Ltd. and National Energy Shipping Co., Ltd.for their unwavering support and encouragement. The presence of distinguished leaders and officials at the conference underscored the importance of this event and highlighted the significance of navigation technology in global development.

Our sincere thanks go to the Organizing Committee and the Academic Committee for their meticulous planning and execution. Their dedication and expertise ensured that the conference ran smoothly and that the academic content was of the highest quality. We also extend our gratitude to the keynote speakers, session chairs, and all the presenters who shared their insights and research, enriching the discussions and fostering a vibrant exchange of ideas.

Special thanks are due to the sponsors and exhibitors who supported the conference and showcased the latest advancements in navigation technology. Their participation added immense value to the event and provided attendees with a glimpse into the future of navigation and related technologies.

We are also grateful to the volunteers and support staff who worked tirelessly behind the scenes to ensure that every aspect of the conference was well-organized and executed. Their hard work and dedication were crucial in creating a positive and memorable experience for all attendees.

Finally, we extend our thanks to all the participants from around the world who attended the IAIN World Congress 2024. Your active engagement, insightful questions, and collaborative spirit made this event a true success. We hope that the knowledge shared and the connections made during this conference will continue to inspire innovation and collaboration in the field of navigation.

As we look forward to the next IAIN World Congress in Poland in 2027, we remain committed to advancing the field of navigation and fostering international cooperation. Thank you all for your contributions and for being part of this remarkable journey.

Baochen Zhang
President, IAIN
November 2024

Proceedings of the International Association of Institutes of Navigation World Congress 2024 – Chen et al. (Eds)
© 2026 The Editor(s), ISBN: 978-1-041-03597-8
Open Access: www.taylorfrancis.com, CC BY-NC-ND 4.0 license

Editor Biographies

Chen Shijun is Director of the Editorial Department of the Chinese Society of Navigation. He is responsible for the editing and publication of the journals China Navigation and Navigation Technology, organizing and implementing the high-quality scientific journal grading directory project in the field of navigation, and promoting innovative application of Chinese and English journals in terms of equivalence, scientific achievements and talent. He has also served as the project leader in the development of the National Government Service Platform for Water Transport and Maritime Administrative Approval System and has organized the implementation and drafting of national and industry standards, including the Port and Shipping Data Resource Directory and Electronic Certificates for Waterway Transport.

Dong Siwen holds a master's degree in Human Resource Development Policies from the ISCTE-University Institute of Lisbon, Portugal. She currently serves as an Editor at the Navigation of China and Navigation Technology editorial offices of the China Institute of Navigation, with intermediate qualifications in publishing. Specializing in the publication of scientific journals, she organizes international academic conferences and oversees the publication of conference proceedings, notably as the lead editor and coordinator for the IAIN 2024 World Congress Proceedings. She has contributed to projects such as the China Association for Science and Technology's High-Quality Journal Grading Directory and was a key contributor to the first prize-winning projects in the 2022 and 2023 Navigation Science and Technology Progress Awards of CIN. Her research interests include green shipping development, enhancing the academic impact of international journals, and integrated publishing development.

Guo Weihua was born in 1972. She graduated from the School of Continuing Education of Renmin University of China, majoring in International Economics and Trade. She holds an Intermediate Certificate for Professional Qualifications of Publishing Professionals. Currently, she is working as the Director of the Academic Exchange Department at the China Institute of Navigation. Over the years, she has been committed to providing support and services for marine science and technology workers and building high - level academic exchange platforms. By organizing various international academic conferences and other activities, she has promoted technological innovation and knowledge sharing in the field of navigation, facilitated the dissemination and promotion of academic achievements, and made significant contributions to the development of marine science and technology and the cultivation of talents.

Wang Yan holds a master's degree in English Translation from Dalian Maritime University and currently serves as the Special Project Director of the Academic Exchange Department at the China Institute of Navigation. Specializing in international exchange and cooperation in maritime technology, she has extensive experience in organizing international academic conferences. As the coordinator of international review experts for IAIN World Congress 2024, she oversees the entire paper review process, including initial and final review organization, score compilation, expert confirmation, and the coordination of paper presentation sessions. Her research interests focus on English translation and international academic exchange mechanisms.

Zhang Feifei is Associate Editor of the Editorial Department at the Chinese Society of Navigation. She is responsible for the editing and publication of the journals China

Navigation and Navigation Technology. Zhang has organized and implemented several provincial and ministerial-level projects, including the "National Society-hosted Journal Publishing Capacity Improvement Plan" and the "High-Quality Scientific Journal Grading Directory in the Field of Navigation." She is dedicated to researching ways to enhance the academic influence of scientific journals in the field of navigation and supports technological innovation in the maritime sector.

Liu Yang is currently an editor for China Navigation and Navigation Technology. His research focuses on maritime culture, smart shipping, maritime technology, and green shipping. He has participated in the implementation of the high-quality scientific journal grading directory project in the field of navigation and promoting innovative application of Chinese and English journals in terms of equivalence, scientific achievements and talent. Additionally, he contributed to the Research on the Construction of the Scientific and Technological Popularization System in the Transportation Industry project. The research outcomes were adopted by the Ministry of Transport's Guiding Opinions on Strengthening the Popularization of Science and Technology in Transportation, and the results won the "Second Prize for Scientific and Technological Progress." He also participated in writing and publishing the book Research and Practice on the Construction of the Scientific and Technological Popularization System in the Transportation Industry.

Wang Yuxiang has been working in the publishing industry for five years. He holds an intermediate-level certification for publishing professionals and is a certified dive master. He currently serves as Editor at the Chinese Society of Navigation. Over the years, he has been dedicated to publishing books in the field of marine science. He provided technical support for this book.

Ma Erdan holds a master's degree in finance and graduated from the University of Plymouth, UK. She is an intermediate economist and has established a good reputation in the workplace for his professionalism and patience. Currently, she is involved in organizing activities such as international academic conferences at the China Institute of Navigation.

Proceedings of the International Association of Institutes of Navigation World Congress 2024 – Chen et al. (Eds)

Organizing Committee

Chair:
Krzysztof Czaplewski *President of IAIN, Former Presient of Polish Navigation Forum*

Co-chair:
ZHANG Baochen *Senior Vice-President of IAIN, Executive President of CIN*
Hesham Helal *Secretary General of IAIN, President of Arab Institute of Navigation*
LI Shixin *Vice President of CIN*
ZHAO Jicheng *Executive Vice President of GNSS and LBS Association of China*

Member:
QIN Jiuyi *Deputy Director of China Centre for International Science and Technology Exchange*
LIU Dake *Vice President and Secretary General of GNSS and LBS Association of China*
YAO Junchen *Vice President and Secretary General of Chinese Society of Aeronautics and Astronautics*
ZHU Yaochun *Deputy Secretary General of China Air Transport Association*
JIANG Jun *Deputy Secretary General of Chinese Society of Astronautics*
WANG Dapeng *Deputy Secretary General of China Highway and Transportation Society and Director of WTC Operation Center*
WANG Yingkuan *Secretary General of Chinese Society of Agricultural Engineering*
SHI Xin *Vice President of Shanghai Maritime University*
WANG Jun *Deputy General Manager of China Energy Shipping Co., Ltd.*
WANG Qi *General Manager of Science and Technology Department of CCCC-Tianjin Dredging Co., Ltd.*

Secretary General:
YAN Xiaobo *Secretary General of CIN*

Proceedings of the International Association of Institutes of Navigation World Congress 2024 – Chen et al. (Eds)
© 2026 The Editor(s), ISBN: 978-1-041-03597-8
Open Access: www.taylorfrancis.com, CC BY-NC-ND 4.0 license

Academic Committee

Chair:
Krzysztof Czaplewski *President of IAIN, Former Presient of Polish Navigation Forum*

Co-chair:
ZHANG Baochen *Senior Vice-President of IAIN, Executive President of CIN*

Members (without particular order):
John Pottle *Director of the Royal Institute of Navigation, Former President of IAIN*
Hesham Helal *Secretary General of IAIN, President of Arab Institute of Navigation*
Dana Goward *President of the Resilient Navigation and Timing Foundation*
Adam Weintrit *Rector of Gdynia Maritime University*
SHI Chaojian *Professor, Shanghai Maritime University*
Yasuyuki NIWA *Principal Researcher of National Maritime Research Institute, Director of Japan Institute of Navigation*
HU Qing *Director of the National Engineering Research Center of Ship Navigation System of Dalian Maritime University, Member of the Communication Navigation Committee of the CIN*
ZHANG Bingqi *Deputy Director and Senior Engineer of Transportation Beidou Application Development Center*
Tom Willems Officer, *Treasurer of IAIN, Radio Navigation Engineer of the Netherlands Institute of Navigation*
Daisuke Terada *Professor of National Defense Academy, Director of Japan Institute of Navigation*
Jiri Svaton *Doctor and Scientist of Czech Technical University in Prague*
Vaclav Navratil *Ph.D. and Associate Professor, Department of Radio Engineering, Czech Technical University in Prague*

Proceedings of the International Association of Institutes of Navigation World Congress 2024 – Chen et al. (Eds)

Navigation requirements for marine autonomous surface ships

B.C. Zhang
China Institute of Navigation, Beijing, China

J.N. Luo
China Waterborne Transport Research Institute, Beijing, China

ABSTRACT: Maritime Autonomous Surface Ship is a hot topic and clear development direction in the shipping industry. In response to the unclear demand for Maritime Autonomous Surface Ship navigation, the scope and basic elements of ship navigation are first defined. Then, by comparing and analyzing machine driving modes and combining with the verification practice of "Zhifei" ships, three typical navigation requirements for Maritime autonomous surface ship are proposed. Finally, the development path of navigation technology for Maritime autonomous surface ship is proposed.

1 INTRODUCTION

The development of intelligent ships can be traced back 70 years. From the 1940s to the 1960s, the development and application of reconnaissance-type remote-controlled ships were carried out in the United States and the Soviet Union. In the 1970s and 1980s, research focused on the development of automatic control systems. Perception and obstacle avoidance navigation technology has undergone several iterations in the two decades before and after this century. Since 2010, we have seen the rapid advancement of technologies such as communication, navigation, computing, and artificial intelligence, at the same time, the global shipping industry has been accelerating its transformation strategy, focusing on cost reduction, efficiency improvement, energy conservation, emissions reduction, and enhancing intrinsic safety. Therefore, automation, remote operation, reduced-crew, and even unmanned shipping have become key development directions, and the era of smart shipping is rapidly approaching.

In 2017, at MSC 98, MASS was officially defined by the IMO. It is a ship that can operate with varying levels of human intervention. Besides, the IMO has initiated work to define the scope of MASS legislation. In 2018 at MSC 99, the IMO proposed four levels of MASS. In 2020, at MSC 101, the IMO adopted the 'Interim Guidelines for the Trial Operations of Autonomous Ships'. This supports the international community and stakeholders in conducting trials of MASS systems and infrastructure.

In 2024, at MSC 108, the IMO decided to finalize and adopt the non-mandatory MASS CODE at the 110th session in 2025. An experience-gathering phase will begin in 2026, with the development of the mandatory MASS Code starting in 2028. The mandatory code is expected to be adopted at a meeting before July 1, 2030, and to come into effect on January 1, 2032. MASS CODE will establish a comprehensive security regime for MASS to ensure the safety of lives at sea, as well as the safety of cargo on board and of MASS itself.

Colleagues who are developing global MASS navigation legislation rules at IMO, the international community has already had many practices in MASS construction and operation. Norway, South Korea, Japan, Russia and other countries have conducted MASS

DOI: 10.1201/9781003624424-1

navigation tests, and some ships have been put into operation. Navigation technology is the fundamental core technology of MASS. Although there is a good foundation, the current MASS navigation is mainly in open waters. For more complex scenarios, existing navigation technology cannot fully meet the needs of MASS navigation. The analysis and demonstration of MASS navigation requirements internationally are also extremely limited. In order to promote MASS to accumulate more experience before 2032 and achieve true commercial operation, it is urgent to carry out MASS navigation feature analysis to provide support for the research and development of MASS perception, decision-making, and control systems.

2 SCOPE AND DRIVING MODES OF MASS NAVIGATION

2.1 *Scope of MASS navigation*

Navigation is the process of determining a location, planning a route, and guiding a moving object from one point to another. This basic paradigm reveals the core of ship navigation, which is the collection of ship positioning, route planning, and guidance information during the process of a ship moving from one berth to another. (Navigation support and security information are provided onshore, while the ship can obtain perception and decision-making assistance information. The coordination and interaction between the ship and shore can better guide the vessel's navigation.) Therefore, technical equipment related to these three elements should be considered as part of the "navigation" equipment system. Based on the above definitions, I will talk about my understanding of the demand for MASS navigation from these three aspects.

2.2 *Driving modes of MASS*

We know that the four MASS levels proposed by IMO MSC 99 clearly define four MASS scenarios. MASS Level 1 refers to vessels with automated processes and decision support for assisted navigation. MASS Level 2 is about the remote control of the ship with a crew on board. MASS Level 3 is about the remote control of the ship without a crew on board, while MASS Level 4 describes fully autonomous vessels. The four scenarios actually correspond to three driving modes using artificial intelligence technology, with remote operation divided into crewed and uncrewed vessels. Overall, this has little relation to the levels of MASS. Nowadays, the concept of "unmanned ships" is often used, but among the four scenarios, only MASS 4 truly refers to unmanned ships.

2.3 *Sample of requirement analysis*

From 2018 to 2021, we led the project of "Research on Key Technologies for Intelligent Navigation and Control of Ships Based on Ship-Shore Collaboration," a national key research and development program in China. In this project, the theory of intelligent ship-shore collaboration was studied; a number of standards, such as the "Guidelines for the Construction of an Intelligent Ship-Shore Collaboration System,"were formulated; a number of key technologies were overcome; 16 technical systems, platforms, and equipment were developed; inland intelligent navigation systems and maritime intelligent navigation systems were developed; a ship intelligent navigation test system marked by the Qingdao Sea Experimental Test Site was completed; and the "Zhi Fei" container ship, which combines manual driving and autonomous navigation, was designed and built. Since its commercial operation, the ship has maintained safe operation for 700+ voyages and 70,000+ kilometers, achieving comprehensive intelligent decision support for manual driving, remote control driving from berth to berth, autonomous collision avoidance in open waters with high frequency, and automatic berthing and leaving in the whole process without the participation of the driver on board.

Figure 1. ZHIFEI.

3 MASS NAVIGATION REQUIREMENTS

3.1 *Basic judgment*

Through recent research and the practical experience gained from the "Zhi Fei," we recognize that the biggest difference between MASS and conventional ship navigation is that conventional ship navigation is oriented towards the needs of "human pilots", while MASS navigation is oriented towards the needs of "machine pilots". The reason for this is that in the four scenarios of MASS, assisted driving and remote control driving are still driven by people. Humans have the ability to perceive, acquire, and process information, as well as the ability to make decisions and respond to emergencies. Its main navigation needs are information and support for decision-making, which is consistent with the navigation needs of traditional ships. Existing navigation technology and equipment, route planning, and navigation support and assurance information can generally meet demand. If there is any change, it is mainly in the immediacy of information, the authenticity of the audiovisual interface, and the credibility of decision-making suggestions. These new needs have been basically solved in technology research and development and are no longer bottleneck obstacles affecting the application of auxiliary driving and remote driving technologies. However, for fully autonomous navigation, that is, unmanned navigation, the main demand for navigation changes from humans to machines, generating a series of new demands that exceed those of humans.

3.2 *Overall requirement*

Route planning at the macroscale does not reflect new navigation requirements. Dynamic optimization of navigation at the meso- and microscale in specific scenarios requires fast response and flexible adjustment.

Autonomous route planning is a form of macro or large-scale navigation. Whether piloted or machine-driven, it is determined based on the geographical and environmental conditions of the route, the mission of the voyage, costs, meteorological and oceanographic conditions, etc., and basically does not require new navigation.

Dynamic optimization of routes includes navigation at large, medium, and small scales. Adjustments to planned routes on a large scale, such as changes in destination ports or severe weather events, can be classified as issues related to autonomous route planning. Medium- and small-scale route dynamic optimization is specifically reflected in the adjustment of course, speed, and route planning based on guidance information such as port scheduling, navigation warnings, and distant ships that may be encountered. The system needs to be able to predict the position and distance of obstacles and the movement trajectory of other ships in order to assess the risk of collision, automatically optimize the navigation route and provide collision avoidance strategies. Because it involves collision avoidance, the system needs to be able to respond quickly and adjust flexible.

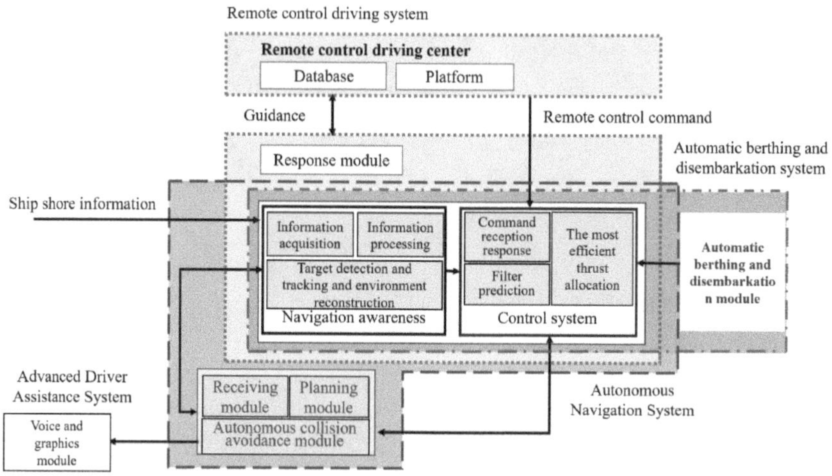

Figure 2. Design of MASS navigation system and relevant modules.

3.3 *Navigation requirements for autonomous collision avoidance*

According to previous research and experimental observations, autonomous navigation is more dependent on and has higher requirements for the navigation system.

Basic argument 1: In terms of relative positioning in scenarios such as complex waters, narrow waterways and severe weather conditions, the navigation requirements for autonomous collision avoidance are reflected in high precision and high timeliness.

Using positioning technologies such as the Global Navigation Satellite System (GNSS), coastal ground-based augmentation systems, and PPP-RTK, it is now possible to achieve decimeter or even centimeter-level positioning, which can meet the spatial positioning accuracy requirements of autonomous sailing ships in open waters and their relative positions to other ships. However, in scenarios such as complex waters, narrow waterways, berthing, collision avoidance, and severe weather conditions, the positioning accuracy of GNSS is subject to certain limitations. For example, it is difficult for existing technology to provide an effective solution for positioning in shaded areas. Collision avoidance occurs more frequently and environmental conditions are more critical. Autonomous sailing ships cannot completely rely on absolute spatial position positioning technology. Therefore, MASS must rely on the surrounding environment and references to achieve positioning, and it needs to accurately identify the direction, distance, status, etc. Otherwise, the autonomous navigation system cannot make quick and accurate judgments, and it is impossible to make precise collision avoidance decisions. In other words, it is necessary to optimize the location discrimination method in the relative space and subdivided scenarios within the existing GNSS navigation framework and give the autonomous navigation system the ability to judge and analyze spatial location like a human being, such as the ability to judge relative location and quickly integrate and judge external information, so as to achieve high-precision and timely navigation in all scenarios, including open waters, complex waters, and restricted waters. So, the MASS navigation system must integrate multiple sensors, such as radar, lidar, sonar, and cameras, and process and analyze the collected data using advanced perception algorithms to form a comprehensive understanding of the surrounding environment.

3.4 *Guidance information*

For autonomous driving, the new demand for guidance information (navigation information) is characterized by high sufficiency and uniqueness.

Collision avoidance decisions for fully autonomous navigation are made independently by the autonomous navigation system. The information provided by the aforementioned ship-end perception system alone is not sufficient. It is also necessary to fully grasp diverse information such as meteorological and sea conditions, navigation security, port scheduling, shipping companies, shipowners' scheduling and maritime supervision. As mentioned above, in scenarios such as complex waters, narrow waterways, and adverse weather conditions, the positioning accuracy of GNSS is subject to certain limitations, and the end-of-ship perception system may also have limitations. As mentioned above, in scenarios such as complex waters, narrow waterways, and adverse weather conditions, the positioning accuracy of GNSS is subject to certain limitations, and there may also be limitations with the ship-end perception system. Therefore, the problem of sufficient and timely supply of guidance information should be solved through ship-shore cooperation. This involves four questions. The first is the channel—a reliable information and communication link.

Figure 3. Shore based operation and control center.

The second is the standard—a structured method, specification, and mechanism for transmitting information. The third is the "eyes"—a perception system that integrates and complements the ship and shore. The last is the "brain"—a higher level of situational awareness of the navigation environment, cognitive capabilities, and collision avoidance algorithms with higher reliability and robustness. Currently, most MASS is at the "open water autonomous collision avoidance" level. The reason is that perception methods, cognitive algorithms, and collision avoidance algorithms still need to be upgraded. For example, shore-based navigation systems for complex waters and narrow waterways have not yet been integrated into public maritime safety facilities. If such a public facility is available, the perception system of the autonomous sailing ship will be more economical.

4 DISCUSSION

4.1 *The synergy between MASS navigation and infrastructure*

Absolute and relative spatial positioning and navigation are inseparable from the construction and collaborative application of an integrated space-air-ground navigation infrastructure. The positioning and navigation of the relative spatial position of the MASS propulsion require investments in the navigation industry's infrastructure. GNSS, ground-based augmentation systems, unmanned high-frequency surveying and mapping, and other infrastructure will remain the focus of future development. ROC is also an important infrastructure for MASS's remote driving. The construction of a high-precision geographic information navigation framework based on infrastructure will be an important direction for future development.

4.2 *The synergy between MASS navigation and regulatory standards*

It is not difficult to see from the IMO's process of promoting MASS that the legal compliance status definition of MASS, the implementation of technical guidelines, and the formulation and implementation of the MASS CODE have been given top priority. Advancing the development of MASS navigation technology requires accelerating the improvement of relevant laws and regulations and the construction of the standard system, as well as increasing government strategic planning and policy guidance. MASS navigation equipment, systems, and the interaction and collaboration with infrastructure all require a high degree of standardization, including hardware interfaces, software protocols, and many other aspects. Regulatory standards are essential for the effective operation, maintenance, and management of intelligent ships by the ship's crew team, the ROC management team, and the shore-based third-party service team. The new generation of information standards represented by S-100 will bring new technological changes to MASS navigation.

4.3 *The synergy between MASS navigation and communication technologies*

Integrated communication and navigation technology will continue to play an important role in ship navigation and will have a significant impact on MASS navigation. Synergies with communication technology are mainly expressed in two aspects. On the one hand, MASS can quickly transmit relative spatial location information, including transmission between MASS's own navigation system and decision-making system and transmission of location information between MASS and other ships, so as to achieve instant determination and updating of relative spatial location information. On the other hand, it is the long-distance, efficient, stable, and reliable transmission of guidance information, including the exchange of ship-shore cooperative information and the exchange of information between the infrastructure and MASS. There is no doubt that 6G will bring about new technological changes in future MASS navigation application scenarios, integrating MASS sensing, computing, and communications to provide high-precision positioning and sensing capabilities.

4.4 *The synergy between MASS navigation and talent training*

Compared with traditional marine navigation disciplines, MASS navigation has put forward new demands for talent training. There is a clear shortage of digital and information literacy, thinking, and ability in this area in the international community. Knowledge is more complex, comprehensive abilities and the required technical level are higher. This has broken through the traditional discipline of marine navigation. At this stage, it is urgent to construct a knowledge map of MASS navigation technology and disciplines, clarify the key technologies and the connection between them, and propose a responsive technology application system and technical equipment system.

4.5 *The synergy between MASS navigation and the Intelligent shipping system and its industrialization*

The development of MASS navigation technology equipment and infrastructure as part of the intelligent shipping system and its industrialization needs to be included in the overall thinking of the intelligent shipping system and its industrialization. In China, some achievements have been made in building an intelligent shipping industry chain. However, it is also necessary to further integrate resources to promote multifunctional integration and mutual assistance in theoretical research, design and construction, equipment development, testing and inspection, demonstration application and legislative research.

5 CONCLUSION

MASS presents a new set of navigation, which is distinct from the navigation needs of manned vessels. This has not only led to new directions in navigation technology, but also led to local reconfigurations of navigation systems, such as the need for land-based navigation systems in complex waters with heavy traffic.

The next five to eight years will be an important stage in the development of MASS and related technologies, facing multiple tasks of restructuring the system of technologies, standards, and rules. The IAIN faces an important development opportunity to better serve humanity, as well as the challenge of how to play its role.

REFERENCES

China Classification Society. (2022). Intelligent Ship Development Outlook 2022. Released at *the 2022 North Bund International Shipping Forum*.

China Classification Society. (2024). *Guide for Intelligent Energy Efficiency Management Inspection of Ships 2024*.

Lin S.Y. and Zhong T. (2018). Discussion on the relationship between intelligent ship and intelligent aviation security. *Science and Technology Innovation Herald* (25): 120–122.

Yan X.P. (2016). Research status and development trend of intelligent ships. *Public Utilities*, 3(1): 25–28.

Zang B.C. The Development of Intelligent Shipping Technology Will Bring the Era of Unmanned Ships. (2020-01-06)[2024-06-28].

Zhang B.C. A Brief Analysis of the Ship Intelligent Navigation Driving Mode and External Information Interaction Coordination. (2020-05-29)[2024-06-28].

Zhang B.C. How Far Away is the Intelligent Navigation of Ships from Us?. (2018-12-27)[2024-06-28].

Zhang B.C., Geng X.F., Li Y.B., *et al.* (2022). Research Progress on the Development of Ship Intelligent Navigation Technology. *Science & Technology Review* 40(14).

Proceedings of the International Association of Institutes of Navigation World Congress 2024 – Chen et al. (Eds)

Exploiting AIS data for vessel traffic flow prediction: Online recurrent extreme learning machine

Guanya Wang, Bin Lin & Miyuan Zhang
Dalian Maritime University, Dalian, Liaoning, China

ABSTRACT: Accurate prediction of vessel traffic flow is essential for maritime navigation departments to develop waterway planning. However, machine learning algorithms have been widely applied in marine traffic flow prediction. Existing researches still exhibit limitations in various applications, such as slow convergence speed and low prediction accuracy. To solve the problems, this paper proposes an offshore traffic flow prediction model based on an Online Recurrent Extreme Learning Machine. Firstly, we preprocess the vessel AIS (Automatic Identification System) data, and input AIS data to the model for training; Secondly, we continuously and recursively update the weight matrix by using recursive least squares to enhance its overall prediction performance; lastly, we employ the trained network to predict the future short-term vessel traffic flow. The experimental results demonstrate that the method proposed in this paper exhibits a high prediction accuracy compared to the Extreme Learning Machine and Online Sequential Extreme Learning Machine methods.

Keywords: traffic flow prediction, machine learning, AIS data

1 INTRODUCTION

With the acceleration of the 21st Century Maritime Silk Road, the demand for maritime transportation [1] has entered a new normal of stable growth. The development of the maritime industry has resulted in a dramatically increasing number of vessels in offshore waters, leading to a continuous rise in maritime traffic accidents. Therefore, proposing an efficient and accurate method for predicting vessel traffic flow is crucial to ensure maritime navigation safety.

Currently, vessel traffic flow prediction methods [2,3] can be mainly categorized into two types: statistically based on prediction methods and machine learning prediction methods based on vessel traffic flow.

(1) Statistical prediction methods: This method is based on mathematical or statistical theory, combined with traditional forecasting models for prediction. It mainly includes the grey theory model, theoretical combination model, and autoregressive moving average model. Reference [4] established a prediction model combining Markov chain and grey theory to predict traffic flow in specific water areas. Reference [5] considered the impacts of port geographical features on vessel traffic flow and established a short-term forecast combination model network based on historical vessel traffic flow data. Reference [6], considers the influence of noise information on the prediction results and performed denoising on AIS data for vessels. The missing data were reconstructed using cubic spline interpolation and established an autoregressive moving average model to enhance the precision of predicting vessel traffic flow.
(2) Machine learning algorithms mainly include Support Vector Machine (SVM), Back Propagation Neural Network (BPNN), and Convolutional Neural Network (CNN) for predicting vessel traffic flow. For example, in reference [7], researchers considered the

 DOI: 10.1201/9781003624424-2

issues of non-linear and seasonal features of traffic flow and used an adaptive algorithm based on SVM for traffic flow prediction. Reference [8] established a BP neural network prediction model for the traffic flow prediction problem of Tianjin Port through analysis of AIS data and assessed the reliability and applicability of the model through residual analysis. Reference [9] proposes an advanced deep learning method for predicting vessel traffic network patterns, which encodes all vessel-related information into a two-dimensional channel. By utilizing the spatial feature extraction function of neural networks, researchers incorporated five fully connected layers into the model for feature extraction to predict future vessel traffic flow and vessel behavior characteristics. Therefore, to address the limitations of existing models, this paper delineates a methodology for forecasting offshore vessel traffic flow by using the Online Recurrent Extreme Learning Machine.

2 PREDICTION METHODOLOGY

2.1 *Extreme Learning Machine*

Extreme Learning Machine (ELM) [10] is classified as a pre-feedback neural network learning method, where the input weights and corresponding bias values are randomly generated and remain consistent throughout the learning process. The output weights are determined using the pseudo-inverse parsing technique. Consequently, in comparison to traditional models such as BP neural networks, this approach demonstrates the potential to achieve an optimal global solution.

The conventional ELM network model, as depicted in Figure 1, is partitioned into input, output, and hidden layers based on its structure [11]. In the ELM prediction model, the input layer consists of the existence of N sets of training samples (X_i, Y_i), where the variable i represents the dimension of the target vector of inputs, and X_i Y_i represent the input and output vectors, respectively. The hidden layer consists of a regression function that seeks the input to the target output, the function is denoted as:

$$O_j = \sum_{i=1}^{L} \beta_i \times g(\omega_i \times x_j + b_i) \tag{1}$$

Where j represents the output weight of the network; g(x) is the activation function; ω_i represents the network output weight, and β_i represents the i hidden layer unit. The final output of the model can be simplified as a linear combination of the hidden layer output and the input weights β. As shown in Equation (2):

$$Y = H\beta \tag{2}$$

The aim of the ELM is to minimize the difference between the model output and the actual theoretically desirable output. To achieve more accurate prediction or classification results. The expression (3) is as follows:

$$\min_{\beta} \|H\beta - Y\|_2 \tag{3}$$

Due to the stochastic allocation of input weights and thresholds, Iterative adjustment of the output weights between the hidden and output layers is necessary. The weight β of the output layer can be obtained by solving the least squares problem, and the specific expression is:

$$\beta = H^{\Psi} Y \tag{4}$$

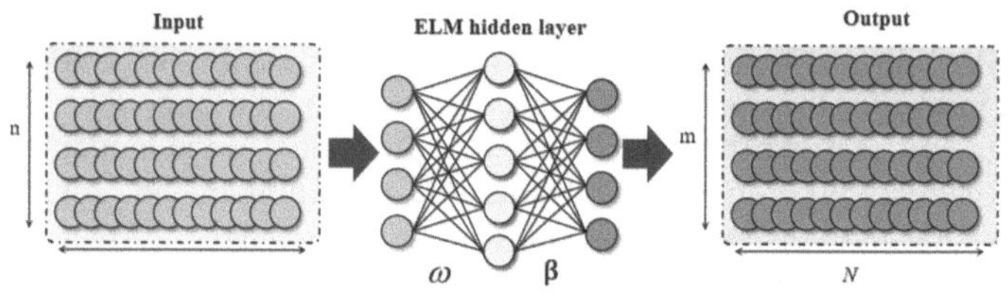

Figure 1. The structure of ELM.

2.2 *Online sequential Extreme Learning Machine*

Although ELM has good results in terms of training speed, generalizability, and obtaining optimal solutions. However, the ELM network may affect the network model output results because it needs to manually determine the number of neurons in the hidden layer of the network, which increases the complexity of the prediction model; and the ELM is unable to adjust the structure of the network parameters with the update of the data during the training process.

Therefore, current research studies have proposed an Online Sequential Extreme Learning Machine (OS-ELM) algorithm [11] based on ELM, which has been an algorithm designed for online learning and integrates the concepts of incremental learning and online sequential learning. By dynamically adjusting the network parameters and structure to gradually adapt to data changes, OS-ELM can continuously update the weights and change the network architecture compared to ELM.

The training process of OS-ELM is divided into two phases: (1) First phase is to compute and initialize the output weights β_i by a small number of training samples using the ELM algorithm. The hidden layer output matrix H_{t+1} is calculated as shown in Equation (5).

$$H_{t+1} = \begin{pmatrix} g\left(\omega_1^T \times X_1^{(t+1)} + b_1\right) & L & g(\omega_L^T \times X_L + b_L) \\ M & 0 & M \\ g\left(\omega_1^T \times X_{Nt+1}^{(t+1)} + b_L\right) & L & g\left(\omega_L^T \times X_{Nt+1}^{(t+1)} + b_L\right) \end{pmatrix}_{N*L} \cdot \tag{5}$$

At this stage, the covariance matrix Q_0 is also initialized as:

$$Q_0 = \left(H_0^T H_0\right)^{-1} \tag{6}$$

(2) After the initial training, the OS-ELM moves into the online learning phase, where new data samples are incrementally processed. When new input data x_t with corresponding output. Y_t becomes available at time step t, the hidden layer output matrix H_t is updated as follows:

$$H_t = g(WX_t + B) \tag{7}$$

where W and B are the input weights and biases initially determined during the first phase. The key here is that the input weights and biases do not change, ensuring that only the output weights β_t are updated. The update of the output weights β_{t+1} is computed as:

$$\beta_{t+1} = \beta_t + Q_{t+1} H_{t+1} \left(R_{t+1}^T - H_{t+1}^T \beta_t\right) \tag{8}$$

The covariance matrix Q_t is also updated at each time step to account for the new data. The Q_t function is expressed as (9). This recursive update ensures that the covariance matrix accurately reflects the model's current state based on all data seen up to time t.

$$Q_{t+1} = Q_t - \frac{Q_t H_{t+1} H_{t+1}^T Q_t}{1 + H_{t+1}^T Q_t H_{t+1}} \tag{9}$$

2.3 *Online recurrent Extreme Learning Machine*

Compared to ELM, OS-ELM has the advantage of continuously updating the weights and network architecture. However, when dealing with large-scale datasets, OS-ELM input weights are randomly updated during the training process, which may lead to an overall performance degradation in the subsequent prediction process. To address this problem, this paper establishes an Online Recurrent Extreme Learning Machine (OR-ELM) for vessel traffic flow prediction.

OR-ELM is an improvement based on OS-ELM, aimed at addressing the limitations of OS-ELM in predicting short-term traffic flow fluctuations. The structure of the OR-ELM network, as shown in Figure 2, consists of two main components:

(1) Recurrent Neural Network (RNN): This is the core component of OR-ELM [12], responsible for handling the prediction tasks. RNN utilizes its feedback structure to effectively capture dynamic changes in time series data, thereby enhancing prediction accuracy.
(2) Single-hidden Layer Feedforward Neural Network (SLFN): SLFN as an autoencoder network, used to learn and update the input and hidden weights of the RNN. When new input data arrives, SLFN adjusts the number of hidden layer nodes to convert the input features into sparse or compressed representations, dynamically updating the model's weights to adapt to new data and improve online prediction performance.

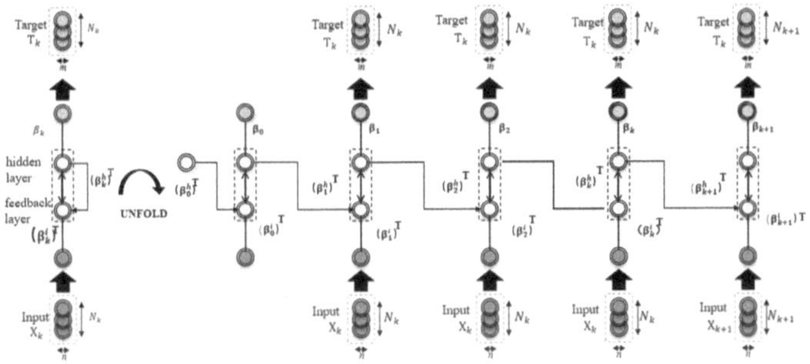

Figure 2. The structure of the OR-ELM.

This paper is based on the OR-ELM prediction model, utilizing recursive least squares normalization to optimize the weights and biases between the input layer and the hidden layer. The model parameters are gradually optimized to improve overall prediction performance and ensure that the model can effectively handle short-term traffic flow mutations. The specific steps are as follows:

Step 1: Initialize network parameters
Before starting training, the parameters of the OR-ELM model need to be initialized, including the feedforward neural network input weights, the number of hidden neurons, the activation function, and the recurrent neural network parameters.

Step 2: Data Input
The historical data undergoes preprocessing before being used as input for the model's hidden layer during training. This involves determining the optimal number of neurons in the hidden layer, selecting an appropriate activation function, and configuring parameters for the recurrent neural network. The dataset contains both input features and target variables, which are subsequently fed into the model sequentially for training purposes.

Step 3: Normalization
Through the forward propagation algorithm, the input data is fed forward in the feedforward neural network is propagated to get the output of the hidden and output layers.

Step 4: Update weights
Update the parameter weights in the feed-forward neural network through the recursive least squares algorithm to adjust the predictive ability of the model.

Step 5: Cyclic Training
At each time step, memory and learning are performed by recurrent neural networks to retain historical information about the time-series data, and training iterations are performed on the entire dataset.

Step 6: Model Evaluation
Following the completion of model training, evaluate the model. using cross-validation or other evaluation metrics to assess the performance and accuracy of the model.

Recursive Least Squares Normalization is an algorithm used for online sequence learning. This algorithm achieves good results in incremental learning tasks by continuously recursively updating the weight matrix. In the RLSN algorithm, normalization can play a good optimization role. Specifically, if there is a significant variance difference in the input vector, then while updating the weight matrix, it is also necessary to continuously adjust the normalized statistics between the input vector and the output vector. Normalization can effectively suppress the scaling effect between features, reinforce the algorithm's capacity to enhance precision in predictions, and facilitate broader inference. The normalization method can be achieved through the following steps: (1) For each input vector, calculate the mean and variance of each feature. (2) The scaling factors necessary for normalization can be derived by utilizing the estimated sample mean and variance. (3) Normalize the input vector according to scale coefficients. (4) Update the weight matrix in the RLSN algorithm and simultaneously update the normalized statistics.

3 DATA SOURCES AND PROCESSING

3.1 *Data sources*

AIS data mainly includes Maritime Mobile Service Identity (MMSI), speed, position, heading, name, ground speed, voyage time, and other relevant information. By mining this data, macro traffic flow parameters can be obtained.

The Port of Rizhao is a busy shipping area in China with significant variations in vessel traffic flow, making long-term prediction of vessel traffic flow necessary. This study selects the AIS dataset of vessels at Rizhao Port from January 31, 2023, to April 30, 2023. Considering the impact of upstream and downstream traffic flow on the current waterways, the study constructs multi-segment data for four selected channels in the port. The research area is a rectangular region with the top-left coordinate (119.85°E, 35.00°N) and the bottom-right coordinate (119.89°E, 35.05°N).

3.2 *Model evaluation*

In this study, to measure the difference between the model prediction results and the actual observations. Mean Absolute Error (MAE), Root Mean Square Error (RMSE) and Fitting Coefficient are introduced to assess the model prediction performance.

The study evaluates predictive accuracy through the application of metrics such as MAE, RMSE and R^2. MAE represents the mean absolute error, quantifying the average deviation between predicted and true values. RMSE, or root mean square error, is particularly attuned to larger discrepancies between predicted and true values. Meanwhile, the coefficient of determination is employed to gauge how effectively a model aligns with observed data; its values range from 0 to 1, with closer proximity to 1 signifying superior model fitting. The expressions (10-12) are defined as follows:

$$RMSE = \sqrt{\frac{1}{n}\sum_{j=1}^{n}(p_i - o_i)^2} \tag{10}$$

$$R^2 = \frac{\left(n\sum_{i=1}^{n} p_i o_i - \sum_{i=1}^{n} p_i \sum_{i=1}^{n} o_i\right)^2}{\left(n\sum_{i=1}^{n} (o_i)^2 - \sum_{i=1}^{n} (o_i)^2\right)\left(n\sum_{i=1}^{n} (p_i)^2 - \sum_{i=1}^{n} (p_i)^2\right)} \tag{11}$$

$$MAE = \frac{\sum_{i=1}^{n} (p_i - o_i)}{n} \tag{12}$$

3.3 *Data preprocessing*

Considering that neural networks are sensitive to input data and that different units and ranges of input data can affect prediction results, the following preprocessing steps are necessary:

(1) Detect anomalies in the data. For missing data, use the average value of adjacent periods to mitigate the impact of random factors on the prediction results.
(2) Normalize the data. To improve the accuracy of network input parameters, each dimension of the data is normalized to the range of [-1, 1]. The mapping function is represented as follows:

$$y = \frac{x - x_{\min}}{x_{\max} - x_{\min}} \tag{13}$$

where x is the initial input data, $x_{\min}$ is the minimum initial input value, $x_{\max}$ is the maximum initial input value, and y is the normalized value. Part of the vessel traffic data after normalization is shown in Table 1.

Table 1. Vessel traffic data initialization balues.

Contents	Sampling Time	Initialization Values
1	1/31 2:12	0.345527
2	1/31 2:27	0.476488
3	1/31 2:42	0.214566
4	1/31 2:57	−0.178318
5	1/31 3:12	1.000333

3.4 *Experimental parameters*

The experimental data is partitioned into two segments, with 70% initially designated for training and the remaining 30% allocated for testing to evaluate the predictive efficacy of the proposed model.

The efficacy of a neural network is significantly impacted by the configuration of its topology. Before the application of OR-ELM to the data, it is imperative to appropriately fine-tune specific hyperparameters. In Table 2, the input represents the number of characteristics in the neural network's input, which amounts to 1625, output represents the output dimension as 1; neurons quantity determines the depth of the model, with each neuron reflecting the width; the incorporation of lamb is intended to address the complexity of the model and mitigate against overfitting; and output weight represents the weight of the feed-forward part of the neural network as 0.92.

Table 2. Hyperparameters of OR-ELM.

Category	Hyper-parameters	Values
1	Learning rate	0.01
2	Iterations	100
3	Neurons	10
4	lamb	0.1
5	Inweight	1.0
6	Outweight	0.92

4 RESULTS AND DISCUSSIONS

Figure 3 compares the performance of the OS-ELM and OR-ELM prediction models, illustrating their predictive effects on the same dataset. Each data point in the figure represents 15 minutes, with a total prediction duration of four days. The OS-ELM model effectively captures the overall traffic flow trends, especially the upward trends between time steps 0 to 120 and 200 to 250, as well as the downward trend from 120 to 180. However, at step 370, where the flow sharply increases to its peak, the OS-ELM model falls into local optima due to its inability to update the output weights in a timely manner, resulting in significant local prediction errors. In contrast, the OR-ELM model demonstrates stronger predictive capability. By continuously updating and optimizing the network output weights using the least squares method during training, OR-ELM not only accurately predicts the overall trend but also approximates actual traffic during local peak periods. This effectively reduces prediction errors across the entire network, further enhancing the model's overall predictive performance. The comparison shows that OR-ELM exhibits better adaptability and accuracy than OS-ELM when handling large-scale datasets and dynamic traffic flow changes.

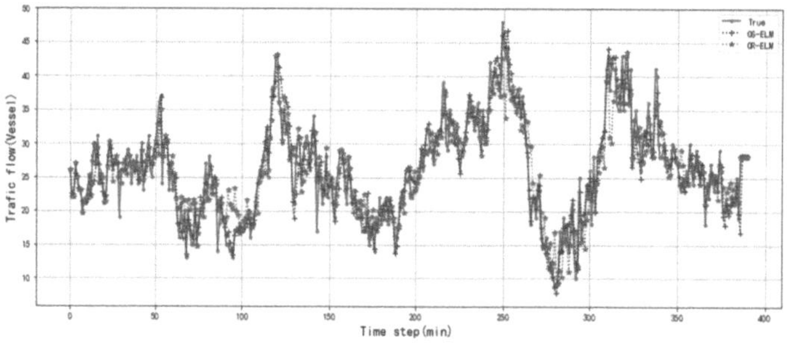

Figure 3. Different traffic flow predictions.

The performance of three prediction models, ELM, OR-ELM, and OS-ELM, is compared in Figure 4. It can be observed that ELM exhibits a relatively higher overall error when compared to the OR-ELM and OS-ELM models. Conversely, both OR-ELM and OS-ELM demonstrate lower overall prediction errors with a similar trend depicted in the Figure 4.

Figure 4 compares the errors of OR-ELM and OS-ELM. The figure illustrates that compared with OS-ELM, OR-ELM has a more stable overall error prediction change and can also show accurate prediction results in the face of sudden changes. We compared the MAE, RMSE and R^2 of the three models in Table 3. The figure illustrates that the OR-ELM model has a decrease in RMSE and MAE. Compared with the OS-ELM model, the MAE has increased by 0.18, about 10.1%, and the RMSE has increased by 0.87, about 14.7%, showing good predictive performance.

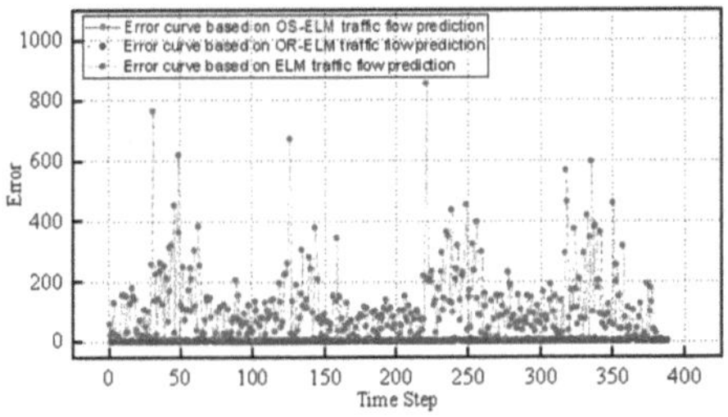

Figure 4. Comparison of OS-ELM, ELM and OR-ELM prediction.

Table 3. Comparison of error results.

Model	MAE	RMSE	R^2
ELM	1.906	5.952	0.965
OS-ELM	1.774	5.903	0.892
OR-ELM	1.592	4.934	0.901

5 CONCLUSION

In this paper, the research focuses on predicting offshore vessel traffic by using a machine learning algorithm. However, during the training process, the network parameters are set randomly and cannot adapt to real-time data changes, leading to poor prediction performance. To address this issue, we propose the OR-ELM model, which continuously and recursively updates the weight matrix of the network using recursive least squares to improve overall prediction performance. Experimental results demonstrate that the proposed method achieves high prediction accuracy with faster convergence speed and smaller model loss. This algorithm can serve as a valuable reference for the maritime sector. In future research, we aim to incorporate factors such as maritime characteristic behaviors affecting vessel traffic flow into our algorithm to further enhance the overall prediction performance of the network.

AUTHOR'S BIOGRAPHY

Guanya Wang: received the B.S. degree in Information Countermeasure Technology from the School of Equipment Engineering, Shenyang Ligong University, Shenyang, China, in 2018. He is currently pursuing the M.S. degree at the Department of Information Science and Technology, Dalian Maritime University, Dalian, China, under the supervision of Prof. B. Lin. His main research interests include maritime communication networks and network deployment optimization. E-mail: 982236730@qq.com

Bin Lin: (Senior Member, IEEE) received the B.S. and M.S. degrees from Dalian Maritime University, Dalian, China, in 1999 and 2003, respectively, and the Ph.D. degree from the Broadband Communications Research Group, Department of Electrical and Computer Engineering, University of Waterloo, Waterloo, ON, Canada, in 2009.

She is currently a Full Professor and the Dean of Communication Engineering Department, School of Information Science and Technology, Dalian Maritime University. She has been a Visiting Scholar with George Washington University, Washington, DC, USA, from 2015 to 2016. Her current research interests include wireless communications, network dimensioning and optimization, resource allocation, artificial intelligence, maritime communication networks, edge/cloud computing, wireless sensor networks, and Internet of Things.

Dr. Lin is an Associate Editor of IEEE TRANSACTION ON VEHICULAR TECHNOLOGY and *IET Communications.*

Minyuan Zhang: received the B.S. degree in Communication Engineering from Kexin College, Hebei University of Engineering, Handan, China, in 2019, and the M.S. degree in Information and Communication Engineering from Liaoning University of Technology, Jinzhou, China, in 2023. He is currently working toward the Ph.D. degree at the Department of Information Science and Technology, Dalian Maritime University, Dalian, China, under the supervision of Prof. B. Lin. His main research topics include maritime communication networks, deep learning-based communications and semantic communications. E-mail: zhangmiruan@163.com

REFERENCES

[1] Sahdev S.L., Chotai N. and Banerjee S. (2023). A Study on Analyzing the Importance of Factors Impacting Shipping in Maritime Industry in the Context of the Indian Economy-An AHP Approach [C], *2023 4th International Conference on Computation, Automation and Knowledge Management (ICCAKM)*. IEEE, 01–05.

[2] Bai X., Zuo Y., Li T., *et al.* (2021). A systemic method of traffic flow velocity prediction in narrow waterways using ais data[C], *2021 International Conference on Security, Pattern Analysis, and Cybernetics (SPAC)*. IEEE, 308–312.

[3] Xiao H., Xiao J., Deng X., *et al.* (2021). Traffic flow prediction based on traffic and meteorological data fusion in non-stationary environments[C], *2021 International Conference on Electronic Information Technology and Smart Agriculture (ICEITSA)*. IEEE, 154–158.

[4] Xiaopeng T., Xu C., Lingzhi S., *et al.* (2015). Vessel trajectory prediction in curving channel of inland river[C], *2015 International Conference on Transportation Information and Safety (ICTIS)*. IEEE, 706–714.

[5] Li X., Xinghua Z. and Shi H.J. (2010). Short-term traffic flow time series forecasting based on grey interval forecasts method[C], *2010 International Conference on Educational and Information Technology*. IEEE, 1: V1-384–V1-389.

[6] Hang W., Xu M., Chen X., *et al.* (2015). Vessel traffic flow prediction model based on complex network [C], *2015 International Conference on Transportation Information and Safety (ICTIS)*. IEEE, 473–476.

[7] Suyono H., Santoso H., Hasanah R.N., *et al.* (2018). Prediction of solar radiation intensity using extreme learning machine[J]. *Indonesian Journal of Electrical Engineering and Computer Science*, 12(2): 691–698.

[8] Kim K.I. and Lee K.M. (2018). Deep learning-based caution area traffic prediction with automatic identification system sensor data[J]. *Sensors*, 18(9): 3172.

[9] Wang X., Li J. and Zhang T. (2019), A machine-learning model for zonal ship flow prediction using ais data: A case study in the south atlantic states region[J]. *Journal of Marine Science and Engineering*, 7 (12): 463.

[10] Huang G.B., Zhu Q.Y. and Siew C.K., (2004). Extreme learning machine: a new learning scheme of feedforward neural networks[C], *2004 IEEE international joint conference on neural networks* (IEEE Cat. No. 04CH37541). IEEE, 2: 985–990.

[11] Peng X., Zheng W., Zhang D., *et al.* (2017), A novel probabilistic wind speed forecasting based on combination of the adaptive ensemble of on-line sequential ORELM (Outlier Robust Extreme Learning Machine) and TVMCF (time-varying mixture copula function)[J]. *Energy Conversion and Management*, 138: 587–602.

[12] Abrar S., Loo C.K., Kubota N. and Tahir G.A., (2020), "A Personalised Blood Pressure Prediction System using Gaussian Mixture Regression and Online Recurrent Extreme Learning Machine, *2020 International Symposium on Community-centric Systems (CcS)*, Tokyo, Japan, pp. 1–7.

*Proceedings of the International Association of Institutes of Navigation World Congress 2024 –
Chen et al. (Eds)*

Data-driven approach of harbour emergency traffic networks

Di Cui
Safety and Emergency Protection Division, China Waterborne Transportation Research Institute, Beijing, China

ABSTRACT: Seaport road transit networks are among the most congested traffic areas, particularly under emergencies conditions. For example, during the Tianjin Port explosion on August 12, 2015, and the Ningbo seaport closure due to the COVID-19 pandemic in 2019, the criticality of this infrastructure was highlighted. This paper proposes a novel real-time traffic estimation and prediction system based on data-driven approaches. It simulates the movement of stretch trucks within the dynamic harbor district. Additionally, enhanced microscopic emergency traffic flow models for the harbor district roads are developed, including the static queuing model and the vehicle emergency movement model.

Keywords: Tianjin seaport 8·12 explosion, Emergency traffic networks, Microscopic traffic model, Emergency traffic control, Seaport emergency traffic simulation

1 INTRODUCTION

With the worldwide rapid growth in the number of seaport vehicles, a series of social and environmental issues have emerged, along with harbour emergency traffic congestion, especially in the large-scale major seaport. Consequently, several traffic management technologies have been proposed to tackle these problems, including those triggered by emergencies such as the Tianjin Port 8·12 explosion and the closure of the Ningbo seaport during the COVID-19 pandemic in 2019.

Unfortunately, most of the proposed traffic management strategies come with certain limitations. To overcome these drawbacks, supplementary traffic control strategies have been proposed [10,11]. However, the majority of these strategies are based on theoretical approaches, making them difficult to implement in large - scale real - world scenarios. The Max-Min Variable Speed Limits (MMVSL) is a mainstream control method that is widely used in dynamic traffic systems. MMVSL messages can effectively display real-time traffic conditions, which in turn affect the route-choice behavior of road users. Additionally, the MMVSL strategy has proven effective in alleviating traffic congestion [9]. Studies have demonstrated that emergency traffic jams on harbor emergency highways can be reduced by 20% to 30% [1,2]. Furthermore, the MMVSL model has the potential to mitigate the occurence of rear-end accidents [13].

The content of traffic optimization varies with different situations. Shi *et al.* [2] proposed an optimization approach for one-way traffic organization in the urban micro-circulation transportation network, with road saturation as a constraint. Long *et al.* [3] proposed different models and algorithms for the same problem. Dussutour *et al.* [4] delved into the optimization behavior of ants, and their research findings were published in *Nature*. Jia [5] systematically studied the problem of traffic organization optimization on urban main roads. Xu *et al.* [6] conducted traffic organization optimization using simulation methods. Yao *et al.* [7] proposed a more practical information-based method . Yang *et al.* [8,9] proposed regional traffic organization methods integrating time periods. Sun *et al.* [10,11] also

DOI: 10.1201/9781003624424-3

provided some valuable suggestions for urban passenger transport hubs through simulation methods. Zhengyan [12] addressed some technical methods for emergency evacuation traffic organization. Recently, Carlson *et al.* [2010a; 2010b] proposed a mainstream traffic flow control method on freeways by applying MMVSL [6,7]. Carlson(2011) designed a simple yet highly efficient controller [5]. Some studies indicate that the cooperative Multi-Modal Variable Speed Limit Strategy (MMVSLS) has the potential to contribute to traffic flow harmonization [8]. Hegyi [9] studied the benefits of implementing the MMVSL.

Based on the analysis of the aforementioned documents, most studies focus on traffic flow within the realm of traffic organization. To ensure the rationality of traffic organization, it is essential to evaluate both its strengths and limitations. In light of the drawbacks of static methods, this paper puts forward a dynamic approach. To establish a scientific traffic management system, it is necessary to analyze the relationship between traffic demand and traffic supply. By forecasting the interaction of traffic demand, it becomes possible to simulate the response behavior of drivers, and analyze traffic control schemes. The self-developed software system, the Developed Intelligent Traffic System (DITS), was utilized to perform specific tasks in this regard

2 THE MODELS AND ALGORITHM

Several studies have demonstrated that the cooperative MMVSL (Multi-Modal Variable Speed Limit) strategy shows great potential in mitigating the impacts of traffic congestion [8]. Hegyi [9] delved into the advantages of implementing MMVSL for the real-time reduction of crash risks. However, the control of MMVSL Model constitutes a nonlinear and time-varying system. In this study, a self-developed dynamic traffic simulation software is adopted. Additionally, this research focuses on enhancing the emergency response capacity. This simulation-based approach represents a component of the Developed Intelligent Traffic System (DITS), which is designed to estimate and predict traffic conditions in real-time.

Some studies have shown that the cooperative MMVSL (Variable speed limits strategy) has the potential to reduce the impacts of traffic jams [8]. Unfortunately, the MMVSL control system for the harbour expressway mainline is a nonlinear and time-varying one. It ramains a challenge to describe the basic rules of the MMVSL control strategy in a deterministic model. The purpose is to examine the effectiveness of the MMVSL strategy. The network under consideration contains an overpass and some nearby harbour road sections. This is a simulation-based real-time module within the Developed Intelligent Traffic System (DITS). DITS is capable of estimating and predicting traffic conditions. with the aim of realizing the short-term transportation planning for special events such as fires and exploration-related activities. Moreover, it can provide the information of averaged speed, averaged density and averaged flow.

(1) Anisotropic Supply Simulator: The Developed Intelligent Traffic System (DITS) employs two types of simulation tools: a demand simulator and a supply simulator. The demand simulator, functioning as a microscopic traffic simulation tool, primarily focuse on estimating time-dependent origin-destination (OD) flows.
(2) Availability of Harbour Emergency Floating Traffic Data (HEFCD): The Developed Intelligent Traffic System (DITS) utilizes standard loop detector data to calibrate model parameters [3].
(3) Modeling Mixed Traffic Flow in a Mixed-Traffic System, including Trucks, Ambulances, Container Lorries, Emergency Evacuation Buses, and Fire Engines.
(4) The DITS system is expected to incorporate a wide range of supplementary functionalities, including harbor emergency transport simulation, parallel computing technology, advanced traffic management strategies, and harbor emergency event detection capabilities, among others.

3 THE DATA-DRIVEN ANALYSIS

The data-driven analysis primarily revolves around the development of advanced computational techniques. Firstly, these approaches are based on the assumption that all physical systems are well-characterized. Secondly, these models tend to be computationally inefficient, demanding significant computational resources. Consequently, data-driven methods have emerged as a novel paradigm in the context of harbor emergency management.

The general concepts and features of any data-driven approach are summarized in Figure 1. In recent years, remarkable advancements have been made in computational intelligence techniques, including artificial neural networks, probabilistic reasoning, fuzzy logic, genetic algorithms, and chaos theories. These cutting-edge techniques significantly contributed to the evolution of data-driven methodologies.

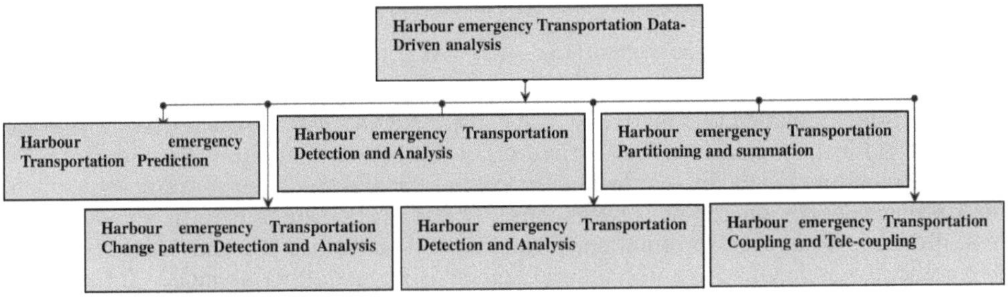

Figure 1. Harbour emergency transportation data-driven analysis.

4 THE MICROSCOPIC TRAFFIC SIMULATION MODELS AND ALGORITHM

A variety of models have been developed within the Developed Intelligent Traffic System (DITS), including the harbor emergency traffic road network model, traveler behavior model, path generation algorithm, and capacity models. Among these, the microscopic traffic simulation model and the dynamic OD estimation model are regarded as the most critical components. The proposed simulation process has been designed with high efficiency in mind. The subsequent discussion will focus on these models and algorithms.

4.1 *The microscopic traffic simulation models*

The microscopic simulation process involves two key models. One of them is the speed-density relationship model, which needs data on the flow, speed, or density of harbor emergency roads to more accurately reproduce the current road traffic conditions. Typically, one or more parameters must be calibrated in the following function:

$$V_u = \max\left\{V_{\min,} V_{\max}\left\{1 - \left\{\frac{k - k_{\max}}{k_{jam} - R}\right\}^b\right\}^a\right\} \tag{1}$$

Where the parameter R is the safety risk.

Where V is the vehicle speed, k is the density, k_{jam} is the jam density, $k_{\max}$ is the critical density, $V_{\min}$ and $V_{\max}$are the minimum and maximum speed designed. In this paper, traffic simulation technology is utilized on multiple occasions. The minimum speed is set to prevent the simulation system from failing to initialize. To achieve greater realism, the scenario can be described as follows: the upstream of a long road maintains a constant speed, while the downstream adheres to the speed-density relationship. The mathematical formulation is presented below:

$$V_u = \begin{cases} V_f \mathrm{k} <= \mathrm{k}_{max} \\ \max\left\{ V_{\min,} V_{\max}\left\{ 1 - \left\{ \frac{k - k_{\max}}{k_{jam} - R} \right\}^b \right\}^a \right\} \mathrm{k} > k_{\max} \end{cases} \tag{2}$$

where R and V are parameters. Otherwise, the R is the safety risk.

4.2 *The estimation method of dynamic OD demand*

The real-time OD estimation model is needed. In this section, the generalized least square estimation model is used to resolve the problem.

$$\check{x}_h = argmin\left[\left(x_h - x_h^a \right)^T \right] W_h^{-1} \left(x_h - x_h^a \right) \tag{3}$$

where $\check{x}_h$ is the estimated OD matrix; x_h^a is the historical OD matrix which refers to a priori values of x_h for the time period h; n_l is the section number with monitoring equipment; $yh \in R^{nl}$ is the monitored value on section n_l in time period h; $a_h^p$is the dynamic traffic assignment matrix which defines a mapping between the OD vector x_p in p and the traffic volume vector in h; if there is no a priori information, $W_h$and R_h can be defined as a unit matrix.

Another dynamic OD demand estimation model is Kalman filtering model. It includes a state and space model. The basic idea was proposed by Ashok and Ben-Akiva in 1933. Unlike other approaches in the literature, the approach is based on deviations from historical values. For a network with OD pairs and link counts, let us assume the following notation: xh+1 is the matrix representing the number of vehicles between each OD pair that depart from their origins during time interval h and h is the corresponding historical estimated matrix. The transition equation can be expressed as

$$\begin{aligned} \check{x}_h = argmin\Big[& \left(x_h - x_h^a \right)^T w_h^{-1} \left(x^h - x_h^a \right) \\ + & \left(y_h - \sum_{p=h-p'}^{h-1} a_h^p \check{x}_p - a_h^h x_h \right)^T . R_h^{-1} \left(y_h - \sum_{p=h-p'}^{h-1} a_h^p \check{x}_p - a_h^h x_h \right) \Big] \end{aligned} \tag{4}$$

Where $\check{x}_h$ is the estimated OD matrix; x_h^a is the historical OD matrix which refers to a priori values of x_h in the time period h; n_l is the section number with monitoring equipment; R^{nl} is the monitored value on section in time period h; $a_h^p (p = h, h-1, \ldots, h-p)$ is the dynamic traffic assignment matrix which defines a mapping between the OD vector x_p in p and the traffic volume vector in h. The W_h and R_h can be defined as a unit matrix.

4.3 *Simulation process*

The designed simulation process plays a crucial role in ensuring simulation accuracy. The simulation time is divided into multiple update phases of equal length. Each update phase is further subdivided into several advance intervals of equal length. During these advance intervals, vehicles are reclocated to their new positions. The update phase serves to update the traffic dynamics parameters （e.g., densities, speeds, and flow） used in each simulation cycle. If the simulation horizon is T, without loss of generality, we assume that the time horizon starts at time $t = 0$ and finishes at time $t = T$. Update phase is Δt_{update}; Advance phase is $\Delta t_{advance}$. So,

$$T = k_u \Delta tupdate \tag{5}$$

$$\Delta tupdate = k_A \Delta t_{advanc} \tag{6}$$

k_u and k_A are the positive integers; the specific simulation process is as follows.

Step 1.

Describe the network and the simulation horizon. Document details of all vehicles at the start of the interval, including their locations on the network, destinations, and current routes. Account for all vehicles in the network at time t= 0 and the loaded positions.

Step 2.

Set up network nodes, sections, and other parameters; execute the following program.

Step 2.1. Update the definition of each segment based on the incident description. Refresh the output capacities of all segments.

Step 2.2. For each segment, compute the upstream-speed and the density.

Step 2.3. Calculate the speed of the downstream segment.

Step 2.4. Execute the advanced phase for each vehicle.

Step 2.5. i= 0.

Step 3. Terminate the algorithm. $i = k_A$.

4.4 *The case*

The selected data consist of one-day records from 2022. First, all invalid data are systematically removed. Although the dataset includes various types of vehicles, such as buses and trucks, these are uniformly converted into equivalent standard car units for analysis. The model parameters are defined as follows: v_{free}=120 km/h, ρ_{jam}=0.1150, α is 1.9420, β is 0.5040, and the AMS length is 150 m.

The total simulation period is 15 hours and is divided into 60 time periods with each lasting 15 minutes. If there is no any network incident, the simulation can output the average density, average flow and average speed for every segment, as shown in Figures 2 and 3.

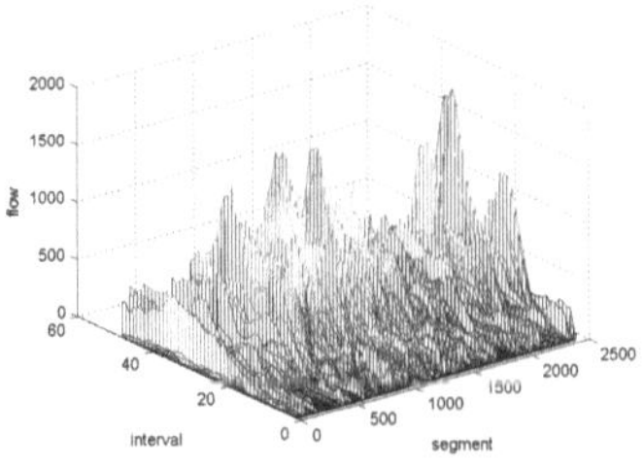

Figure 2. The flow on segment per interval.

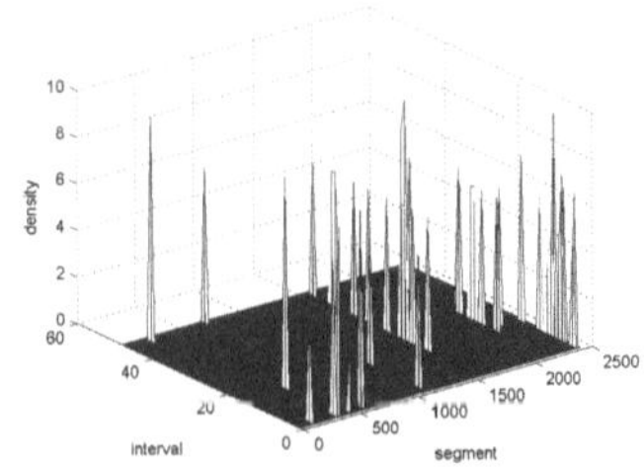

Figure 3. The density on segment per interval.

Regarding the effectiveness of information guidance within the transportation network, we analyzed the average flow, average density, and average speed of each road segment when dynamic guidance information was generated. The spatio-temporal maps of traffic flows are presented in Figures 4 to 7.

To assess the effectiveness of information guidance within the transportation network of the port area, we analyzed the average traffic flow, density, and speed of each road segment in two distinct scenarios. The spatio-temporal maps of traffic flows are depicted in Figures 4 to 7, showcasing the average flow, density, and speed for each segment.

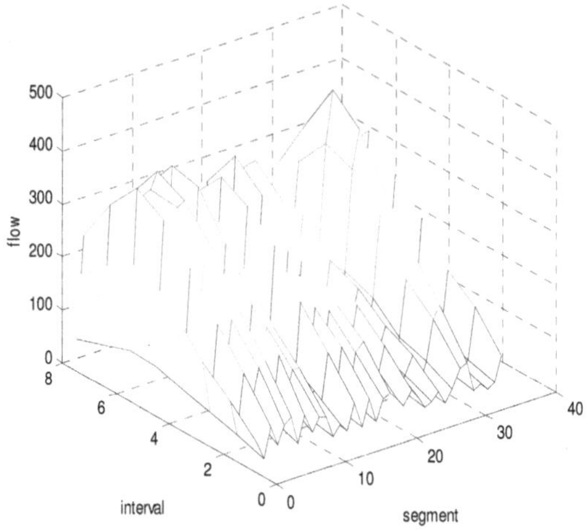

Figure 4. Flow during the pre-induction segment.

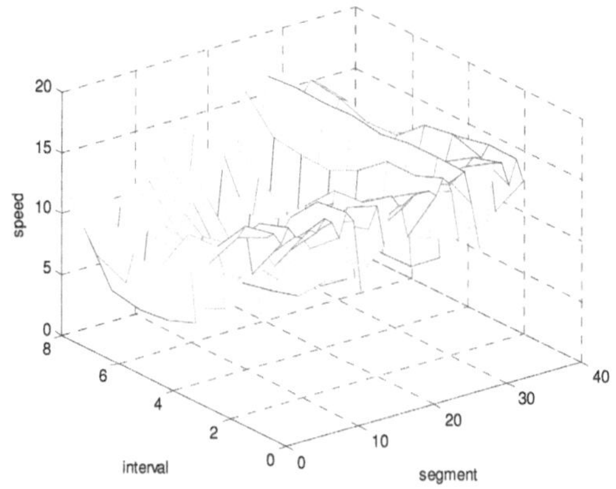

Figure 5. Speed during the pre-induction segment.

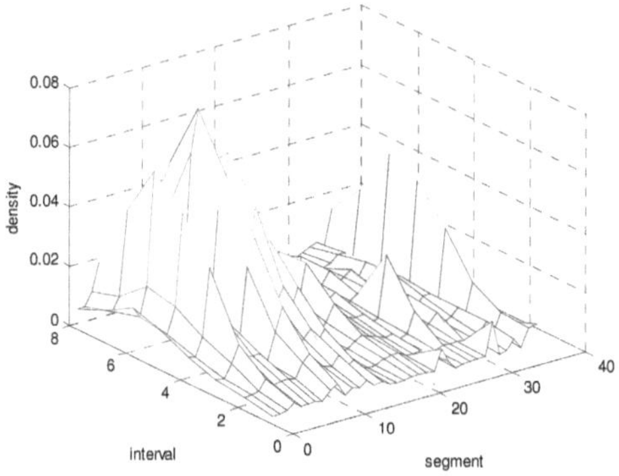

Figure 6. Density of segmental periods after-induction.

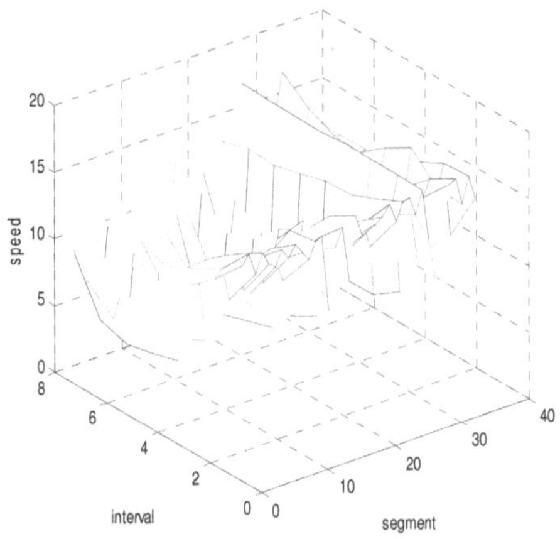

Figure 7. Speed during the segment after induction.

This paper focuses on the mixed traffic flow in harbor road areas, including trucks and tank trucks, and explores methods to curb the spread of traffic congestion in harbor regions. To alleviate traffic congestion, a three-dimensional traffic management guidance system is developed.

5 CONCLUSION

From the above spatial-temporal graph, it can be seen that the density of harbour traffic vehicles in the harbour area varies. It shows the different levels of smoothness at each segment during different time periods, indicating that the induced information causes some vehicles to change their original paths. Due to the knowledge of harbour congested road network, some travelers choose a relatively smooth route for travelling. When comparing several sections of the intersection, with dynamic information guidance in place, the speed of vehicles in the port area is significantly improved, and the overall traffic flow is relatively higher. This indicates that the traffic situation in the harbour road area has significantly improved. That is to say, dynamic information induction can effectively divert traffic jam in the harbour area, effectively preventing the occurrence of traffic congestion in the harbour road, and facilitate the evacuation of vehicles for personnel in the seaport area.

ACKNOWLEDGMENTS

This work is supported by the National Natural Science Foundation of China (Grant nos. 72242102,Safe operation and resilience recovery of urban systems in the face of extreme disasters and 72271021, Collaborative optimization of intelligent connected vehicle trajectories and smart intersection signals in a new hybrid transportation environment).

REFERENCES

[1] Zhai Z., (2004), *Road traffic Organization Optimization*, China Communication Press, Beijing, China.

[2] Shi F., Huang E.-H., Chen Q., and Wang Y.-Z., (2009), "Optimization of one-way traffic organization for urban micro-circulation transportation network," *Journal of Transportation Systems Engineering and Information Technology*, vol. 9, no. 4, pp. 30–35.

[3] Long D.-F., Shi F., and Wang Y.-Z., (2010), "One-way traffic organization based on traffic load and road equity," *Journal of Transportation Systems Engineering and Information Technology*, vol. 10, no. 6, pp. 109–114.

[4] Dussutour A., Fourcassie V., Heibing D., and Deneubourg J.-L., (2004), "Optimal traffic organization in ants under crowded conditions," *Nature*, vol. 428, no. 6978, pp. 70–73.

[5] Jia B., (2010), *TTe study on traffic organization and optimization of urban arterial road* [M.S. thesis], Chang'an University, Chang'an, China.

[6] Xu H., Tan Z., Long K., and Zhong Z., (2009), "Operational optimization method and simulation of bottleneck section based on urban arterial road," *Computer and Communications*, supplement 1, pp. 4–8.

[7] Yao G.-Z., Wu K.-Q., and Wu H.-J., (2011), "Empirical analysis of the traffic Micro-circulation organization in heavy traffic networks," *Journal of Transportation Systems Engineering and Information Technology*, vol. 11, no. 1, pp. 160–166.

[8] Yang Y.-Z., Mao B.-H., Zhang X.-J., Huang R., and Feng X.-J., (2010), "Study on traffic micro-circulation organization in heavy traffic networks based on route guidance," *Journal of System Simulation*, vol. 22, no. 7, pp. 1580–1584.

[9] Mao B. H., Sun Z. Z., Jia S. P. *et al.*, (2007), *Regional traffic Organizational Optimization with Practice*, China Communications Press, Beijing, China.

[10] Sun L. and Ren F., (2007), "Optimize of the transfer organization in urban transportation terminal," *Road Traffic and Safety*, vol. 7, no. 3, pp. 18–21 (Chinese).

[11] Sun L.-S., Yao L.-Y., Liu B.-H., Wu Y.-Y., and Rong J., (2012), "traffic optimization of transportation terminal based on dynamic simulation technology," *Journal of Beijing University of Technology*, vol. 38, no. 4, pp. 570–574.

[12] Zhengyan W., (2011), *Study on the methods of traffic organization optimization for emergency evacuation* [Ph.D. thesis], University of Jilin.

[13] Wang J., Kim Y.H., He X. and Peeta S., (2017). Analytical model for information flow propagation wave under an information relay control strategy in a congested vehicle-to-vehicle communication environment, *Transportation Research Part C*, 23: 738–757.

[14] Xie D.F., Zhao X.M. and He Z.B., (2019). Heterogeneous traffic mixing regular and connected vehicles: modeling and stabilization. *IEEE Transactions on Intelligent Transportation Systems*, 20(6): 2060–2071.

[15] Xie D.F., Gao Z.Y., Zhao X.M., David Z.W. Wang, (2012). Cellular automaton modeling of the interaction between vehicles and pedestrians at signalized crosswalk, *Journal of Transportation Engineering*, 138(12): 1442–1452.

[16] Xie D.F., Gao Z.Y., Zhao X.M., David Z.W. Wang, (2012). Agitated behavior and elastic characteristics of pedestrians in an alternative floor field model for pedestrian dynamics, *Physica A*, 391(7): 2390–2400.

[17] Zhou G.J., Xie D.F., Zhao X.M., Lu C., Zhou C. D., Zhang C. Y. (2024). Collaborative optimization of vehicle scheduling and speed regulation for autonomous customized bus systems. *Applied Mathematical Modelling*, 128: 410–430.

[18] Xie D.F., Yu Y.P., Zhou G.J., Zhao X.M. and Chen Y.J. (2023). Collaborative optimization of electric bus line scheduling with multiple charging modes. *Transportation Research Part D*, 114: 103551.

Proceedings of the International Association of Institutes of Navigation World Congress 2024 – Chen et al. (Eds)

Research on the collision governance platform for merchant and fishing vessels based on E-navigation platform

C. Chen
Zhejiang Ocean University, Zhoushan, Zhejiang, China
Zhejiang E-navigation Zhoushan Municipal Doctoral Innovation Station, Zhoushan, Zhejiang, China

Y.H. Luo
Shandong Maritime Safety Administration, Qingdao, Shandong, China

D.R. Wu
Zhejiang E-navigation Information Technology Co., Ltd, Zhoushan, Zhejiang, China

ABSTRACT: With the rapid development of the shipping industry, collisions between merchant and fishing vessels frequently occur, posing a serious threat to maritime safety and the lives and property of those involved. This paper aims to research and develop a governance platform for collisions between merchant and fishing vessels based on the E-navigation Platform, utilizing innovative technologies to address these challenges. The platform consists of a "merchant vessel version" smart end installed on merchant ships, a "fishing vessel version" smart end installed on fishing vessels, the E-navigation platform, and a management end used by maritime and fisheries authorities.It can issue collision warnings in advance, facilitate coordinated collision avoidance through one-on-one audio and video channels during incidents, and support one-click distress calls and coordinated rescue efforts organized by the platform after incidents, thus fully addressing the visibility and communication issues during encounters between merchant and fishing vessels. By equipping the vessels with smart ends connected to a three-in-one communication module (Beidou RNSS positioning + 4G communication + AIS signal reception) and utilizing 4G/5G communication technology and the Beidou navigation system, the platform enables real-time position sharing and communication between vessels up to 20 nautical miles offshore. Leveraging its robust data processing and information dissemination capabilities, the platform offers a new technical solution for managing collisions between merchant and fishing vessels, aiming to effectively reduce such incidents, enhance maritime navigation safety, and ensure maritime traffic safety.

1 INTRODUCTION

With the rapid development of China's shipping industry, maritime traffic volume has been continuously increasing. A large number of merchant and fishing vessels navigate and operate in the same waters, frequently crossing paths at sea, leading to frequent collision accidents between these vessels. Such collisions often result in significant casualties and have become one of the largest sources of maritime risk in coastal areas(Feng 2022), severely threatening maritime traffic safety as well as the lives and property of the crew.

According to statistics on maritime traffic accidents, compared to the "Twelfth Five-Year Plan" period, the number of maritime traffic accidents and the number of deaths and missing persons involving Chinese-flagged vessels during the "Thirteenth Five-Year Plan" period decreased by 34.4% and 25.5%, respectively. However, the number of accidents and the

DOI: 10.1201/9781003624424-4

number of deaths and missing persons involving both maritime transport vessels and fishing vessels increased by 21.6% and 36.2%, respectively, during the same period. Additionally, the proportion of such accidents in the total maritime traffic accidents nationwide increased from 8.2% to 8.5%, and the proportion of deaths and missing persons rose from 18% to 25.3%. This shows that the collision accidents and losses that have occurred between merchant vessels and fishing boats at sea in the past 10 years are contrary to the overall improvement trend of maritime traffic safety and show a trend of reverse growth (Yang 2022). In particular, in 2020, there were three major collisions between merchant and fishing vessels, resulting in 32 deaths or missing persons.

Collisions between merchant and fishing vessels have always been a "long-standing" problem in maritime safety. According to extensive statistical analysis by the Ministry of Agriculture and Rural Affairs, the large number of fishing vessels and concentrated operating areas, the neglect of collision avoidance rules by fishing vessel crews, unpredictable operating courses, and poor communication between merchant and fishing vessels are all significant causes of such collisions (Du *et al.* 2023).

This study focuses on the entire process of collisions between merchant and fishing vessels, aiming to achieve "effective prevention before the incident, efficient communication during the incident, and timely rescue after the incident." Utilizing the E-navigation platform based on internet cloud computing and artificial intelligence technology, the study proposes a solution to comprehensively address the visibility and communication challenges during encounters between merchant and fishing vessels. This includes pushing traffic information and collision warnings before the incident, ensuring effective communication through one-on-one audio-video channels during the incident to achieve coordinated collision avoidance, and supporting one-click distress calls and organized coordinated rescue efforts by the platform when the incident happens. These measures are expected to significantly enhance maritime navigation safety, reduce the occurrence of collisions between merchant and fishing vessels, and protect the lives and property of the crew.

2 DEVELOPMENT STATUS OF INTERNATIONAL PLATFORM-LEVEL VESSEL COLLISION GOVERNANCE PLATFORMS

Internationally, many countries and regions have developed and implemented various platform-level vessel safety monitoring solutions to enhance maritime traffic safety. For example, Europe's SafeSeaNet and Sea Traffic Management (STM) systems have achieved significant success in vessel safety monitoring. SafeSeaNet is a centralized information exchange platform that integrates maritime traffic information from member states, primarily focusing on vessel safety and security information. STM, on the other hand, focuses on voyage plan sharing and dynamic traffic management, enhancing navigation efficiency and safety through real-time data exchange and optimized routing.

The United States' National Automatic Identification System (NAIS) integrates AIS data, radar data, and satellite monitoring to achieve real-time tracking and management of vessel activities, providing collision warnings and incident reporting functionalities. Other global vessel tracking and monitoring platforms such as MarineTraffic, BigOceanData, and FleetMon also play crucial roles in enhancing vessel navigation safety by utilizing AIS data and advanced technologies.

However, these platforms are primarily designed for merchant vessels. While some features and applications can be applied to fishing vessels, they often lack considerations specific to the needs of fishing vessels and do not provide targeted measures for visibility and communication between merchant and fishing vessels. Therefore, it is essential to develop a platform specifically designed for managing collisions between merchant and fishing vessels, addressing their visibility and communication issues, and improving the navigation safety for both types of vessels.

3 MAIN CAUSES OF MERCHANT AND FISHING VESSEL COLLISIONS IN CHINA AND GOVERNMENT GOVERNANCE

3.1 *Main causes of merchant and fishing vessel collisions in China*

China possesses an extensive coastline and abundant marine resources, leading to exceptionally busy maritime economic activities in coastal areas, with a large number of merchant vessels such as cargo ships, oil tankers, and container ships arriving and departing ports daily. These merchant vessels are large, fast, and have significant inertia, requiring substantial safe water areas for navigation. On the other hand, China is also the world's largest fishing nation, with a vast number of fishing vessels, including trawlers, seine netters, and other fishing boats, distributed in the Bohai Sea, Yellow Sea, East China Sea, and South China Sea. Fishing vessels are smaller, slower, often operate within fixed fishing zones, and are relatively maneuverable, but also prone to disorderly operations.

High-incidence areas for merchant and fishing vessel collisions are mainly concentrated in offshore waters with intensive fishing activities and major busy shipping lanes. The main issues include:

(1) Overlapping Navigation Areas. Merchant and fishing vessels often navigate and operate in the same waters, especially in coastal fishing areas and major shipping lanes, leading to overlapping navigation areas and increased collision risks.
(2) Unclear Boundaries Between Navigation Routes and Fishing Zones(Du *et al.* 2023). Important navigation areas are generally planned for merchant or passenger ships, with few dedicated routes for fishing vessels. The boundaries between navigation routes and fishing zones are often not clearly defined.
(3) Disorderly Use of Fishing Gear Positioning Instruments. Fishermen's extensive use of positioning instruments and issues like arbitrary deployment and unauthorized encoding lead to merchant vessels' communication and navigation equipment becoming overwhelmed by numerous fishing vessel markers, making it difficult to distinguish between fishing vessels and positioning instruments, significantly increasing collision risks.
(4) Inadequate Communication Equipment. Some fishing vessels lack proper communication equipment or do not use it correctly, resulting in ineffective communication with merchant vessels during emergencies. Additionally, language barriers and insufficient operational skills hinder effective communication(Liu & Han 2022).
(5) Weak Collision Avoidance Awareness. Some fishing vessel crew members lack strong safety awareness and necessary collision avoidance knowledge and skills. They may fail to detect and avoid merchant vessels in time, increasing collision risks. Some even deliberately cross in front of larger vessels, seeking to take the lead.
(6) Difficulty in Supervision. The complex maritime operating environment makes maritime supervision challenging. Despite existing laws, regulations, and management measures, actual implementation faces difficulties, unable to cover and effectively regulate the navigation behavior of all vessels.
(7) Insufficient Information Sharing. Existing management systems and platforms have shortcomings in data processing and information sharing, unable to provide accurate real-time navigation safety information, affecting coordination and collision avoidance between merchant and fishing vessels(Chen *et al.* 2023).
(8) Lack of Familiarity Between Merchant and Fishing Vessels. Both types of vessels are unfamiliar with each other's communication methods, vessel handling characteristics, and navigation operations, leading to a lack of mutual understanding and consensus on collision avoidance during crises.
(9) Need for Improved Platform Data Analysis and Information Push Capabilities. Existing platforms struggle to efficiently analyze actual data and provide effective warnings and decision support. In urgent situations, merchant and fishing vessels cannot timely access critical information, impacting collision avoidance and rescue efficiency.

3.2 *Government governance for merchant and fishing vessel collisions in China*

The Chinese government places great importance on governing merchant and fishing vessel collisions. In 2021, the Ministry of Transport, Ministry of Agriculture and Rural Affairs, and Ministry of Emergency Management jointly formulated a series of policies and measures, including the "Opinions on Strengthening Maritime Transport and Fishing Vessel Safety Risk Prevention and Control" and the "Implementation Plan for the 2021 Special Action on Merchant-Fishing Vessel Co-Governance." In 2022 and 2023, these ministries continued to issue the "2022 Special Action Plan on Merchant-Fishing Vessel Co-Governance" and the "2023 Special Action Plan on Merchant-Fishing Vessel Co-Governance," reinforcing joint governance actions against collisions. Since 2020, the total number of collisions and the number of deaths and missing persons have continuously decreased, and the frequent occurrence of major accidents has been initially curbed. In 2022, the number of general-level and above merchant and fishing vessel collisions and the number of deaths and missing persons in China's coastal areas decreased by 31.8% and 25.4%, respectively, compared to 2021(Lin & Li 2023).

Although special actions have shown some effectiveness, these results are achieved through short-term, high-intensity governance at the safety supervision level. They serve as a means to raise industry safety awareness and deter illegal activities through short-term campaigns, accumulating experience and seeking fundamental solutions. However, to achieve long-term effective prevention of merchant and fishing vessel collisions, relying solely on long-term governance is insufficient. It is necessary to establish a systematic, regularized management mechanism.

Building on the favorable safety shipping situation created by special governance activities, the development of a governance platform for merchant and fishing vessels based on the E-navigation platform has become an inevitable choice.. The E-navigation platform leverages internet cloud computing and artificial intelligence technologies to replace labor-intensive offline inspections and monitoring activities with advanced technology, providing dedicated navigation safety services for merchant and fishing vessels in the navigation area.

4 ARCHITECTURE OF THE MERCHANT AND FISHING VESSEL COLLISION GOVERNANCE PLATFORM

The frequent collisions between merchant and fishing vessels have posed a serious threat to maritime traffic safety. Traditional safety management methods have become inadequate in dealing with the complex maritime traffic environment and the increasing number of vessels. To effectively reduce the occurrence of collisions between merchant and fishing vessels and enhance the safety of vessel navigation, it is necessary to develop a platform-level solution that integrates multiple technological methods. The governance platform for merchant and fishing vessel collisions based on the E-navigation platform is designed to meet this need, providing comprehensive navigation safety assurance through advanced communication and data processing technologies.

The Merchant and Fishing Vessel Collision Governance Platform consists of three main components: the onboard smart end(merchant/fishing vessels), the E-navigation platform, and the management end(maritime/fishery authorities).

The onboard smart end is equipped with the corresponding version of the 'Hai e xing'App (merchant version/fishing version), which displays official chart data and dedicated fishing vessel routes. It collects nearby AIS information, facilitates one-on-one audio and video coordinated collision avoidance, and provides one-click alarm and detailed incident and distress reporting functions. The E-navigation platform is deployed on cloud servers, utilizing big data storage and processing technologies. It displays the overall operational status of merchant and fishing vessels within the area, analyzes traffic data using artificial intelligence algorithms, sends relevant weather notifications, traffic information, and navigation warnings to vessels in different areas, and issues collision warnings to merchant and fishing vessels at potential risk of collision.

The management end provides a dedicated web-based platform, refers to the rescue suggestions given by the E-Navigation end through artificial intelligence algorithms, and implements a hierarchical emergency response mechanism based on the maritime search and rescue plan. That is, the provincial maritime search and rescue center, the municipal maritime search and rescue center, and the county (city, district) level maritime search and rescue center can all receive and display alarm information. The county (city, district) level maritime search and rescue center will organize and carry out search and rescue according to the first response situation. The onboard end connects to a three-in-one communication module (Beidou+AIS+4G/5G), using the Beidou system for positioning, 4G/5G signal modules for connecting to shore-based networks, and AIS antenna module for receiving nearby AIS signals. The received signals are processed in the end application and uploaded to the platform for collision risk calculation and other analyses.

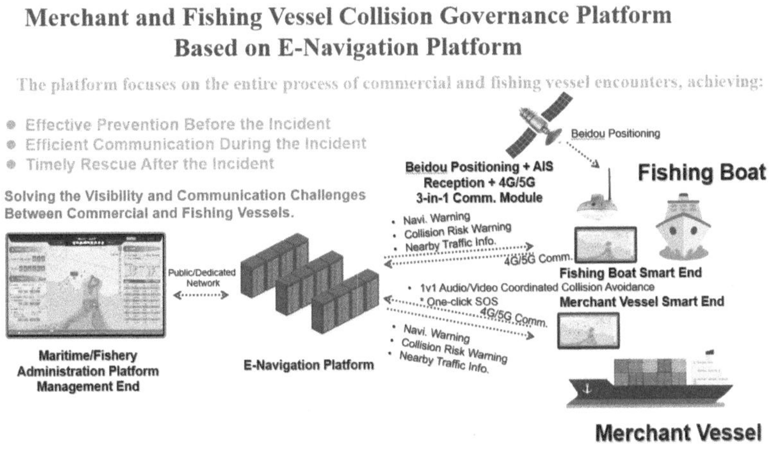

Figure 1. Architecture of the merchant and fishing vessel collision governance platform.

5 IMPLEMENTATION OF THE MERCHANT AND FISHING VESSEL COLLISION GOVERNANCE PLATFORM

Figure 2. Onboard smart end for merchant and fishing vessels.

5.1 *Installing the "Hai e Xing" app to build the onboard end*

The "Hai e Xing" app currently has nearly 2 million users distributed across coastal provinces in China. Considering the different usage needs and scenarios of merchant and fishing vessels during encounters, the "Hai e Xing" app used for the collision governance platform is divided into two versions: the "Merchant Version" and the "Fishing Version."

The "Merchant Version" focuses solely on monitoring fishing vessels (small vessels), providing collision warnings related only to fishing vessels, while collisions between merchant vessels are handled as before. Similarly, the "Fishing Version" focuses only on monitoring merchant vessels, providing warnings related only to merchant vessels, with collisions between fishing vessels handled as before.

The merchant and fishing versions of the app are designed specifically for merchant and fishing vessels, simplifying the user interface and operation processes. Crew members only need to pay attention to warning information relevant to their vessel type, reducing the risk of information overload and increasing operational convenience and efficiency. By focusing on specific collision risks, the system's burden is reduced, improving performance and response speed, thus providing faster and more accurate warnings to enhance overall safety. Additionally, separate processing of collision warnings for different types of ships avoids unnecessary information interference and redundancy, and reduces the possibility of misjudgment and misoperation.

The onboard smart end integrates a Beidou three-in-one communication module (Beidou RNSS positioning, 4G/5G communication, and AIS signal reception), enabling real-time position sharing and communication between merchant and fishing vessels within 20 nautical miles offshore. This design enables merchant and fishing vessels to maintain mutual visibility and communication over a larger range, thereby improving the safety and efficiency of navigation.

5.2 *Information prompts from the onboard smart end*

The onboard smart end prominently displays important navigation information related to merchant and fishing vessel collisions, such as collision danger zones published by management authorities, locations of incidents, and sunken ship points. In areas with dedicated fishing vessel routes, these can also be published through the platform. This helps crew members identify high-risk areas and take preventive measures in advance to avoid entering dangerous zones, reducing collision incidents. Clear markings of dedicated fishing routes help both merchant and fishing vessels plan their navigation paths reasonably, reducing the probability of navigation intersections and collision risks.

To achieve more precise positioning and information matching, the E-navigation platform merges user location data provided by the Beidou system with the Maritime Mobile Service Identity (MMSI) trajectory data of vessels. By matching user location trajectories with vessel trajectories, if multiple segments of user trajectories successfully match vessel trajectories, the user and vessel data from the Automatic Identification System (AIS) are merged, matching with the vessel names in the maritime database, and displaying the real vessel names directly on the search and rescue center command end. This technology significantly improves the accuracy and reliability of information, ensuring efficient emergency responses.

Additionally, upon receiving warning signals from the platform, the onboard smart end immediately pops up a warning window and sounds an alarm through high-power speakers. This ensures that warning information is quickly noticed by the crew and necessary measures are taken, further enhancing accident prevention and emergency response efficiency. Through multi-layered technological integration and innovation, the platform provides more comprehensive and precise navigation safety assurance.

5.3 *Providing various information push notifications related to merchant and fishing vessel navigation*

The platform can push weather notices, traffic information, and navigation warnings regularly, irregularly, or in real-time. When a pressing situation with collision risk is detected between merchant and fishing vessels, the platform sends warning messages to the relevant vessels. The onboard smart end receives the warning messages, pops up an alert window, and activates high-power speakers to ensure the warning information is heard by the on-duty crews, even if they are not at the station, prompting a quick return to the control station. This information push mechanism enhances crew response speed and warning capability to potential risks, optimizing navigation decisions, ensuring timely transmission of all important information, thus reducing collision incidents and improving overall navigation safety for merchant and fishing vessels.

5.4 *Building a coordinated collision avoidance communication channel through audio and video technology*

Typically, communication between merchant and fishing vessels relies on VHF calls, but establishing contact is often difficult due to incomplete or incorrect use of communication equipment on fishing vessels. The onboard smart end allows for one-on-one calls using the vessel's MMSI code, establishing a communication channel through "video + audio" or audio only, depending on the communication bandwidth. The system optimizes call quality to ensure effective coordinated collision avoidance.

5.5 *One-click alarm for quick and accurate maritime rescue location and response*

In the event of a collision, the onboard smart end supports a one-click alarm function, allowing crew members to report an alarm immediately and communicate with the platform management end via audio, video, or other forms to further report the situation. The management end can quickly locate the accident vessel based on the onboard end's navigation trajectory, refer to rescue suggestions provided by the E-navigation platform, and arrange rescue according to the rescue plan. Additionally, the onboard end features a Beidou short message communication function, ensuring one-click alarms and command dispatch information transmission when other communication methods are unavailable.

The onboard smart end receives positioning data via the built-in Beidou navigation positioning module, with a positioning frequency in milliseconds, receiving multiple positioning returns per second from approximately 15 Beidou satellites. In areas without public communication network coverage, the Beidou III short message communication function establishes a reliable alarm communication link, ensuring the timeliness and reliability of emergency rescue and command dispatch information transmission.

5.6 *False alarm prevention and convenient cancellation*

Verifying false alarms is a challenging issue in current maritime search and rescue operations. To minimize false alarms on the maritime mutual rescue assistance alarm platform, three preventive measures are adopted during the design process:

(1) Providing an alarm test mode during the alarm process to confirm the alarm link is smooth, avoiding false alarms triggered by user curiosity.
(2) Implementing a reconfirmation mechanism during real alarms, with a confirmation dialog box reminding users that false alarms will bear legal responsibility.

(3) Offering a cancel alarm button to cancel the alarm promptly after a false alarm. Additionally, platform users are authenticated with mobile phone numbers, allowing further verification of the distress situation by phone.

6 CONCLUSION

Addressing the issue of collisions between merchant and fishing vessels requires a concerted effort from both government management and technological solution providers to achieve the desired outcomes. Through effective government management and technical support, the E-navigation platform-based collision governance platform for merchant and fishing vessels can achieve effective prevention before incidents, efficient communication during incidents, and timely rescue after incidents, effectively addressing the visibility and communication challenges between merchant and fishing vessels.

As the interaction navigation data between merchant and fishing vessels accumulate, and with the iteration of artificial intelligence algorithms and the expansion of application scenarios, the future can bring even smarter, more instructive, and efficient navigation services. For instance, incorporating automatic translation functions (including dialect translation) into the one-on-one communication process, providing text prompts alongside voice communication, can completely solve the problem of language barriers. These innovations and improvements will further enhance the practicality and reliability of the collision governance platform, providing more comprehensive and efficient protection for the safe navigation of merchant and fishing vessels.

In the future, with continuous technological advancements and deeper applications, it is hoped that the collision governance platform for merchant and fishing vessels can be promoted globally, contributing to the improvement of international maritime traffic safety management level.

REFERENCES

Chen, G., Bai, J., Yu, Y.Q. and Chen, K. (2023). Construction ideas for the Ningde demonstration area to prevent merchant and fishing vessel collisions. *China Maritime* (09): 12–16.

Du, W.C., Wang, C., Ni, C. and Xiang, Y. (2023). Research on the co-governance of merchant and fishing vessels in Tongzhou Bay waters, Nantong City. *China Maritime* (10): 45–48.

Fishery Administration Department of Hebei Provincial Department of Agriculture and Rural Affairs. (2022). Hebei Province conducts the "2022 Special Action on Merchant-Fishing Co-Governance" along the coast. *Hebei Agriculture* (05): 8–9.

Feng, S. (2022). Digital empowerment and intelligent control of intrinsic safety on the water - Introduction to the AIS anti-collision system for commercial and fishing vessels. *China Emergency Management* (05): 76–79.

Liu, J.Y. and Han, C. (2022). Analysis of the causes of collisions between merchant and fishing vessels due to VHF communication failures in coastal waters of China. *China Maritime* (03): 33–35.

Lin, Y.X. and Li, R. (2023). Ministry of Transport and Ministry of Agriculture and Rural Affairs jointly launch the "2023 Special Action on Merchant-Fishing Co-Governance". *Farmers' Daily*, 2023-06-03(002).

Yang, X.Z. (2022). Interpretation: Deepening merchant-fishing co-governance to firmly prevent major collisions-Interview with Yang Xinzhai, Deputy Director of the Maritime Safety Administration of the Ministry of Transport on the "2022 Special Action on Merchant-Fishing Co-Governance". *China Maritime* (06): 57–58.

Proceedings of the International Association of Institutes of Navigation World Congress 2024 –
Chen et al. (Eds)

An approach for determining the safe distance between offshore oil platforms and shipping routes based on fuzzy logic

Junhui Zhang, Wenkai Wang, Yuxin Liu & Liang Yue
China Waterborne Transport Research Institute, Beijing, China

ABSTRACT: The rapid development of offshore oil fields has led to increasingly prominent conflicts between offshore oil platform facilities and shipping routes, especially in areas where oil field resources are abundant along individual ship routes. To standardize the site selection for offshore oil platforms, ensure the safety of ship navigation, and promote the coordinated development of offshore oil field exploitation and the shipping economy, it is urgent to study the safe distance between offshore oil platforms and shipping routes. However, determining the safe distance between offshore oil platforms and shipping routes involves multi-attribute decision-making issues such as the natural environment, route environment, and platform characteristics, which pose significant challenges. In response to this issue, this paper innovatively constructs a safe distance model between offshore oil platforms and shipping routes based on fuzzy logic. The model is developed through field investigation of safety emergency data from offshore oil platforms, analysis of AIS data from passing ships, and determination of safety distance impact factors via literature review and expert consultation. According to the model analysis results, the proposed method has significant advantages in analyzing the problem of fuzzy safety distance boundaries between offshore oil platforms and shipping routes. Furthermore, it can also categorize the safety distance into four intervals based on the emergency response capabilities of offshore oil platforms: the strict control zone, control zone, buffer zone, and open zone. This paper provides a technical approach for achieving safety supervision of offshore oil platforms and optimizing the allocation of marine resources.

Keywords: offshore oil platform, shipping routes, safe distance, fuzzy logic

1 INTRODUCTION

Energy is the foundation and driving force of human civilization and progress, vital to the national economy, people's livelihood, and national security; while shipping is an important link in the development of international trade, playing an extremely important role in the global economy. However, with the development of offshore oil and gas resources, the contradictions between offshore oil platforms and shipping routes have become increasingly prominent, leading to increasingly scarce maritime resources for ship navigation and gradually compressed navigational waters for ships [1]. If the planning and development of maritime resources are not scientific and coordinated, it will pose a serious threat to maritime traffic resources. For example, the dense construction of offshore oil and gas field platforms near shipping routes will change the original navigation habits of ships, increase the difficulty of avoidance and the risk of collision, and also increase fuel and time costs; the construction of oil and gas field platforms on both sides of the routes will reduce the width of the navigable waters, and even artificially form "narrow channels" increasing the density of

DOI: 10.1201/9781003624424-5

ship traffic and the difficulty of navigation; the construction of oil platforms in the waters where ships are accustomed to entering and leaving ports will hinder the navigation of ships entering and leaving ports, increasing the risk of ship collisions. According to the analysis of ship types involved in platform collision accidents in the global maritime accident database, nearly 23% of collision accidents involve passing ships, 36% of collisions are related to supply ships, and the probability of collisions between ships and platforms occurs on average once every two years [2].

Offshore oil platforms come in two types: mobile and fixed, and the construction of any structural form will have a certain impact on the passage of surrounding vessels [3]. To better maintain maritime traffic order and the external safety of offshore oil platforms, a lot of scientific research has been conducted between offshore oil platforms and shipping routes. In order to ensure the safety of ship navigation and the safety of the drilling platform itself, based on the platform's pool fire theory and ship maneuverability theory, Zhuang *et al.* constructed a comprehensive calculation model for the safe distance between ships and fixed drilling platforms from two aspects: the platform's own safety and the safety of extreme conditions of ship navigation (emergency situations and out-of-control situations) [4]. To provide passing vessels with a safe and economical distance reference value for the drilling platform, Xu *et al.* used AIS data to statistically analyze the distances of a large number of ships passing through the drilling platform, obtaining the distances of ships of different types (bulk carriers, container ships, tankers) and different tonnages when passing the platform. Combining the safety zone of the drilling platform and the wind and current elements, Li *et al.* revised the Fujii ship domain model and proposed a calculation model for determining the safe distance between the drilling platform and the customary shipping route. The research model indicates that the safe distance between the drilling platform and the customary shipping route mainly depends on the radius of the drilling platform's safety zone, the scale of the ship's domain, and natural factors such as wind and current in the water area [5].

According to maritime customs, under conditions of good visibility where the shipping route and nearby prominent landmarks are available for positioning and avoiding hazards, the distance should be maintained at no less than 1 nautical mile or more. However, with the vigorous development of the offshore oil and gas industry, the number of offshore oil platforms continues to increase. If the safety distance is set according to the aforementioned conditions, it will inevitably lead to the waste of spatial marine resources and the waste of resources caused by ship detours. Current research on safety distances mainly considers single-factor spatial issues, ignoring the impact of the platform's own safety protection capabilities, wind, currents, visibility, traffic volume, route width, route tonnage, and the layout characteristics of the platform on the safety distance. Therefore, there is an urgent need to adopt a multi-attribute analysis decision-making method that can comprehensively assess the degree of influence of different factors on the safety distance between offshore oil platforms and shipping routes to obtain the optimal solution for the safety distance. Fuzzy logic has a great advantage in analyzing problems with vague boundaries of the object and has been widely applied to various engineering risk management aspects, such as ship-bridge collision [6], risk assessment [7], etc., and can be used to solve multi-attribute problems. Therefore, This paper intends to use Fuzzy logic to solve the safety distance between offshore oil platforms and shipping routes.

2 DEVELOPMENT OF THE SAFETY DISTANCE MODEL

2.1 *Overview of fuzzy logic*

Fuzzy logic is an approximate inference method based on imitative behavior, and the essence of the method is to use fuzzy logic to define a mapping process from input to output. The inference process of this method is implemented based on a fuzzy rule base, which is generally composed of fuzzy rules summarized from expert knowledge and is mainly used to analyze problems with vague boundaries of the object. Generally speaking, fuzzy logic consists of four components that make up the decision-making framework: fuzzification, knowledge base, fuzzy inference, and defuzzification [8]. A fuzzy inference system is composed of a fuzzy controller and a controlled object, as shown in Figure 1.

(1) Fuzzification. The role of fuzzification is to convert the input precise quantities into fuzzy quantities. The input quantities include external reference inputs, system outputs or states, etc., all of which are precise quantities, while the fuzzy inference machine processes fuzzy quantities. Therefore, the input quantities must first be processed to become the inputs required for fuzzy inference. Secondly, the processed input quantities are scaled to transform them into their respective universes of discourse. Finally, the input quantities that have been transformed into the universe of discourse are subjected to fuzzification processing, turning the originally precise input quantities into fuzzy quantities, represented by corresponding fuzzy sets.
(2) Knowledge Base. The knowledge base contains knowledge within a specific application domain and the control objectives required. It is usually composed of two parts: the database and the fuzzy rule base. The database mainly includes the membership functions of each linguistic variable, scaling transformation factors, and the grading numbers of the fuzzy space, etc.; the rule base includes a series of control rules expressed by fuzzy linguistic variables.
(3) Fuzzy Inference. Fuzzy inference is the core of the fuzzy controller, which has the ability to simulate human inference based on fuzzy concepts. This inference process is carried out based on the implication relations and fuzzy rules in fuzzy logic.
(4) Defuzzification. The role of defuzzification is to transform the control quantities obtained from fuzzy inference into crisp quantities actually used for control.

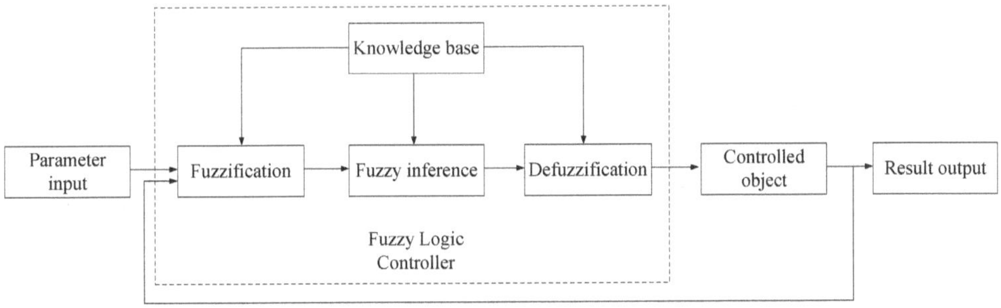

Figure 1. Fuzzy logic inference framework.

2.2 *Risk factor identification*

The safe distance between offshore oil platforms and shipping routes is a multi-factor decision-making issue. Different scholars have different hierarchical divisions of risk factors for offshore oil platforms and routes, and the risk factors they choose vary. These risk factors and evaluation systems have played a certain role in promoting the research and application of the safety of offshore oil platforms and ship navigation.

The determination of risk factors for offshore oil platforms and routes should be both operable and practical, while avoiding the selection of too many factors, which could make the risk factors not prominent and affect the rationality of the evaluation results. Overall, the safe distance between offshore oil platforms and ship routes is mainly influenced by the following factors: natural environment, route environment, and platform characteristics. Among them, the natural environment includes three indicators: visibility, number of windy days, and current speed; the route environment includes three indicators: traffic volume, route width, and route tonnage; platform characteristics include two indicators: platform distance and platform layout.

2.3 *Risk factor fuzzification*

The input of a fuzzy logic inference system is some precise information, and during the inference process, it is necessary to describe these precise pieces of information in a fuzzy manner. The membership function can achieve the above function, so constructing a reasonable and scientific membership function for each risk factor is of great significance. This paper adopts triangular, trapezoidal, and Gaussian membership functions for different risk factors to match the differences in various risks. To facilitate the description of the evaluation level of risk factors, a 5-level risk hierarchy is used, as shown in Table 1.

(1) Visibility. Visibility is one of the important factors that determine the safety of navigation for ships nearby the platform. The quality of visibility is directly related to the crew's ship maneuvering skills and the safety of navigation. This paper uses "days with visibility <1000m/year" as a quantitative standard and takes the multi-year average value of a certain sea area.
(2) Strong winds. In strong wind or typhoon weather, the safety and traffic efficiency of ships are adversely affected. Strong winds can cause ships to deviate from their course, limiting their maneuverability, and also cause the ships to sway, affecting the regular lookout by the crew.
(3) Current speed. For ship maneuvering, current speed can affect the ship's speed, course, and steering effectiveness. Excessive current speed can lead to turbulent flow, affecting the ship's ability to maintain course. Especially in narrow waters, faster current speeds are more likely to lead to maritime accidents. Therefore, the current speed is an important indicator of the safety level of the navigation environment (This paper does not consider the impact of the current direction on ship navigation, and the current speed refers to the average current speed).
(4) Traffic volume. Traffic volume is a direct reflection of the actual maritime traffic situation in a certain water area, indicating the number of all ships passing through a certain cross-section in the water area within a unit of time. This indicator can directly reflect the congestion and danger level of ship traffic in the water area.
(5) Route width. Route width refers to the width available for safe navigation of ships and is an important indicator for measuring the environment of the shipping route. The wider the route, the greater the area available for ships to navigate or take evasive actions; the narrower the route width, the greater the safety hazards of ship navigation.
(6) Route tonnage. The route tonnage represents the tonnage of ships that can navigate the route, which can usually be represented by the representative ship tonnage of the route. According to convention, shipping routes are generally determined by the representative ship tonnage based on 90% of the cumulative traffic volume of all merchant ships in the study area, thereby determining the tonnage of the shipping routes.
(7) Platform layout. Platform layout is a characteristic manifestation of the distribution of offshore oil platforms, and different layout forms have different impacts on the driving pressure of passing ships. This paper, combined with research and expert consultation, divides the layout forms of offshore oil platforms into single-sided, interlocking, parallel, and corresponding combined forms.

Table 1. Membership degree of risk factor.

Risk factors	Very Low	Low	Moderate	High	Very High
Visibility (d)	(0 0 5 10)	(5 15 22)	(16 25 33)	(27 35 45)	(40 45 60 60)
Strong winds (d)	(0 0 20 40)	(20 45 70)	(50 80 110)	(90 125 160)	(140 150 180 180)
Current speed (kn)	(0 0 0.4 0.6)	(0.3 0.9 1.5)	(1 1.7 2.4)	(2 2.7 3.4)	(3 3.2 4 4)
Traffic volume (vessels/day)	(0 0 6)	(1.5 7.5 13.5)	(9 15 21)	(16.5 22.5 28.5)	(24 30 30)
Route width (km)	(5.2 6 6)	(4.2 5 5.8)	(3.2 4 4.8)	(2.2 3 3.8)	(2 2 2.8)
Route tonnage (t)	(0 0 1)	(0.25 1.25 2.25)	(1.5 2.5 3.5)	(2.75 3.75 4.75)	(4 5 10 10)
Platform layout	Single-sided	Parallel	Interlocking or multiple parallel	Parallel/multiple parallel and interlocking	Multiple parallel and multiple interlocking

2.4 *Establish the fuzzy rule base*

The safety distance model between offshore oil platforms and shipping routes mainly carries out a series of analyses on factors such as the natural environment, route environment, and platform layout. Based on these influencing factors, the model further refines the indicators and establishes fuzzy rule base for the risk factors affecting the platform. At the same time, using dual-input/single-output or triple-input/single-output hybrid models in Simulink, various sub-modules are constructed at different levels. These sub-modules include meteorological conditions module, route environment module, natural environment module, environmental conditions module, comprehensive conditions module, and platform distance module, all of which together form a safety distance inference machine, as shown in Figure 2. By integrating the inputs and outputs of various modules, the model obtains a quantitative risk value for the offshore oil platform and the shipping route. Subsequently, different intervals are defined for different risk values to serve as the safety distance between the offshore oil platform and the shipping route.

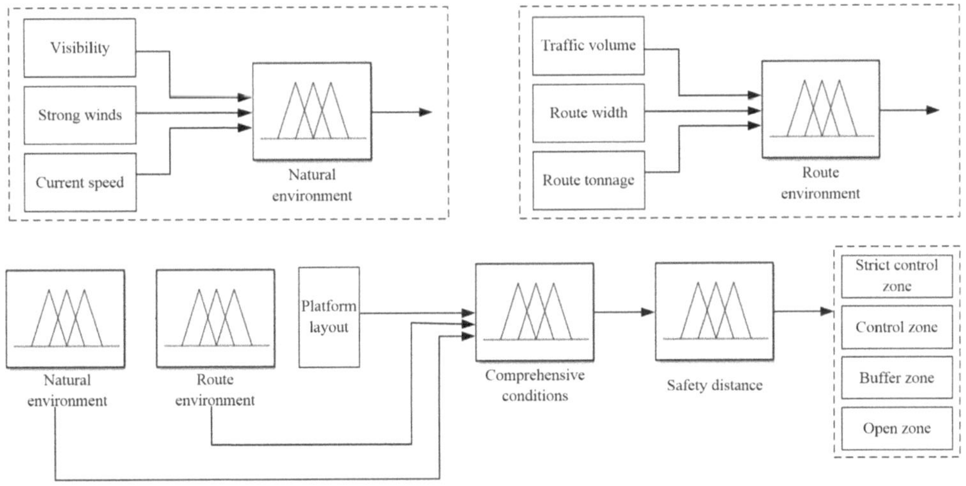

Figure 2. Layout of the fuzzy inference machine for safety distance.

2.5 *Fuzzy inference of safety distance*

For each module, fuzzy inference rules are established separately, with the output being considered as the risk level. A triangular membership function is used to establish a 5-level risk classification, represented by numerical values from 1 to 5, to indicate the level of risk, as shown in Figure 3.

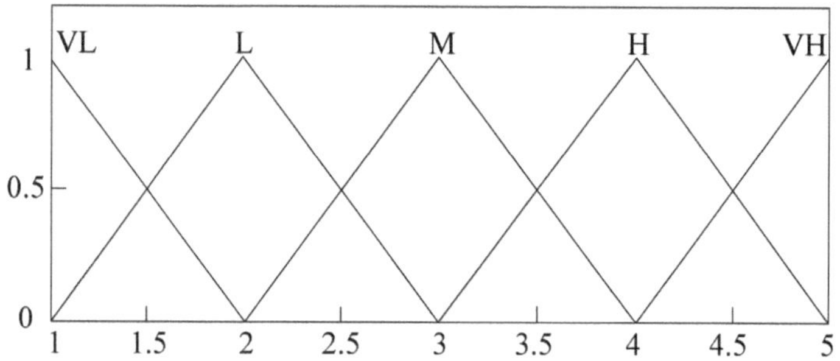

Figure 3. Triangular membership function.

Using the shipping route environment as an example, the process of fuzzy inference is explained. Traffic volume, route width, and route tonnage are factors that characterize the shipping route environment. By establishing a corresponding fuzzy rule base, fuzzy surfaces for the shipping route environment are generated for the pairs traffic volume-route width, traffic volume-route grade, and route width-route grade, as shown in Figure 4.

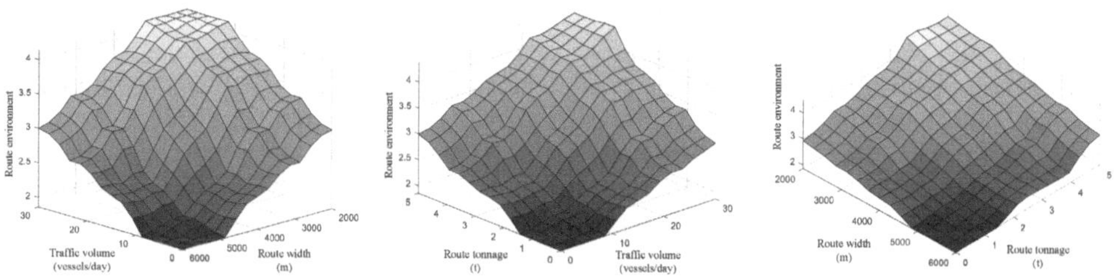

Figure 4. Surface figure of fuzzy inference rules.

2.6 *Defuzzification of safety distance*

Defuzzification is the process of mapping a fuzzy set to a crisp set in fuzzy logic. Common defuzzification methods include the centroid method, the bisector method, and the maximum membership degree method. Due to the intuitive and reasonable characteristics of the centroid method [9], this paper adopts centroid method to solve the safe distance between offshore oil platforms and ship routes.

3 APPLICATION OF OFFSHORE PLATFORM AND ROUTE DISTANCE MODEL

3.1 *Description of safety distance scenario*

The Bohai Sea is rich in oil and gas resources and has relatively shallow waters, making it one of the most important offshore oil and gas development areas in China. By the end of

2022, more than 80 platforms had been constructed in the south of the Bohai Sea, and there are still oil field platforms under planning and construction. At the same time, there are many ports in the south of the Bohai Sea, forming multiple shipping routes such as Changshan Waterway—Tianjin Port, Changshan Waterway—Huanghua Port, Binzhou Port—Changshan Waterway, Dongying Port—Changshan Waterway, etc., with a relatively busy shipping traffic. A large number of offshore oil platforms and ship routes are concentrated in the southern waters of the Bohai Sea, making the navigation environment of this sea area relatively complex. Therefore, selecting this water area as the research object has strong representativeness.

Within the scope of the study, the platforms that have been built or are under construction are mainly distributed near the Binzhou Port-Changshan Waterway route, Dongying Port-Changshan Waterway route, and Changshan Waterway-Tianjin route. Platforms arranged in the form of single-sided, parallel, and interlocking water areas near the above three routes were selected for model verification. The specific selected locations are shown in the following Table 2 and Figure 5.

Table 2. The relationship between the platforms and the nearby shipping routes.

No.	Shipping route	Platform layout
1	Binzhou Port—Changshan Waterway	Single-sided
2	Dongying Port—Changshan Waterway	Single-sided
3	Changshan Waterway—Tianjin Port	Single-sided
4	Binzhou Port—Changshan Waterway	Parallel
5	Dongying Port—Changshan Waterway	Parallel
6	Changshan Waterway—Tianjin Port	Parallel
7	Binzhou Port—Changshan Waterway	Interlocking
8	Dongying Port—Changshan Waterway	Interlocking
9	Changshan Waterway—Tianjin Port	Interlocking
10	Binzhou Port—Changshan Waterway	Parallel/multiple parallel and interlocking
11	Dongying Port—Changshan Waterway	Parallel/multiple parallel and interlocking
12	Changshan Waterway—Tianjin Port	Parallel/multiple parallel and interlocking

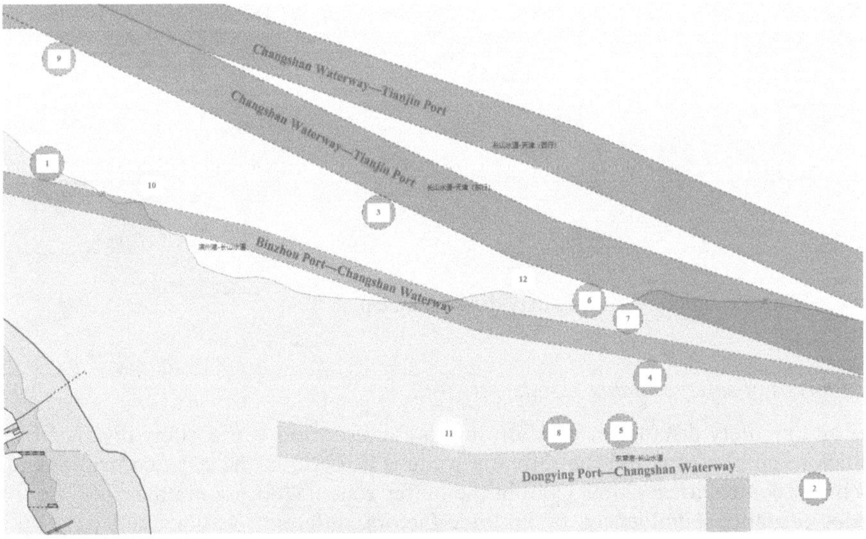

Figure 5. The layout between the platforms and the nearby shipping routes.

3.2 *Calculation of safety distance model*

The 12 selected locations differ only in the choice of parameters, but the method for calculating the safe distance between offshore oil platforms and shipping routes is consistent. Therefore, one location can be chosen as a case for illustration. Location 1 is in the west section of the Binzhou Port-Changshan Waterway and is arranged in a single-sided layout. The parameter values for the offshore oil platform and shipping route at this location, as well as the results of the safety distance calculation, are shown in the following Table 3 and Figure 6.

Table 3. Risk factors parameter input.

No.	Risk factors		Parameter input
1	Visibility (d)		35.6
2	Strong winds (d)		95
3	Current speed (kn)		1.5
4	Traffic volume (vessels/day)		25.41
5	Route width (km)		2.4
6	Route tonnage (t)		1
7	Platform layout		1
8	Calculation results	Strict control zone (m)	<1320
		Control zone (m)	1320~2460
		Buffer zone (m)	2460~3200
		Open zone (m)	>3200

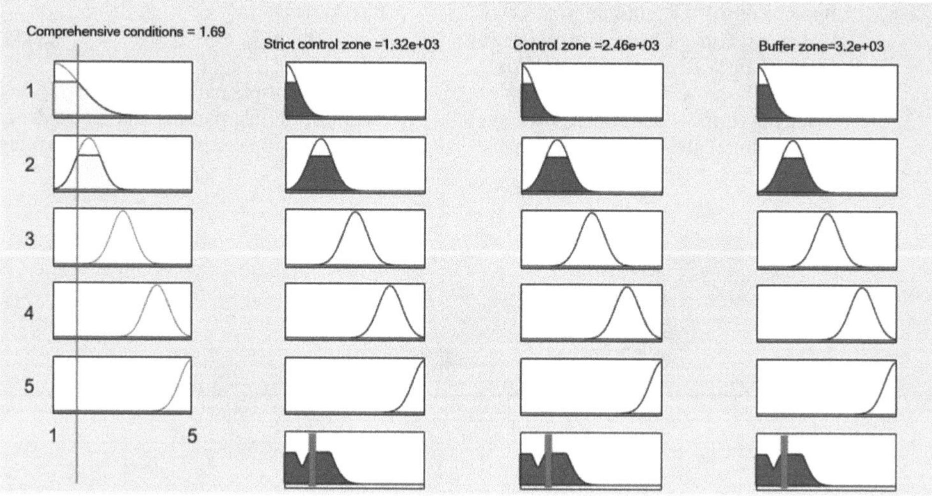

Figure 6. The calculation results of safety distance model.

3.3 *Analysis of safety distance results*

Based on the safety distance calculation results for Location 1, the safety distance between the offshore oil platform and the shipping route is [0 1320) in the strict control area; [1320 2460) in the control area; [2460 3200) in the buffer zone; [3200 $+\infty$) in the open area. Since the safety distance is influenced by multiple factors, different locations will yield different results. To enhance the universality of this method, allowing the use of the same safety distance standard on the same shipping route, the maximum value of different parameters on the same route can be adopted to solve for the safety distance. Furthermore, it is recommended that offshore oil platform construction should not be carried out within the strictly controlled area. If construction is indeed necessary, the safety distance can be appropriately reduced after comprehensive deliberation, taking into account the complexity of the navigation environment of the route, the corresponding protective measures of the platform, and the importance of the oil and gas development project. When offshore oil platform construction is required within the control zone and buffer zone, corresponding levels of protective measures should be taken to ensure the safety of the offshore oil platform.

4 CONCLUSIONS

In response to the issue of safety distance between offshore oil platforms and ship routes, this paper innovatively develops a fuzzy logic model to calculate the safety distance, and proposes a method to obtain safety distance standards for a specific route, enhancing the universality and applicability of the safety distance model. The paper fully considers the impact of natural environment, route environment, and platform characteristics on the safety distance, and can flexibly solve the safety distance with shipping routes when selecting the location of offshore oil platforms.

It should be noted that the input parameters of the model need to be adjusted according to different sea conditions to avoid situations where the safety distance violates the requirements for the location of offshore oil platforms. In order to further optimize the allocation of marine resources and improve the utilization rate of marine resources, the safety distance between offshore oil platforms and shipping routes can be moderately reduced in the future according to the degree of intelligence of offshore oil platforms or passing ships.

REFERENCES

[1] Xie H., Liu Z., Xu X. and Zhang J., (2020), Research on the safe distance between passing ship and offshore drilling platform based on theory and statistics, *Proceedings of the Institution of Mechanical Engineers, Part M: Journal of Engineering for the Maritime Environment*, vol. 234, no. 3, pp. 642–650, doi: 10.1177/1475090220902305.

[2] Mujeeb-Ahmed M.P., Seo J.K. and Paik J.K., (2018), Probabilistic approach for collision risk analysis of powered vessel with Offshore oil platforms, *Ocean Eng.*, vol. 151, pp. 206–221, doi: 10.1016/j.oceaneng.2018.01.008.

[3] Quan S. and Shanxin L., (2023), Research on safe distance from ship to offshore drilling rig, *Port, Waterway and Offhore Engineering*, vol. 60, no. 5, pp. 23–26, doi: 10.16403/j.cnki.ggjs20230506.

[4] Yuan C. Mengxue Z. and Yujiang G., (2021), Calculationon the safety distance between ship and drilling platform based on comprehensive model, *Ship & Ocean Engineering*, vol. 50, no. 05, pp. 132–137.

[5] Li G., Zhang W. and Yin J., (2016), Determination of the safety distance between drilling platform and customary route, *Ship & Ocean Engineering*, vol. 45, no. 01, pp. 186–190.

[6] Wu B., Yip T.L., Yan X. and Guedes Soares C., (2019), Fuzzy logic based approach for ship-bridge collision alert system, *Ocean Eng.*, vol. 187, pp. 106152, doi: 10.1016/j.oceaneng.2019.106152.

[7] Hu Y. and Park G., (2020), "Collision risk assessment based on the vulnerability of marine accidents using fuzzy logic," *Int. J. Nav. Arch. Ocean*, vol. 12, pp. 541–551, doi: 10.1016/j.ijnaoe.2020.06.005.

[8] Wu B., Cheng T., Yip T.L. and Wang Y., (2020), "Fuzzy logic based dynamic decision-making system for intelligent navigation strategy within inland traffic separation schemes," *Ocean Eng.*, vol. 197, pp. 106909, doi: 10.1016/j.oceaneng.2019.106909.

[9] Yin H., Liu C., Wu W., Song K., Dan Y. and Cheng G., (2021), An integrated framework for criticality evaluation of oil & gas pipelines based on fuzzy logic inference and machine learning, *J. Nat. Gas Sci. Eng.*, vol. 96, pp. 104264, doi: 10.1016/j.jngse.2021.104264.

Proceedings of the International Association of Institutes of Navigation World Congress 2024 – Chen et al. (Eds)

Construction of autonomous navigation system test scenarios with typical accident characteristics

C.X. Jiang & X.Y. Zhang
Department of Navigation College, Maritime Intelligent Transportation Research Team, Dalian Maritime University, Dalian, China

ABSTRACT: Responding to the demand for high-risk test scenarios in the safety evaluation of autonomous navigation systems (ANS) in maritime autonomous surface ships (MASS), this paper proposes a method for constructing test scenarios based on typical accident characteristics. Through the analysis of maritime traffic environments and fundamental vessel data at the time of ship collisions, we extracted scenario elements such as the incident time, relative heading, speed, distance, and collision location from 337 accident reports. Utilizing a clustering algorithm that combines hierarchical clustering and K-means, we identified typical accident characteristics across various scenario elements. Subsequently, we reorganized these typical accident characteristic parameters to construct seven test scenarios that reflect real maritime traffic conditions. The findings provide high-risk simulation test scenarios for the safety evaluation of MASS navigation systems, ensuring the authenticity and comprehensiveness of these tests.

1 INTRODUCTION

Recently, the rapid advancement of artificial intelligence, the Internet of Things, and digital technologies has shifted the global maritime industry's focus toward autonomous surface vessels. These Maritime Autonomous Surface Ships (MASS) are designed to enhance navigational safety, reduce crew workload, and optimize traffic efficiency through intelligent automation (Zhang *et al.* 2021). According to the International Maritime Organization (IMO), MASS are vessels capable of operating independently of human intervention to varying degrees, undertaking perception and decision-making tasks traditionally performed by human operators (IMO 2018). While this shift reduces human error, it introduces significant challenges in ensuring navigational safety.

A critical component in ensuring the safety of MASS is the Autonomous Navigation System (ANS). The ANS must not only follow predefined navigation plans but also adapt flexibly to the dynamic and complex maritime environment to avoid collisions (Felski *et al.* 2020). Given the increasing complexity of maritime traffic, even experienced officers require extensive training and practice to master navigational skills. Collision risks are widely regarded as one of the most severe challenges in MASS operations (Puisa *et al.* 2018). Therefore, comprehensive and in-depth training and evaluation of ANS are essential to validate their performance under various scenarios, ensuring they meet or exceed human capabilities. Traditionally, ANS solutions have involved quantifying navigational rules and incorporating them into algorithmic logic to achieve autonomous collision avoidance in specific waters (Zheng *et al.* 2024). However, due to the highly dynamic and unpredictable nature of the maritime environment, the adaptability of these methods is severely limited. Consequently, addressing the uncertainties and instabilities introduced by new technologies has become an urgent issue.

DOI: 10.1201/9781003624424-6

Testing and validation involve manual or automated methods to assess the functionality and safety characteristics of the system under test (Ding *et al.* 2023). The goal is to determine whether the system meets expected requirements or to identify discrepancies between actual and expected results. To expedite the development of autonomous collision avoidance technology and promote the deployment of MASS, countries worldwide are actively engaging in ANS testing and validation research. Although some countries have established testing platforms for MASS (OffshoreEnergy 2017), there remains a lack of specific testing technologies and systems, as well as comprehensive testing standards and guidelines. Particularly when compared to the testing methods for autonomous road vehicles (Li *et al.* 2016), research on ANS testing methods is relatively underdeveloped and in urgent need of breakthroughs. Traditional physical testing methods (Safety 4 Sea 2019), such as model testing and full-scale ship testing, rely on identifying safety issues from actual navigation processes that could cause ANS failure. However, due to the near-infinite input domain of ANS, theoretically, an enormous number of scenarios need to be tested to fully validate the system's safety. These methods are inherently limited by long testing cycles, high costs, and significant safety risks, severely constraining their application in advanced ANS testing.

To address this need, several countries and companies have developed extensive scenario libraries for virtual simulation testing (Xiao *et al.* 2021), containing a wide range of test scenarios. Drawing from the classification of test scenarios for autonomous road vehicles, these libraries should include both natural navigation scenarios and high-risk scenarios. High-risk scenarios are those that are likely to pose significant safety issues for MASS, and thus are crucial for testing the safety of ANS. However, such scenarios are rare in routine ship navigation. Currently, the scenarios in various MASS test libraries are either derived from expert knowledge abstraction (Arendt *et al.* 2011) or collected from real-world natural navigation data (Bolbot *et al.* 2022). Due to limitations in data collection technology, most natural datasets capture vessels in safe operating conditions, making it challenging for researchers to extract comprehensive high-risk scenarios from a large volume of actual data. Consequently, there is a notable scarcity of high-risk scenarios in existing scenario libraries.

In response to these challenges, this paper proposes a novel method for constructing high-risk test scenarios for ANS. The researchers combined hierarchical clustering with the K-means clustering method to perform a fusion clustering analysis on data from 337 vessel collision accident reports. This analysis identified seven ANS test scenarios that exhibit typical accident characteristics. These scenarios effectively depict high-risk collision conditions in real-world settings, providing a realistic and representative testing environment for the virtual simulation of autonomous navigation systems. Consequently, this approach enhances the safety and reliability of these systems.

2 METHDOLOGY

2.1 *Clustering methods for ship collision accident reports*

Text clustering organizes textual data based on feature similarity, facilitating the extraction of representative instances from large datasets. In this study, we convert accident reports into binary vectors using one-hot encoding, making the data suitable for analysis. We then propose a hybrid clustering algorithm that combines hierarchical and K-means clustering, specifically tailored for ship collision reports. The optimal number of clusters is determined by evaluating clustering quality using clustering curves.

2.1.1 *Format conversion methods for incident reports*

A primary challenge in text clustering is transforming the rich information embedded in texts into mathematically analyzable formats. To assess text similarity within clustering algorithms, metrics such as cosine similarity require mapping texts into vector spaces. In preprocessing accident report texts, we employ One-Hot Encoding, a technique that converts discrete textual features into numerical representations suitable for clustering algorithms.

One-Hot Encoding (Zhou 2020), a specific form of the one-of-k encoding strategy, uses an N-bit binary feature register where each bit uniquely identifies a textual feature state. At any given moment, only one bit is activated (set to 1), while the others remain inactive (set to 0). This design ensures the independence and uniqueness of textual features within the dataset. As illustrated in Figure 1, our approach applies One-Hot Encoding to transform accident report texts into unique binary vectors. Each text instance occupies a distinct position in the vector space, resulting in ship collision accident data suitable for clustering analysis. This representation step enhances data processing flexibility, enabling clustering algorithms to capture feature variations in textual data more effectively, and expands the feature space, improving the algorithms' ability to comprehend complex textual data.

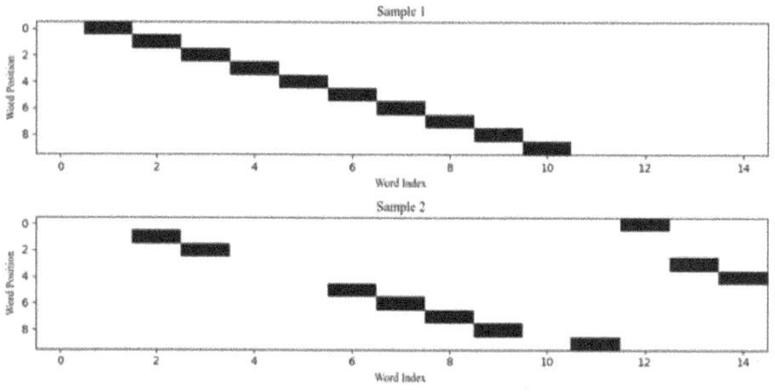

Figure 1. One-hot encoding for ship collision accident report texts.

2.1.2 *Fusion clustering algorithm for ship collision incidents data*

Compared to other text data mining methods, the K-means clustering algorithm has the advantage of not requiring prior knowledge, allowing it to derive features of ship collision accidents based on the natural distribution of the texts. However, in practical applications, the traditional K-means clustering algorithm's performance stability is often affected by outlier samples, the randomness in the selection of initial cluster centers, and the reasonableness of the number of clusters (Fueyo *et al.* 2021). Specifically, when initial cluster centers are located near outlier samples, it can significantly increase the number of algorithm iterations and may cause the algorithm to fall into local optima, thereby compromising both the accuracy and computational efficiency of the clustering results (Song *et al.* 2021). Therefore, targeted improvements to the clustering algorithm are often required to enhance its adaptability to data samples, optimize the clustering analysis process, and better achieve specific analysis goals.

Given the high dimensionality and sparsity characteristics of ship collision accident data and the limitations of the K-means algorithm in this regard, this paper proposes a fusion clustering algorithm suitable for ship collision accident data. This fusion algorithm uses the K-means clustering algorithm as a framework and introduces hierarchical clustering as a preprocessing step to optimize the selection of initial cluster centers. As shown in Figure 2, researchers first perform hierarchical clustering on a randomly selected data sample, then use the hierarchical clustering results as the initial cluster centers for K-means clustering. This approach enhances the scientific and reasonable selection of initial cluster centers, reducing the uncertainty caused by random selection.

The specific implementation process of the algorithm is as follows:

Step 1: Pre-clustering Data Selection: Randomly select a certain number of samples (e.g., 50) from the ship collision accident dataset.
Step 2: Initial Parameter Setting: Set the initial number of clusters K to 2 as the baseline parameter for algorithm initiation.

Step 3: Preset Hierarchical Clustering: Perform hierarchical clustering on the randomly selected samples by calculating the distances between samples and gradually merging the nearest samples until the number of samples is reduced to the preset K.
Step 4: K-means Clustering: Use the K samples obtained from hierarchical clustering as the initial cluster centers for K-means clustering. Calculate the distance from each sample in the overall dataset to each initial cluster center, and assign samples to the corresponding categories based on the nearest distance to establish category labels.
Step 5: Iterative Optimization: Compare the category labels of each sample in consecutive iterations. If there are inconsistencies, update the cluster centers and repeat the K-means clustering process until the category labels of all samples remain stable in consecutive iterations, i.e., the algorithm converges.
Step 6: Convergence Judgment and Parameter Adjustment: Evaluate the algorithm's convergence by plotting clustering curves. If convergence is unsatisfactory, increment the value of K and repeat Steps 2 to 6 until the clustering curve shows good convergence.
Step 7: Clustering Result Selection: Based on the analysis of the clustering curves, select the clustering dataset corresponding to the optimal K value as the final clustering result and output it for subsequent analysis.

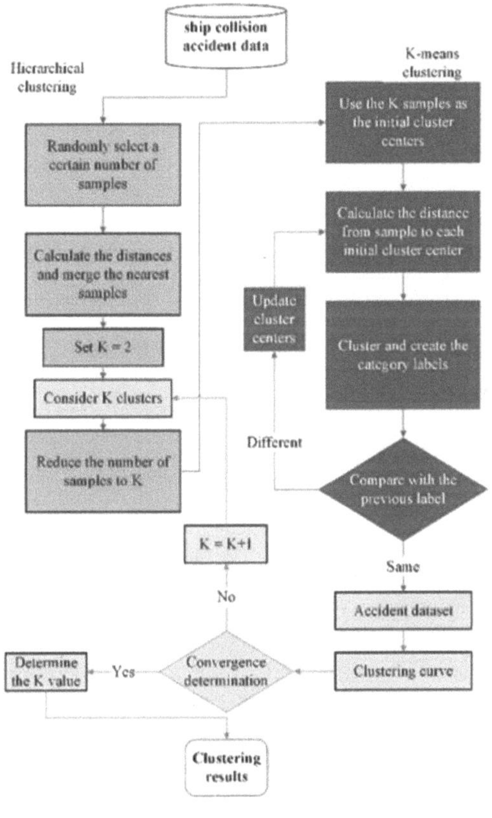

Figure 2. Fusion clustering algorithm process.

2.1.3 *Quality assessment of fusion clustering algorithm*

For the hybrid clustering algorithm applied to ship collision accident data, we evaluate clustering quality by plotting a clustering curve. The clustering curve uses the number of clusters as the x-axis and the sum of squares in cluster (SSC), which represents the sum of the distances of all samples within a cluster to the cluster center, as the y-axis. The SSC is calculated using the following formula:

$$SSC = \sum_{m=1}^{K}\sum_{n=1}^{q_n} dist(X_{mn}, C_m)$$

Where K is the number of cluster, q_n is the number of samples in the n-th cluster, X_{mn} is the n-th sample in the m-th cluster, and C_m is the center of the m-th cluster.

The SSC value is a crucial indicator of the tightness of sample aggregation within clusters. The convergence of the SSC curve serves as a basis for selecting the optimal number of clusters and evaluating the rationality of the clustering. As shown in Figure 3, by gradually increasing the number of clusters K, we observe a decreasing trend in SSC values, indicating an increasing degree of sample aggregation that eventually converges. When K = 7, the clustering curve stabilizes and no longer decreases significantly, demonstrating effective convergence. This indicates that the clustering at this point effectively differentiates between different accident scenarios while avoiding the redundancy of over-segmentation. Therefore, we select 7 as the optimal number of clusters for analyzing the ship collision accident data.

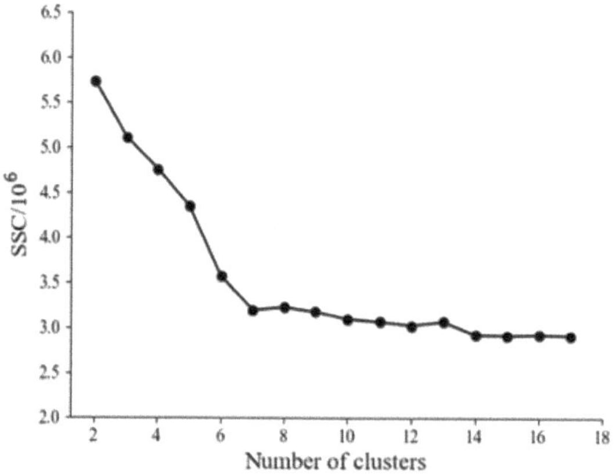

Figure 3. Clustering curve.

2.2 *Strategy for selecting test scenario elements*

In testing Autonomous Navigation Systems (ANS) for intelligent ships, acquiring a sufficient quantity of real-world test scenarios presents a significant challenge, hindering comprehensive testing. Therefore, it is essential to scientifically define the key elements of test scenarios for autonomous navigation systems and to construct diverse, realistic, high-risk test scenarios. This approach enhances algorithm coverage and testing effectiveness, ensuring the validity, safety, and reliability of the tests.

Based on the requirements for test scenarios and the content of ship collision accident reports (Hu *et al.* 2021), our study identifies a series of accident characteristic elements (shown in Figure 4) and uses these as the basis for selecting clustering variables. To precisely define the interaction between two ships, we designate the ship with the collision point at the bow as "striking ship S1" and the ship with the collision point elsewhere as "struck ship S2." Additionally, to scientifically distinguish different encounter types, we use relative bearing intervals, as shown in Figure 5, based on the International Regulations for Preventing Collisions at Sea (COLREGS), to classify the types of ship encounters.

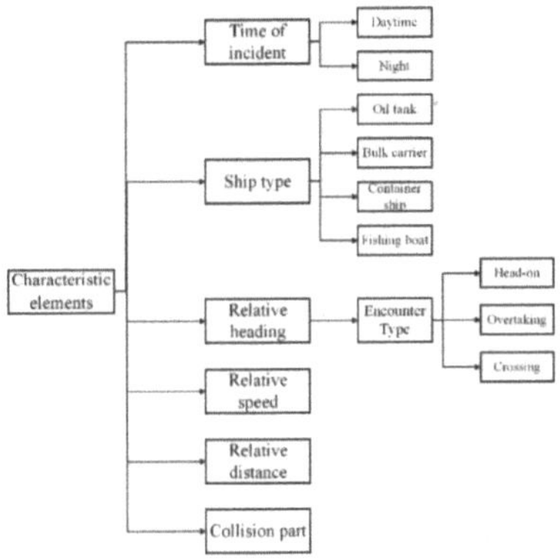

Figure 4. Accident characteristic elements.

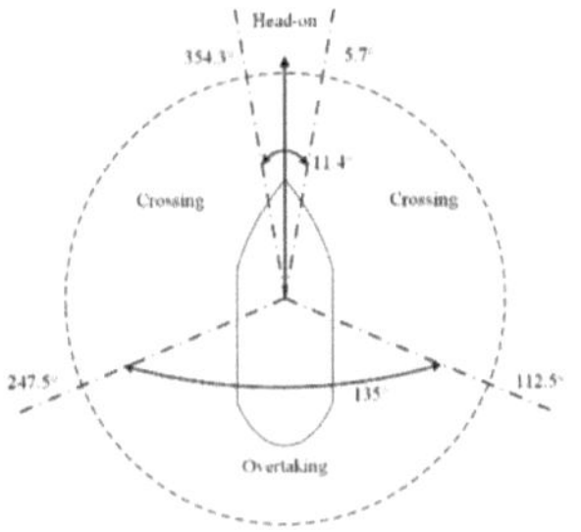

Figure 5. Encounter situations.

2.3 *Cluster analysis of typical incident characteristics*

This study employs the Fusion Clustering Algorithm to categorize ship collision accident data into seven clusters based on selected scenario elements, thereby revealing potential typical accident characteristics. Table 1 presents the clustering results, with bold frequencies highlighting key variable values for each cluster center, which define the typical attributes of each cluster's scenario. We identified seven typical hazardous ship collision scenarios (C1 to C7), each representing specific collision patterns. For instance, scenario C2 describes a collision at night, where the own ship and a fishing boat, within a relative distance of 0-2 nautical miles and approaching at a relative speed of 15-21 knots, collide.

Table 1. Frequency of clustering variables in each cluster for vessel collision accident data.

Clustering variables	Variable value	C1	C2	C3	C4	C5	C6	C7
Number of samples	337	86	41	38	63	56	22	31
Time of incident	Daytime	**62**	0	**34**	0	**56**	0	**29**
	Night	24	**41**	4	**63**	0	**22**	2
Encounter Type	Head-on	0	**29**	0	3	0	1	**24**
	Overtaking	0	0	**31**	**47**	0	1	3
	Crossing	**86**	11	7	13	**56**	**20**	4
Ship type of S2	Oil tank	4	1	6	0	3	0	**19**
	Bulk carrier	**45**	10	**17**	7	14	1	5
	Container ship	16	3	8	11	0	0	7
	Fishing boat	21	**27**	7	**45**	**39**	**21**	0
Relative speed	3-9kn	12	0	**25**	**40**	3	1	6
	9-15kn	**56**	5	10	14	**47**	4	10
	15-21kn	18	**36**	3	9	6	**17**	**15**
Relative distance	0-2nm	**43**	**35**	8	**32**	13	**14**	6
	2-4nm	34	6	**16**	20	**42**	5	**19**
	4-6nm	9	0	7	11	1	3	6

3 RESULTS AND DISCUSSIONS

Through the application of cluster analysis to ship collision data, this study successfully identified and summarized seven representative accident scenarios, each with distinct characteristics. To comprehensively validate the response capabilities of ANS across various collision scenarios, we incorporated the collision points to determine the initial bearing and vessel maneuvering actions of the struck ship(S2). This approach allowed us to construct seven ANS testing scenarios centered on the striking ship (S1) as the test ship and struck ship (S2) as the target ship, reflecting real-world navigation complexities and uncertainties. Detailed descriptions and schematic diagrams of these scenarios are provided in Table 2 and Figure 6.

Table 2. Test scenarios of ANS.

Type of scenario	Time	Encounter Type	Ship type of target ship	Relative speed	Relative distance	Action of target ship	Probability
C1	Daytime	Crossing	Bulk carrier	9-15kn	0-2nm	Starboard	25.52%
C5			Fishing boat	9-15kn	2-4nm	Port	16.61%
C3		Overtaking	Bulk carrier	3-9kn	2-4nm	Steady	11.27%
C7		Head-on	Oil tank	15-21kn	0-2nm	Steady	9.21%
C2	Night	Head-on	Fishing boat	15-21kn	0-2nm	Starboard	12.17%
C4		Overtaking		3-9kn	0-2nm	Starboard	18.70%
C6		Crossing		9-15kn	0-2nm	Port	6.52%

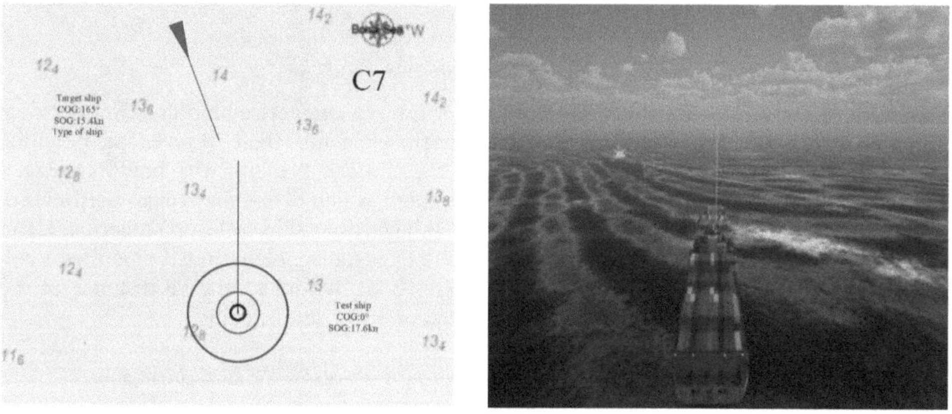

Figure 6. Schematic diagram of partial test scenarios.

This study's test scenarios add a precise dimension to accident data analysis. The comparison of scenarios in Table 2 reveals that fishing boats, due to their smaller size and unpredictable navigation patterns, are the most frequently involved in collisions. Further investigation shows that significant changes in lighting conditions, particularly at night, substantially increase the risk of collisions, as ship operators often fail to notice smaller vessels. The increased relative speed between ships drastically reduces the time and distance available for collision avoidance, thereby elevating the probability of accidents.

Similar to traditional ship operators' decision-making processes, Autonomous Navigation Systems (ANS) use sensors such as radar and cameras to perceive the environment and make collision avoidance decisions based on embedded navigational logic. Consequently, using accident data that includes human errors to train and evaluate ANS can help verify whether ANS can match or exceed the collision avoidance capabilities of human operators under complex and variable conditions.

To enhance the richness and robustness of testing scenarios, future work will incorporate multivariate maritime data, including meteorological data, marine environmental data, and vessel traffic flow data. By leveraging advanced data fusion techniques, we aim to integrate this information into the cluster analysis, thereby identifying more multidimensional accident features. As data volume and complexity increase, we will continue to optimize clustering algorithms to improve efficiency and accuracy in processing large-scale, high-dimensional datasets. Additionally, we plan to adopt a hybrid approach combining real-ship and virtual testing to strengthen the credibility of our results. As autonomous navigation technology evolves, it is essential to continually update and refine the testing scenario library to address emerging navigational risks and technological demands.

REFERENCES

Arendt, R. and Van Uden, E. (2011): A decision-making module for aiding ship system automation design: A knowledge-based approach. *Expert Systems with Applications*, 38(1), 410–416.

Bolbot, V., Gkerekos, C., Theotokatos, G. and Boulougouris, E. (2022): Automatic traffic scenarios generation for autonomous ships collision avoidance system testing. *Ocean Engineering*, 254, 111309.

Ding, W., Xu, C., Arief, M., Lin, H., Li, B. and Zhao, D. (2023): A survey on safety-critical driving scenario generation—A methodological perspective. *IEEE Transactions on Intelligent Transportation Systems*, 24 (7), 6971–6988.

Felski, A. and Zwolak, K. (2020): The ocean-going autonomous ship—Challenges and threats. *Journal of Marine Science and Engineering*, 8(1), 41.9

Hu, L., Bao, X., Lin, M., Yu, C. and Wang, F. (2021): Research on risky driving behavior evaluation model based on CIDAS real data. Proceedings of the Institution of Mechanical Engineers, *Part D: Journal of Automobile Engineering*, 235(8), 2176–2187.

IMO (2018): IMO takes first steps to address autonomous ships.

Li, L., Huang, W. L., Liu, Y., Zheng, N. N. and Wang, F. Y. (2016): Intelligence testing for autonomous vehicles: A new approach. *IEEE Transactions on Intelligent Vehicles*, 1(2), 158–166.

OffshoreEnergy (2017): First test area for autonomous ships opened in Finland.

Puisa, R., Lin, L., Bolbot, V. and Vassalos, D. (2018): Unravelling causal factors of maritime incidents and accidents. *Saf. Sci.*, 110, 124–141.

Safety 4 Sea (2019): Autonomous Ships: Test Areas and Research Centers Making Headlines.

Song, Y., Chitturi, M. V. and Noyce, D. A. (2021): Automated vehicle crash sequences: Patterns and potential uses in safety testing. *Accident Analysis & Prevention*, 153, 106017.

Suarez-del Fueyo, R., Junge, M., Lopez-Valdes, F., Gabler, H. C., Woerner, L. and Hiermaier, S. (2021): Cluster analysis of seriously injured occupants in motor vehicle crashes. *Accident Analysis & Prevention*, 151, 105787.

Xiao, G., Ren, B., Tong, C. and Hong, X. (2021): A quantitative evaluation method for obstacle avoidance performance of unmanned ship. *Journal of Marine Science and Engineering*, 9(10), 1127.

Zhang, X., Wang, C., Jiang, L., An, L. and Yang, R. (2021): Collision-avoidance navigation systems for maritime autonomous surface ships: A state of the art survey. *Ocean Engineering*, 235, 109380.

Zheng, K., Zhang, X., Wang, C., *et al.* (2024): Adaptive collision avoidance decisions in autonomous ship encounter scenarios through rule-guided vision supervised learning. *Ocean Engineering*, 297, 117096.

Zhou, Y. (2020): A review of text classification based on deep learning. In *Proceedings of the 2020 3rd international conference on geoinformatics and data analysis* (pp. 132–136).

Proceedings of the International Association of Institutes of Navigation World Congress 2024 – Chen et al. (Eds)

Study on simulation and optimization measures of ship maneuvering in continuous bends of an artificial canal

Chao Guo & Zhonghua Li
Nanjing Hydraulic Research Institute, Nanjing, China

ABSTRACT: The width of a planned artificial canal channel is relatively narrow and its water depth is shallow, which brings certain difficulties to the navigation of 5000 tons ships. In addition, some canal channels have S-shaped continuous turning sections. In this paper, the maneuverability of a 5000 tons ship passing through the continuous turning section of an artificial canal is simulated on the ship maneuverability simulator, and the maneuverability characteristics of the ship such as drift Angle, and track line are carefully studied, and the influence of different environmental factors on the navigation safety and maneuverability of the ship are analyzed. Some suggestions for optimizing the layout of continuous bends are put forward, such as increasing the turning radius of continuous bends, reducing channel turn angle, and extending the length of straight lines between two bends. The research of this paper will effectively ensure the maneuvering safety and navigation efficiency of large ships under the condition of narrow waterways.

1 INTRODUCTION

During the 14th Five-Year Plan period, China will steadily advance the construction of large-scale cross-water artificial canals, including the Pinglu Canal, Xianggui Canal and Ganyue Canal. This initiative aims to increase the mileage of China's high-grade waterways to 25,000 kilometers, thereby addressing the deficiencies in inland water transportation. Artificial canals typically feature shallow and narrow waterways, where hydraulic fluctuations and changes in flow velocity have a more pronounced impact on ship maneuverability compared to natural waterways. Ships navigating these canals may encounter increased navigational resistance, reduced speed, and diminished rudder efficiency, all of which affect the safety and efficiency of navigation. Some sections of artificial canals are excavated based on existing natural waterways, often characterized by continuous bends. As a special type of waterway, continuous bends present complex three-dimensional turbulence structures, significantly increasing the difficulty of ship maneuvering in channels with multiple turns.

There has been considerable research on the hydrodynamic conditions and navigability of continuous bends. However, studies focusing on ship maneuverability in such environments within China remain limited. Ship maneuvering is influenced by multiple factors, including the vessel's inherent handling characteristics, external environmental conditions such as wind, waves, and currents, as well as the water traffic environment and the judgment of pilots. Among these, the impact of current conditions on ship maneuverability is particularly significant. Currently, both domestic and international research on ship maneuverability primarily employs methods such as numerical simulation, self-propelled ship models, ship maneuvering simulators, and prototype observations.

Using a large-scale ship maneuvering simulator, Zhang Xiufeng *et al.* (2003) conducted a navigation safety and maneuverability test on a challenging channel characterized by

DOI: 10.1201/9781003624424-7

frequent steering, large turning angles, limited turnaround space, and shallow water depth, using the Tugutang channel hub as a case study. Hu Xuyue *et al.* (2010) investigated the drift angle of a thousand-ton motorized barge navigating through continuous curves using a self-propelled ship model, analyzing its influencing factors. Cao Fengshuai *et al.* (2011) examined the key factors affecting the design of channel turns in coastal ports, comparing various design methodologies and providing design parameters for channel turns based on ship maneuvering simulator results. Wang Fei *et al.* (2022) studied the safety of ship navigation in narrow, continuously curved river sections of the navigation hub. Mansuy *et al.* (2022) simulated the maneuverability of ships passing through two sharp bends of the Seine River in Paris, France, using a ship maneuvering simulator, and recommended the maximum dimensions for vessels transiting the Seine River.

The technical specifications of the artificial canal are classified as Inland Waterway Class I, designed to accommodate 5000-ton class inland waterway vessels. The canal section features several continuous S-shaped bends, with a channel width of only 80 meters. Vessels can navigate in both directions; however, the navigation conditions are considerably complex.

In this paper, we simulate the maneuverability of a 5000-ton vessel navigating through the continuous bends of an artificial canal. The simulation analyzes key ship maneuvering characteristics including drift angle, speed, and track line. Based on the simulation results, optimization suggestions for the continuous bend arrangement scheme and corresponding safeguard measures are proposed to provide technical support for enhancing the safety of vessel navigation in this artificial canal.

2 TEST BASIS INFORMATION

2.1 *Wind information*

In this study area, the predominant wind direction is from the North (N), occurring with a frequency of 26%. The second most common wind direction is North-Northeast (NNE), observed at a frequency of 9.2%. Strong winds also originate from the North, reaching speeds as high as 31 m/s. Based on long-term meteorological data collected from stations near the project site, the average annual wind speed is 2 m/s.

2.2 *Design and test vessel type*

The simulation evaluated a 5,000-ton-class inland waterway bulk carrier with the following dimensions: length 90.0 meters, width 15.8 meters, and full load draft of 5.2 meters. The design speed of the vessel in canal navigation is 6.48 knots(3.33m/s).

Figure 1. View of 5000t class inland bulk carrier.

2.3 *Test equipment*

This test is completed on the ship maneuvering simulation test system (NTPro 6500), which can simulate the actual operation process of the ship on the virtual environment, and make a comprehensive test and comparison of the ship navigation conditions before and after the construction of hubs, ports, wharves and other hydraulic buildings, so as to make a scientific assessment of the safety of the newly built hydraulic buildings on the navigable environment and to find out the existing problems. This set of equipment includes ship navigation conditions test hardware platform, ship maneuvering simulator test system simulation software, network-based distributed processing computer system and other components. The three 120° views, ship's bridge (IBS) module, ship control module, navigation instrument module, view system module, sound simulation module, etc. constitute the ship navigation conditions test hardware platform.

2.4 *Test terrain*

Under the original arrangement conditions, the bottom width of the channel is 80m, the slope ratio of both sides of the bank is 1:2~1:3, and the elevation of the bottom of the channel is 27.7 m. There are many continuous bends in the channel, one of which adopts the same width of 80m as that of the flat section, with the upstream turning angle of 61.74° and the downstream turning angle of 74.78°. The radius of consecutive turns of the channel is 360m, and the straight-line distance between two bends is 119.44m (see Figure 2). Between the two bends of the river, a new inland bridge will be built, with a span of 160m and one hole spanning the navigable water.

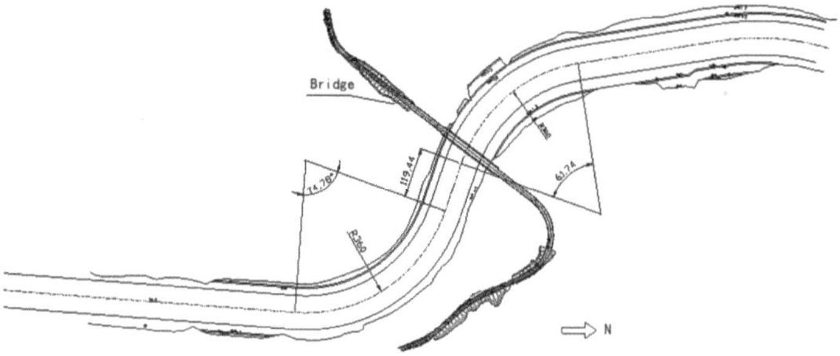

Figure 2. Layout of the channel in the study area (original arrangement).

According to the terrain, a three-dimensional simulation view of ship maneuvering simulation is established, in which the channel time is shown in Figure 3. The simulation scene of ship maneuvering is shown in Figure 4.

Figure 3. Three-dimensional scene of a continuous curve.

Figure 4. Photographs of ship model maneuverability test.

2.5 *Test content*

In this paper, we proposed to carry out ship maneuvering simulation test in continuous reverse bend section, analyzed the ship maneuvering performance of two-lane navigation under hydrostatic and constant wind conditions, and put forward the optimization suggestions of channel width and turning radius in continuous reverse bend section.

3 ANALYSIS AND DISCUSSION OF THE TEST RESULTS OF THE ORIGINAL ARRANGEMENT SCHEME

In the ship maneuvering simulator, the maneuvering characteristics of a 5000-ton vessel navigating through a continuous bend under conditions of a 34-meter water level in the channel, still water, and constant wind direction (wind speed of 5 knots) were simulated to analyze the influence of channel layout and wind direction on navigational safety. Under these conditions, nine sets of maneuvering tests were conducted for upbound vessels and five sets for downbound vessels. Among the nine upbound tests, eight failed due to collisions with the sidewalls when sailing against the wind in the connecting section of the two turns, which was influenced by wind-induced drifting. The drift angle for upbound navigation exceeded 20 degrees. Three out of the five downbound tests were successful, indicating that under the original channel arrangement (turning radius of 360 meters), navigational conditions for downbound vessels are more favorable than those for upbound vessels.

Table 1. Ship navigation results for upstream and downstream passage through continuous bends

No.	Ship type	Ship direction	Wind direction	Note
S01	5000-ton vessel	upward	Northern Wind	Collision, Failure to Pass
S02				Collision, Failure to Pass
S03				Collision, Failure to Pass
S04				Collision, Failure to Pass
S05				Collision, Failure to Pass
S06				Collision, Failure to Pass
S07				Pass
S08				Collision, Failure to Pass
S09				Collision, Failure to Pass
X01	5000-ton vessel	downward	Northern Wind	Pass
X02				Pass
X03				Pass
X04				Collision
X05				Collision

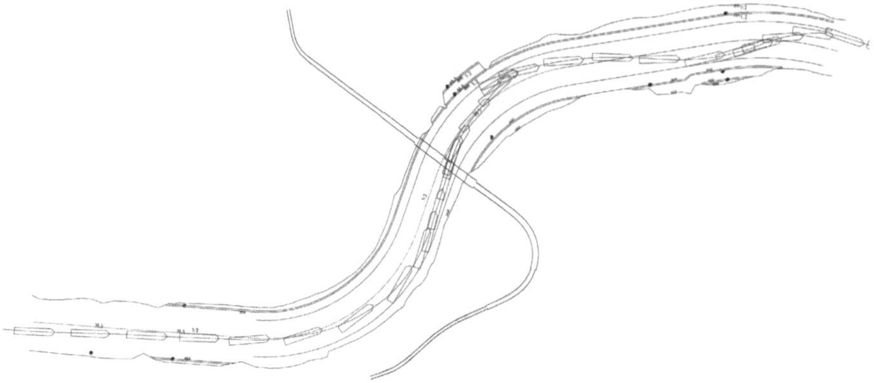

Figure 5. Ship position chart when the vessel is going upward.

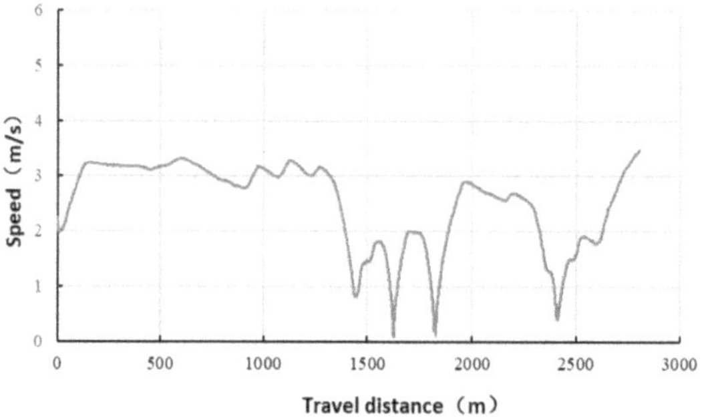

Figure 6. Variation curve of sailing speed with travel distance when the vessel is going upward.

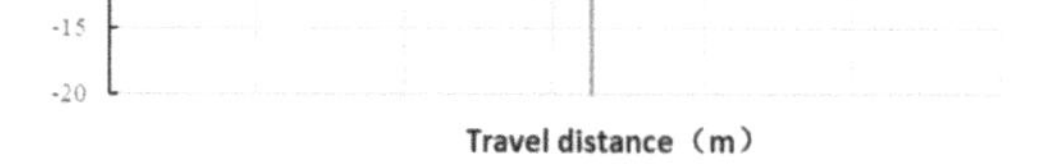

Figure 7. Variation curve of drift angle with travel distance when the vessel is going upward.

From the ship in still water, north wind conditions of the ship up and down through the continuous bends of the ship navigation conditions can be seen, the ship in the transition section of the continuous turns can not be adjusted to the appropriate ship position in order to safely enter the next turn. Unreasonable design of continuous bends leads to the ship entering the next turning segment after leaving the turn without adjusting the ship's position, so the rational design of the turning segment of the channel is crucial to the safe navigation of the ship.

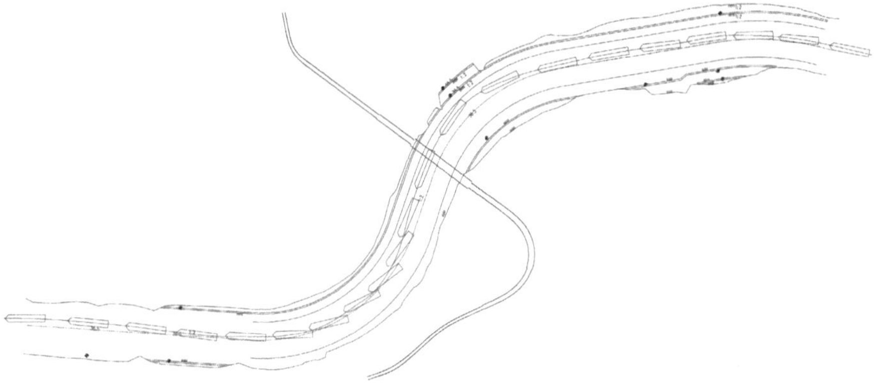

Figure 8. Ship position chart when the vessel is going downward.

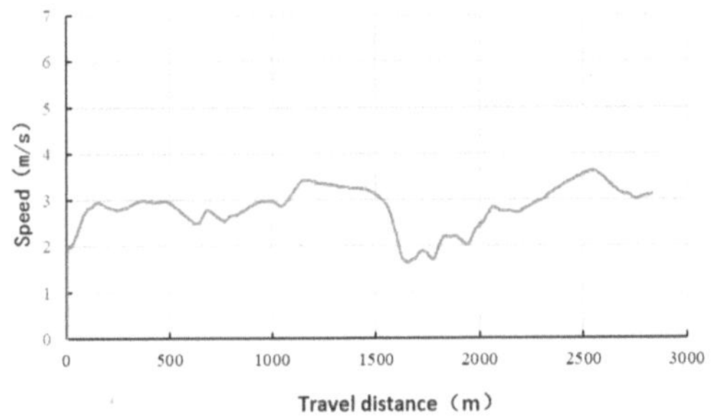

Figure 9. Variation curve of sailing speed with travel distance when the vessel is going downward.

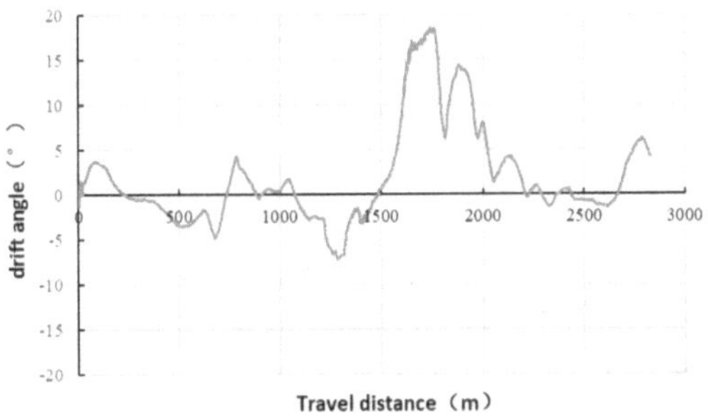

Figure 10. Variation curve of drift angle with travel distance when the vessel is going downward.

The main factors to be considered in the design of the turning section of a channel bend are the turning radius, the width of the turning section, turning angle ,and the length of the straight section between the two bends. Among them, the turning radius of the channel is related to the angle of the channel, the scale of the ship, the ship's speed, the ship's depth-to-draft ratio, the balanced rudder angle, the ship's operating characteristics, the ship's loading condition, the driving technology and other human factors, the natural conditions of the turning section, and the presence or absence of tugboat assistance, and other factors.

The U.S. Army Corps of Engineers' Hydraulic design of deep draft navigation project states that when the steering Angle is greater than 50°, the ratio of turning radius to length is greater than 10. China's "Design Code for Waterway Engineering (JTS 181-2016)" in the explanation of the provisions of the channel steering angle of more than 60° of the project is not too much, practice has proved that the turning angle is large, maneuvering is more difficult to increase the turning radius is necessary. It is proposed that there is the following relationship between the turning radius of the waterway and the ship's speed, ship's length and turning angle:

$$R = \frac{0.5 V_S L}{1 - \sin \frac{\varnothing}{2}} \tag{1}$$

Where R = turning radius (m); V_s = ship speed (knots); L = designed ship length (m): φ = turning angle (°)

The upstream turning angle of this section is 61.74°, and the downstream turning angle is 74.78°, and the turning angles of ship navigation are large. According to the above formula, under the condition of ship design speed 6.48knots, the turning radius of the bend should be around 600m~740m, and the turning radius under the original arrangement condition is only 360m, which can't meet the ship specification requirements, taking into consideration that it is more difficult to maneuver the ship, it is suggested to increase the turning radius of the channel appropriately or (and) to reduce the sailing steering angle.

Whether the ship can safely enter the next bend in the continuous bend lies in whether the fleet arrives at the next bend before the navigation parameters can meet the safety requirements have been adjusted to the appropriate ship position. Waterway Engineering Handbook (China): the length of the transition section of the continuous bend must ensure smooth transition of the water flow, so that the ship can adjust the course in time, from the previous bend out of a straight line, and then turn in the opposite direction of this requirement.

According to the Waterway Engineering Handbook, there are:

$$L \geq \frac{6}{n}\left(1 + \frac{3}{n}\right) L_C \tag{2}$$

$$n = R_C / L_C \tag{3}$$

Where: L = the length of straight line section between reverse curves(m); L_C = the length of the maximum design vessel or fleet(m).

$$R_C = \frac{R_1 + R_2}{2} \tag{4}$$

This project section of the river, two continuous turning section turning radius are 360m, that is, R_C = 360m, n = 4, after calculating $L \geq 2.625 L_c$, thus, between the two curves of the straight line section length should be at least greater than 2.625 times the length of the ship, that is, 236.25m, and the current length of straight line transition section is only 119m, can not satisfy the "Waterway Engineering Handbook" design requirements.

Under the existing arrangement, under the condition that the lowest navigable water level of the river is 34m and there is no tributary convergence, the ship passes through the continuous curved section up and down under the condition of northerly wind, and there are many times that it crosses the route and hits the opposite side of the channel bank or scrapes against the side of the channel bank, which shows the problem of insufficient turning radius of the continuous curved section and the problem that the straight-line connecting section is too short.

Under the conditions of different frequency flow discharge in the canal, ship maneuvering is expected to be more difficult, especially when traveling upstream. It is recommended to appropriately increase the channel turning radius or (and) reduce the turning angle of the bend, adjust the length of the transition section according to the formula, and further verify the maneuverability of the ship through the ship simulator to determine the reasonableness of the design of the channel turning radius and the length of the transition section.

4 OPTIMIZED ARRANGEMENT SCHEME TEST RESULTS

On the basis of the ship maneuverability study on the original arrangement of continuous channel bends, the turning angle and turning radius of the continuous bends were optimized. After optimization, the bottom width of the channel is 80m, the width of the curved section of the channel is the same as that of the straight section, and the elevation of the bottom of the channel is 27.7 m. The turning angle of the upstream is 36.43°, and that of the downstream is 49.48°. The turning radius of the channel is 540m. The straight line distance between two bends is 386 m.

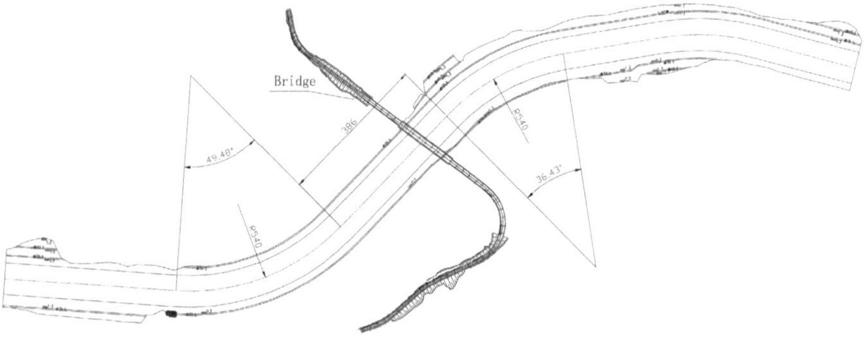

Figure 11. Layout of the channel in the study area (Optimized layout).

Under the optimized arrangement of continuous curve turning radius R = 540m in the study area waterway, ship maneuvering test is carried out at 34m water level, static water, and northern wind conditions, and the up and down track lines of the ship obtained from the test showed that the ship was able to enter the continuous curve smoothly upwards, with the maximal speed of 8.78km/h, the maximal drift angle of 8.68°, and the track belt is within the range of the canal, and the maneuverability of the ship is good.

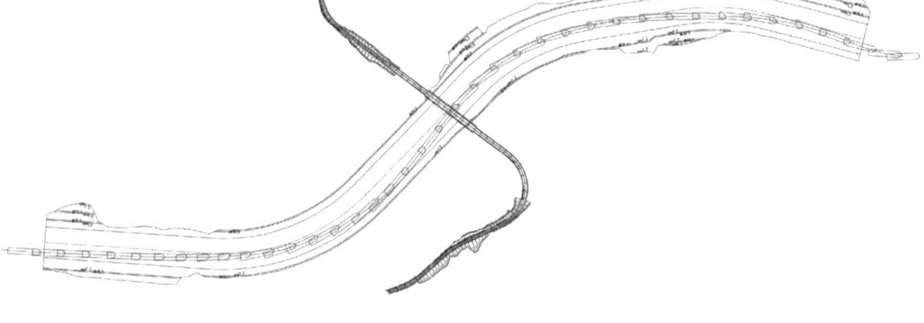

Figure 12. Ship position chart when the vessel is going upward.

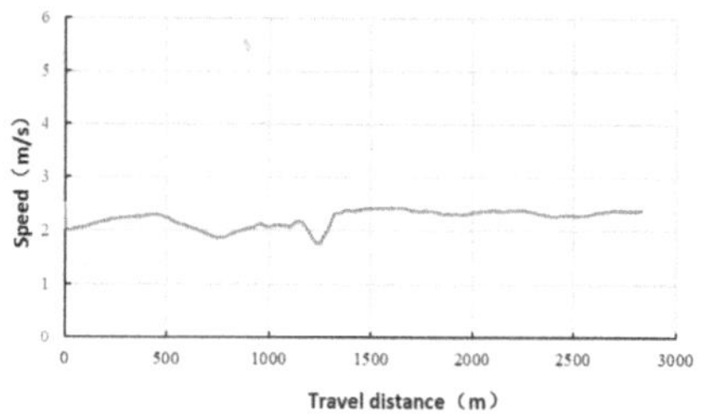

Figure 13. Variation curve of sailing speed with travel distance when the vessel is going upward.

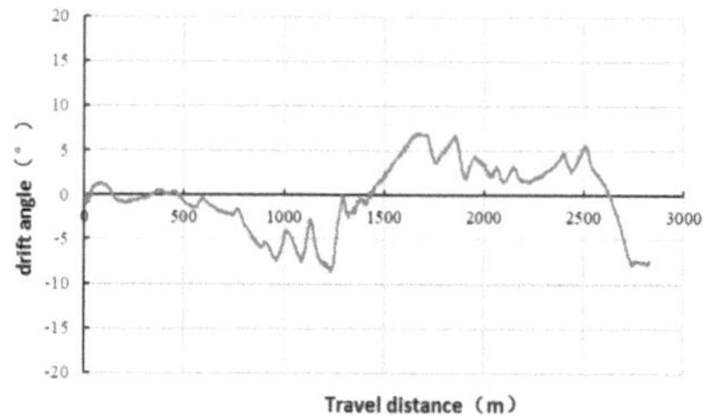

Figure 14. Variation curve of drift angle with travel distance when the vessel is going upward.

Under the optimized arrangement of continuous curve turning radius R = 540m in the channel of the research section, the vessel can smoothly go down into the continuous curve under the condition of 34m water depth static water and north wind grade 2 (constant wind), with the maximum speed of 12.13km/h, the maximum drift angle of 9.75°, and the track belt is within the range of the channel. The maneuverability of the ship when going down is good.

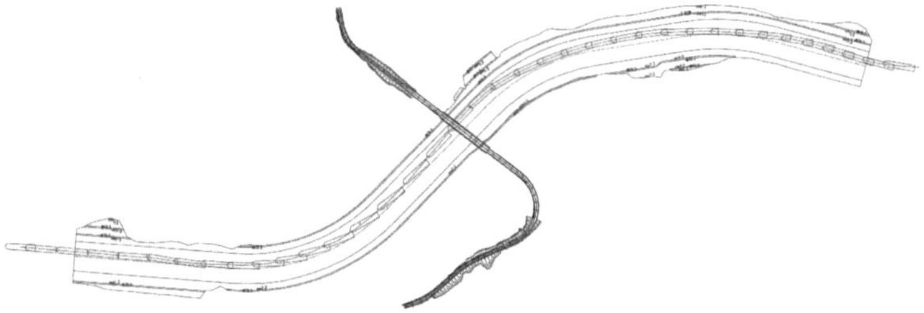

Figure 15. Ship position chart when the ship is going downward.

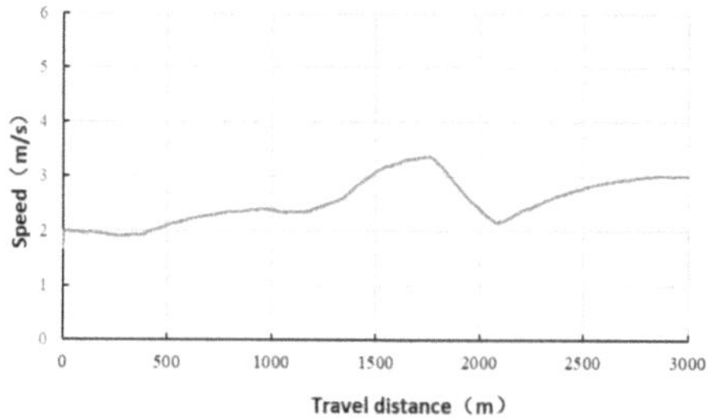

Figure 16. Variation curve of sailing speed with travel distance when the ship is going downward.

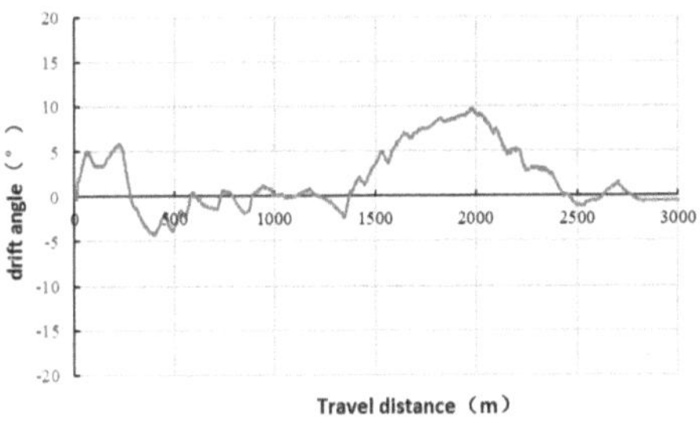

Figure 17. Variation curve of drift angle with travel distance when the ship is going downward.

5 RESULTS AND DISCUSSIONS

(1) In the continuous turning section of the channel with a turning radius of 360 meters and a water level of 34 meters, vessels navigating through this curved section under north wind conditions frequently cross the channel, impacting or rubbing against the opposite bank. This behavior indicates that the turning radius of the continuous curved section is insufficient and highlights issues with the short straight sections connecting the curves.

(2) By appropriately increasing the turning radius of the channel to 540 meters, reducing the steering angle of the bend, and extending the length of the transition section, the vessel can now navigate smoothly into and out of the continuous bend. The trajectory remains within the channel boundaries, significantly enhancing ship maneuverability. Further validation using ship simulators is required to assess maneuverability under varying discharge flow rates in artificial channel locks.

(3) The study area comprises a bi-directional curved channel with a relatively narrow width. Vessels navigating this region are particularly susceptible to shallow water and shore wall effects. During navigation, it is imperative to adhere strictly to established regulations and formal navigation protocols. Given the complexity of the channel environment, avoidance maneuvers and course adjustments occur more frequently, necessitating potential speed reductions or temporary stops. In adverse weather conditions, vessels may also need to anchor and await favorable conditions. Crews should be prepared for such eventualities by ensuring readiness of engines and anchors. Additionally, maintaining a safe distance from other vessels, exercising cautious maneuvering, and enhancing lookout procedures are crucial.

ACKNOWLEDGMENTS

The work described in this paper was fully supported by a grant from the National Key Research and Development Plan of China (No. SQ2023YFC3200166), and the Special Funds for Basic Scientific Research of Nanjing Hydraulic Research Institute (No. Y124018).

REFERENCES

Changjiang waterway bureau (2004). *Waterway Engineering Handbook*. China Communications Press. Beijing. 2004

Cao Fengshuai, Jiang Junjie and Wu Peng (2011). *On design Parameters of Channel Turn in Coastal Ports. Port & Waterway Engineering*. 2011(No.9)

Hu Xuyue, Wang Qiaoyang, Huang Yong (2010). The Analyze of the Drift Angle of the Barge in Inland Waterway. *HUNAN COMMUNICATION SCIENCE AND TECHNOLOGY.* Vol.36 No.2

Marc Mansuy (2022), Maxim Candries, Katrien Eloot, Sebstien Page(2021). Simulation Strudy to Assess The Maximum Dimensions of Inland Ships On The River Seine In Paris. *Proceedings of PIANC Smart Rivers* 2022 October 18-21, 2022, Nanjing, China

Ministry of Transport of the People's Republic of China (2016). *Design Code for Waterway Engineering (JTS 181-2016).* China Communications Press. Beijing. 2016

USACE (2006). *Hydraulic Design of Deep Draft Navigation Project*, Washington: USACE, 2006.

Wang Fei (2022), Zhang Yu-qian, An Xiao-yu, Kong Xian-wei Research on navigation safety of ships in navigation hydropower junction of narrow continuous curved river. *China Harbour Engineering.* Vol.42 No.5

Zhang Xiu-feng, Hong Bi-guang, Jin Yi-cheng (2003). Evaluation Ship Navigation Security with Ship Handling Simulator. *Navigation of China.* No.3. 2003

Proceedings of the International Association of Institutes of Navigation World Congress 2024 – Chen et al. (Eds)

Research on the construction of intelligent navigation brain empowered by digital technology

H.Q. Zhai & J. Zhang
Jiangsu Maritime Institute, Nanjing, China

ABSTRACT: In the era of Big Data, based on the five development concepts of "Digital Empowerment, Intelligent Innovation, Green Safety, Flexible Coordination and Efficient Outstanding " in Intelligent Navigation, a modern "Intelligent Navigation Brain" is established to complete the entire process of "Navigation Situation Awareness, Navigation Cognitive Computing, Digital Decision Control and Network Security Regulations" within Intelligent Navigation. "Navigation Situation Awareness" obtains its own and environmental navigation information. "Navigation Cognitive Computing" extracts and processes relevant navigation elements to "recognize" comprehensively the risks encountered during Intelligent Navigation. "Digital Decision Control" uses the information fed back by "Navigation Situational Awareness" to reaffect the cognition of "Navigation Cognitive Computing" on the situation and itself, and Real-Time feedback the current state to the space for "Navigation Situation Awareness" and "Navigation Cognitive Computing" to further modify the ship navigation situation. Most of all, "Network Security Regulations" is the top priority for Intelligent Navigation in terms of "Green Safety, Efficient and Outstanding". A new "Digital Intelligent + Navigation" era is opening, in which, with Real-Time "Navigation Situation Awareness", Agile "Navigation Cognitive Computing", Intelligent "Digital Decision Control" and Reliable "Network Security Regulations", "Intelligent Navigation Brain" empowered by Digital Technology is ultimately achieving Accurate and Intelligent Navigation Safety.

1 INTRODUCTION

In the era of Big Data, Digital Technology is the key to Intelligent Navigation. In 2017, China officially launched the world's first intelligent merchant ship, the "Intelligent Dolphin" (iDolphin), which was equipped with the world's first self-learning Ship intelligent Operation and Maintenance System (SOMS). The three subsystems, including Intelligent Navigation, intelligent engine room and intelligent energy efficiency management, were integrated into the SOMS, which could enable ship operations to better meet the requirements of safety, environmental protection, and economy in the era [1]. In August 2018, the world's first zero emission fully automatic container ship "Yara Birkeland" was delivered in the first quarter of 2020 after undergoing simulation and testing. "Yara Birkeland" is regarded by industry insiders as a silhouette of highly intelligent technological innovation in the shipping industry [2]. The ship navigation system (Sherpa System for Real, SSR). jointly developed by Japan Mail Ship and Japan Marine Science (JMS), had undergone an autonomous navigation system collision avoidance test for the "Manned Automatic Navigation Ship" on the "Iris Leader" ship in Japan. The test results showed that it could avoid collisions with other ships [3].

Digital transformation, which is an important support for Intelligent Navigation to achieve high-quality development, improves the ability of digital resource allocation and

DOI: 10.1201/9781003624424-8

brings about a great change in navigation in the past century. In the era of big data, Intelligent Navigation should actively increase its investment in digital research and actively try the innovative "Digital + Navigation" mode, thereby giving new momentum to the innovative development of marine transportation.

2 DIGITAL AGE, ADHERING TO THE DEVELOPMENT CONCEPT OF INTELLIGENT NAVIGATION

While Big Data brings hope to the industry, it also leaves everyone at a loss. The advent of the Big Data Era not only illuminates the path ahead like a lighthouse for ships adrift on the vast ocean, but also renders the traditional shipping industry more transparent, intelligent and flexible. At the "SISI International Port and Shipping Development Forum (2019) - Big Data helps Port and Shipping Innovation and Development", which was hosted by Shanghai Maritime University and organized by Shanghai International Shipping Research Center, experts clarified the core competitiveness of the future shipping industry [4].

2.1 *Empowering intelligent navigation with digital technology*

Digitization falls within the scope of informatization and represents a special stage in the process of informatization. On July 15, 2020, the Ship and Ocean Engineering Data Exchange and Sharing Platform CSBC, designed and developed by China Classification Society (CCS), was officially launched, which has laid a solid foundation for building a navigation digital ecological chain, solving the problem of navigation data island, and guiding the industry and technology to frequency [5]. Navigation digitalization is gradually empowering the shipping industry through various means such as ecologicalization, targeting different levels of navigation, Demand-Oriented, and capacity openness.

Navigation digitization contributes to the national strategy of becoming a maritime powerhouse, realizes the objectives of Intelligent Navigation such as transparency, efficiency, environmental-friendliness and safety, and finally integrates the shipping industry into the global digital transformation wave in real time. Digital technology should be systematically and step by step applied to achieve modern Intelligent Navigation, as shown in Figure 1.

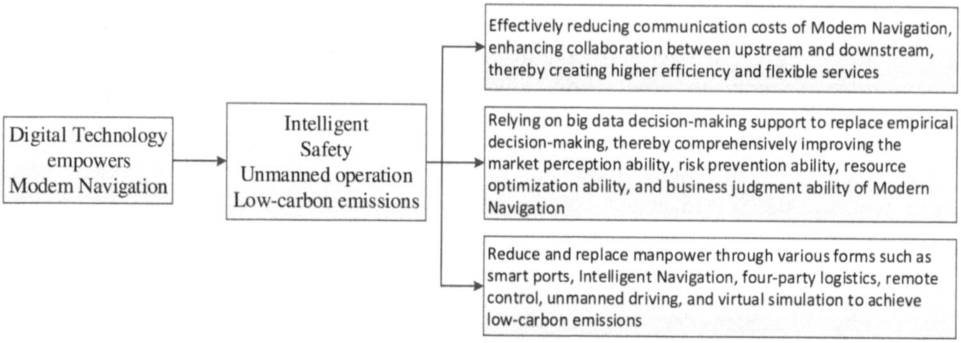

Figure 1. Empowering intelligent navigation with digital technology.

2.2 *Digital age, adhering to the development concept of intelligent navigation*

At present, the shipping industry is still at a low ebb, so the development of intelligent navigation can not only improve the safety of ship navigation, but also reduce operating costs and carbon emissions. At the same time, with the orderly promotion of the Global Carbon Emission Trading Mechanism, the ship Energy Efficiency Design Index (EEDI), and the ship Energy Efficiency Operational Index (EEOI), it has become an inevitable trend for the shipping industry to reduce greenhouse gas emissions. However, Intelligent Navigation should not merely rely on the application of various advanced navigation aids, but also possess the same capabilities as humans in observing, thinking, and handling problems.

With the breakthrough of Artificial Intelligence in the field of natural language processing, it has become possible to use computers to replace humans in reading and understanding vast amounts of data, and to help people make scientific decision-making and formulate schemes. In view of this, cognitive computing technology based on Big Data came into being [6]. In the era of Artificial Intelligence, Big Data and Cloud Computing, Intelligent Navigation should be founded on the five development concepts of Digital Empowerment, Intelligent Innovation, Green Safety, Flexible Coordination and Efficient Outstanding (as shown in Figure 2), so as to make Big Data and Artificial Intelligence Algorithm circulate in the "Shipping Limb" like blood.

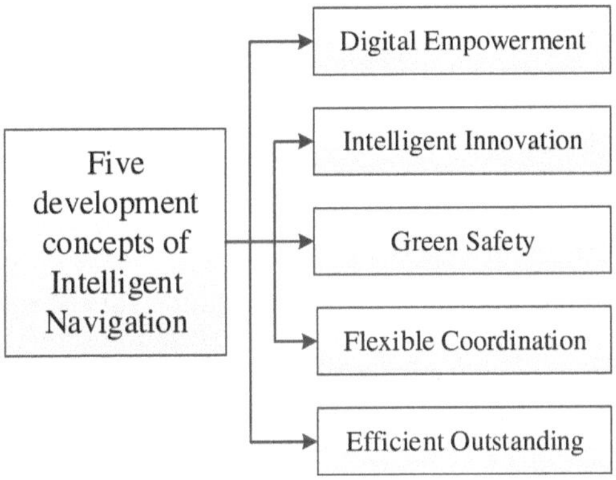

Figure 2. Five development concepts of intelligent navigation.

3 EMPOWERED BY DIGITAL TECHNOLOGY, BUILDING AN "INTELLIGENT NAVIGATION BRAIN"

To foster a profound digital transformation in navigation and achieve the true digital integration of navigation, Intelligent Navigation should enable ships to develop the same capabilities as humans in observing, thinking, and dealing with problems. At the same time, by drawing on the collaborative processing mechanism of the human body's "Brain", "Five Senses" and "Hands and Feet", a modern "Intelligent Navigation Brain" (as shown in Figure 3) for "Cloud-Network-Edge-Teminal Collaborative Computing" [7] can be constructed, so as to complete the entire process of Intelligent Navigation based on "Navigation Situation Awareness, Navigation Cognitive Computing, Digital Decision Control and Network Security Regulations".

After the analysis of Navigation Situation Awareness, Navigation Cognitive Computing extracts and processes relevant navigation elements to comprehensively "recognize" the risks encountered during Intelligent Navigation. Based on this, "Digital Decision Control" uses the information fed back by "Navigation Situation Awareness" to reaffect the "Navigation Cognitive Computing"'s cognition of the situation and itself, and feeds back the current state to the "Navigation Situation Awareness" and "Navigation Cognitive Computing" spaces, so as to further correct the ship navigation situation.

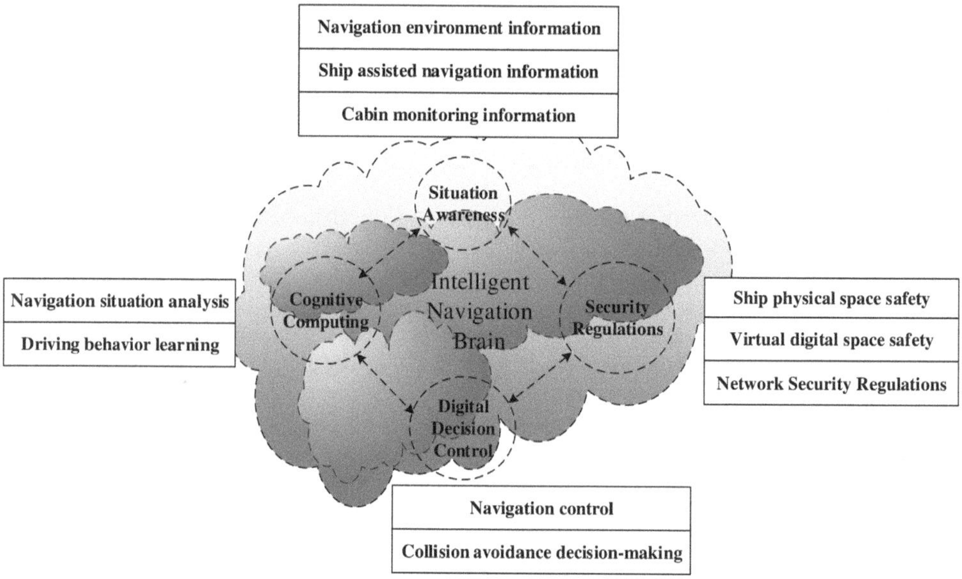

Figure 3. Intelligent navigation brain.

3.1 *Navigation situation awareness*

Navigation Situation Awareness serves as the foundation for the safe navigation of intelligent ships. The Situation Awareness of Intelligent Navigation should not only be capable of perceiving the ship's own state information (including dynamic information such as position, speed and heading and Real-Time monitoring ability of ship structure data), but also be able to perceive the navigation environment information (including Real-Time collection and early warning receiving ability of hydrological and meteorological information such as wind, wave and current). With the deepening of Intelligent Navigation research, new technologies such as the ship Automatic Identification System, marine radar (S-band and X-band), Global Positioning System (GPS) and Artificial Intelligence are widely used in the field of Intelligent Navigation research, which greatly improves the intelligent perception ability of navigation environment.

The "Development of Navigation Situation Intelligent Perception System" project launched in China was highly integrated by new technologies such as radar, panoramic infrared vision, sound recognition, remote sensing and Three-Dimensional reconstruction of digital scene. This integration has completely solved the problems related to the detection, awareness and even cognition of the ship's navigation environment, and finally it has become one of the core equipment for the autonomous navigation of unmanned ships [8]. If the intelligent ship is compared to human beings and the Intelligent Navigation of ships compared to human walking, then Navigation Situation Awareness is equivalent to the comprehensive use of human senses such the eyes, ears, nose to perceive the surrounding environment. Navigation Situation Awareness technology includes Navigation Situation Awareness sensor, Navigation Situation Awareness data fusion and other technologies.

3.2 *Navigation cognitive computing*

With the advent of the Big Data and intelligent era, the research focus of machine learning has begun to shift from the field of Situation Awareness to that of Cognitive Computing. How to improve the cognitive capacity for handling large-scale data has became a research hotspot in intelligent science and technology [9]. Cognition refers to the process through which people acquire knowledge, put knowledge into practice, and process information. Human cognitive activities include feeling, perception, memory, thinking, imagination and language. Cognitive Computing is about exploring uncertain, imprecise, and certain real-world problems within biological systems, with the aim of using computers to simulate the processes of human perception, analysis, thinking, and decision-making.

Combining the data from the Automatic Identification System, Quan Bo *et al.* [10] proposed a Recurrent Neural Networks-Long Short-Term Memory (RNN-LSTM) model by using the method of Deep Learning to predict the ship's navigation trajectory. This model was trained with the ship's navigation data to predict the ship's navigation trajectory. In response to the issue of predicting the impact of ocean meteorology on ship speed, Wang Shengzheng *et al.* [11] confirmed that the prediction results of the Alternating Sparse Auto Encoders (ASAE) network model are more in line with actual sea conditions by using ship navigation data and meteorological data for training and verification, and ultimately provided guidance for navigation optimization strategies such as route planning and speed recommendation.

3.3 *Digital decision control*

The marine environment is complex and the various aspects of the navigation situation are interrelated. As a result, it is becoming increasingly challenging to rely solely on human efforts to understand and predict the navigation situation. In view of this, the introduction of Artificial Intelligence has further improved the ability of Intelligent Navigation Decision-Making and Control. To optimize the dynamic and steady-state performance of ships and realize the intelligent scheduling of ship navigation, Ma Yan [12] proposed an Artificial Intelligence algorithm based on Particle Swarm Optimization for the adaptive optimization of navigation path planning. In the era of Big Data, in order to realize the optimal navigation path, Xie Yi [13] analyzed the genetic characteristics of Big Data related to global path data in navigation. Through this analysis, alternative paths with representative characteristics were obtained. Additionally, the Big Data ant colony optimization algorithm was introduced to optimize the calculation of these alternative paths.

The ship's dynamic positioning first uses sonar to measure the ship's position, and then the ship's automatic control system sends instructions to operate the lateral thrusters installed at the bow and stern, thereby fixing the ship's position. Dynamic positioning technology is one of the core technologies in the field of intelligent ships, boasting the advantages of high positioning accuracy and being unconstrained by water depth. To improve the robustness of ship positioning and resist the interference of external factors on the ship's dynamic positioning system, Wang Hui [14] first established a mathematical model of the ship's dynamic positioning system, and set up a positioning controller control algorithm using an internal disturbance processing mechanism. The simulation results showed that the ship's dynamic positioning system retained a strong anti-interference capacity even under large environmental disturbances. In order to achieve the visualization of parameter settings for various modules of the ship dynamic positioning and the system's output results, Manjiangtao [15] designed a ship dynamic positioning simulation system with a human-computer interaction interface, which allowed users to directly verify the superiority of thrust distribution optimization algorithms and control algorithms, and ultimately obtain an intuitive solution for the ship to maintain its predetermined position under current sea conditions.

Intelligent collision avoidance is a key research object in the field of Intelligent Navigation because waves, currents, and deep-water pressure always interfere with ship navigation. In the research of ship intelligent collision avoidance, the in-depth research on ship intelligent collision avoidance technology based on Big Data and Artificial Intelligence [16–18] is greatly reducing the collision risks for modern sailing ships.

3.4 *Network security regulations*

In the era of network communication, a country's economic development is inseparable from Information and Communication Technology (ICT), so network communication security is the top priority [19]. With the demand for intelligent ships to collect and analyze environmental information during navigation, along with the necessity of communication and information exchange between ships and shore-based systems, the volume of data, privacy, and information involved in network communication are increasing, and therefore the possibility of network attacks is also increasing. In view of the negative impact of the increasing network threats on the safety of Intelligent Navigation, Zhou Xiangyu [20] proposed a STPA-based analysis methodology that Synthesizes Safety and Security (STPA-SynSS). This methodology effectively realized the continuous tracking and closed-loop management of potential hazards. The STPA-SynSS could identify more unsafe control behaviors and loss scenarios, and generate more targeted risk control strategies, thus demonstrating the effectiveness and advanced nature of the method.

Given the fact that legal safety is an important prerequisite for the safety of Intelligent Navigation and the latter must be realized within the boundaries permitted by law, it is extremely necessary to strengthen the revision and improvement of maritime conventions and laws in order to keep up with the development trend of Intelligent Navigation [21]. Although the advantages of Intelligent Navigation, such as safety, efficiency, economy and environmental protection, have brought expectations to people, its rapid development has posed great challenges to the current international legal system. As a specialized agency of the United Nations, the International Maritime Organization (IMO) should gradually incorporate Maritime Autonomous Surface Ships (MASS) into the international regulatory framework and play an important role in the legal supervision of MASS. Li Zhen [22] used the empirical analysis method to briefly introduce the international process of legal supervision over Intelligent Navigation, and further emphasized that Intelligent Navigation should pay high attention to the leading role of IMO in sorting out the MASS laws and regulations.

4 WORKING HARD TOGETHER, OPENING THE ERA OF "DIGITAL INTELLIGENT + NAVIGATION"

As the modern shipping environment is complex, the "Intelligent Navigation Brain" empowered by Digital Technology should have enhanced Real-Time Navigation Situation Awareness, more Agile Navigation Cognitive Computing, more Reliable navigation support, and a Safer navigation environment so as to realize more Intelligent Digital Decision Control. Let's work hard together to usher in the era of "Digital Intelligent + Navigation" (as shown in Figure 4).

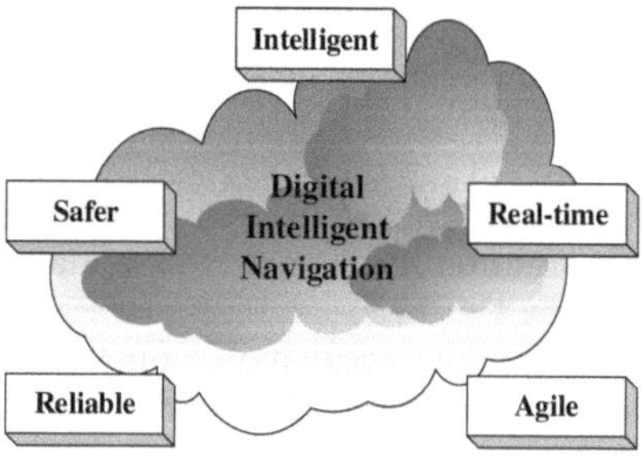

Figure 4. "Digital intelligent + navigation" era.

4.1 *Digital empowerment, real-time and agile navigation awareness and cognition*

"Intelligent Navigation Brain" comprehensively improves the emergency response ability. In the process of ship navigation, the "Intelligent Navigation Brain" based on Big Data should have Real-Time Navigation Situation Awareness under various emergency, abnormal and special circumstances. Based on the multi-channel parallel signal acquisition of FPGA and DSP, Liu Ruqing [23] designed a Real-Time perceptual lidar multi-channel data acquisition system to realize the rapid solution of LIDAR point cloud in the surrounding environment and obstacles.

To realize data sharing, the "Intelligent Navigation Brain" should open up the previous navigation digital island, and use "Digital Technology Empowerment" to realize more Agile Navigation Cognitive Computing. Through Fog Radio Access Networks (F-RAN), Li Zhipeng [24] first integrated cluster nodes such as new navigation aids, IoT devices, and nearby water vessels, and then performed Real-Time Cognitive Computing on a large amount of data from distributed sensors. Finally, a general situational map of the water flow field near the waterway was constructed to ensure navigation safety within the waterway.

4.2 *Big data decision, smart and reliable navigation secure guarantee*

In the era of Big Data, the "Intelligent Navigation Brain" wakes up the sleeping data, enabling the data to speak. It breaks the digital silos and realizes data interconnection through the integration, analysis and application of data. Finally, it realizes the precision and intelligence of modern Digital Decision Control. In order to achieve effective integration of Automatic Identification System (AIS) with technologies such as Dig Data, AI, and Machine Learning, Gao Miao [25] conducted research on "Structured learning of ship intelligent collision avoidance Decision-Making based on Seq-CGAN and ship handling behavior basic splice" to obtain a humanized ship intelligent collision avoidance decision that is in line with conventional maritime practices.

For an industry with high energy consumption such as the shipping industry, the "Intelligent Navigation Brain" based on Big Data can establish a reliable underlying digital capability, thus providing a solid and reliable basic guarantee for modern shipping. At the same time, the "Intelligent Navigation Brain" has established a dynamic Safe protection mechanism, which can issue early warnings, ensuring that the Intelligent Navigation network information will not be leaked and effectively resisting network attacks. In order to achieve energy-saving and emission reduction control of ships with the goal of minimizing operating costs and greenhouse gas emissions (such as CO_2), Guo Qiang [26] established a ship energy-saving control decision support system based on Big Data by mining and analyzing shipping Big Data from the perspective of the entire voyage of ships and predicting BP neural network models.

5 CONCLUSION

Intelligent Navigation is the future development trend of the shipping industry. Based on the development concept of "Green and Safe" Intelligent Navigation, this article refers to the collaborative processing mechanism of the human body's "Brain" and "Five Senses" and "Hands and Feet" to construct a modern "Intelligent Navigation Brain" based on "Cloud-Network-Edge-Teminal Collaborative Computing" that can circulate data and AI algorithms in the "Shipping Limb" like blood. With Real-Time "Navigation Situation Awareness", Agile "Navigation Cognitive Computing", Intelligent "Digital Decision Control" and Reliable "Network Security Regulations", "Intelligent Navigation Brain" empowered by Digital Technology is ultimately achieving Accurate and Intelligent Navigation Safety.

REFERENCES

[1] Wang S.J. (2019): Key to the future of shipping industry, *China Ship Survey*, **10**, 73–77

[2] Zhao B. (2020): these "first" significance is extraordinary, *China Ship Survey*, **02**, 27–31

[3] Wu X.X. (2021): Smart ship approaching and the functional element system needs to be improved, *China Shipping Newspaper*

[4] Wu M.H. (2019): How big data can make the port and shipping industry "digital Nuggets"?, *Navigation*, **06**, 3–4

[5] Wang S.J. (2020): Digital sharing platform enables digital transformation of ship industry, *China Ship Survey*, **07**, 54–57

[6] Du X.T. (2021): Applications of big data cognitive computing in content security governance, *Big Data Research*, 7(**06**), 53–66

[7] Huang X.Q., Wei X.B., Zhu L., *et al.* (2021): Research on Intelligence Distribution Network Based on Cloud-Network-Edge-Teminal Collaborative Computing, *Radio Communications Technology*, **48(01)**, 89–96

[8] Zhang D., Zhao Y.X., Cui Y.F., *et al.* (2021): A Visualization Analysis and Development Trend of Intelligent Ship Studies, *Journal of Transport Information and Safety*, **39(01)**, 7–16+34

[9] Chen W.H., An J.Y., Li R.F., *et al.* (2017): Review on Deep-learning-based Cognitive Computing, *Acta Automatica Sinica*, **43(11)**, 1886–1897

[10] Quan B., Yang B.C., Hu K.Q., Guo C.X., *et al.* (2018): Prediction Model of Ship Trajectory Based on LSTM, *Computer Science*, **45(S2)**, 126–131

[11] Wang S.Z., Shen X.Q., Zhao J.S., *et al.* (2018): Prediction of marine meteorological effect on ship speed based on ASAE deep learning, *Journal of Traffic and Transportation Engineering*, **18(02)**, 139–147

[12] Ma Y. (2017): Ship route planning based on artificial intelligence algorithm. *Ship Science and Technology*, **39(24)**, 22–24

[13] Xie Y. (2019): Research on the optimal route planning algorithms for ship navigation based on large data. *Ship Science and Technology*, **41(20)**, 22–24

[14] Wang H. (2021): Mathematical modeling of ship power system and simulation of positioning controller, *Ship Science and Technology*, **43(02)**, 28–30

[15] Man J.T. (2021): *PID CONTROLLER OPTIMIZATION AND PROGRAMMING OF SHIP DYNAMIC POSITIONING SYSTEM, Harbin Institute of Technology*

[16] Ni S.K. (2020): *Study on Key Technologies for Ship Intelligent Decision Making for Collision Avoidance Based on Rules, Dalian Maritime University*

[17] Nie H.B. (2020): Analysis of ship automatic collision avoidance decision algorithm based on artificial intelligence and soft computing, *China Informatization*, **02**, 80–82

[18] Shi Q.T., Zhu X.Y. (2020): Optimal scheduling algorithm of ship collision avoidance path based on big data and artificial intelligence, *Ship Science and Technology*, **42(02)**, 55–57

[19] Peng S.J., Tu S.P., Yu Y.Q. (2021): Background and Countermeasures of network communication security in information age, *Computer & Network*, **47(05)**, 53

[20] Zhou X.Y. (2020): *Research on Hazard Analysis Method for Autonomous Ships, Dalian Maritime University*

[21] Fan H.G. (2018): Problems and solutions of development of intelligent unmanned ships, *Ship Standardization Engineer*, **51(3)**, 16–19+52

[22] Li Z. (2020): International process of intelligent ship legal supervision, *World Shipping*, **43(03)**, 1–4+15

[23] Liu R.Q., Jiang Y., Li F., *et al.* (2021): Design of multi-channel data acquisition system for real-time perception lidar, *Infrared and Laser Engineering*, **50(05)**, 88–94

[24] Li Z.P., Cui K.T. and Sun W.Q. (2021): Real time situation awareness of channel flow field based on fog calculation, *China Water Transport*, **08**, 70–74

[25] Gao M. (2021): *Structured learning of ship intelligent collision avoidance decision-making based on Seq-CGAN and ship handling behavior basic splice, Dalian Maritime University*

[26] Guo Q. (2020): *Research on Full Range Energy Saving Control of Ship Based on Big Data Technology, Chongqing Jiaotong University*

Proceedings of the International Association of Institutes of Navigation World Congress 2024 – Chen et al. (Eds)

A time window panning filtering method for ship trajectories

Dapeng Jiang & Guoyou Shi
Navigation College, Dalian Maritime University, Dalian, Liaoning, China
Key Laboratory of Navigation Safety Guarantee of Liaoning Province, Dalian, Liaoning, China

ABSTRACT: The rapid development of communication and navigation technologies has made it easier to acquire marine traffic data, which is now widely used for traffic pattern recognition, trajectory prediction, and risk perception in marine transportation studies. Data mining methods, driven by machine learning algorithms, have provided novel approaches to these studies, with high-quality data enhancing the effectiveness of algorithmic models. Ship trajectory data is often affected by factors such as completeness, continuity, spatial coherence, and consistency. This paper addresses these issues by using trajectory segmentation and reconstruction techniques to generate complete voyage trajectories. To improve spatial coherence and consistency, we propose the Time Window Panning Filtering (TWPF) method, which applies a Gaussian kernel within fixed-length time windows to smooth the trajectory features. The time window is then panned along the time axis to maintain consistency across the data. Compared to other filtering methods, TWPF enhances the alignment of ship trajectories with vessel movement, reducing outliers and improving data quality, thus supporting more accurate machine learning models in transportation research.

1 INTRODUCTION

With the advancement of navigation and positioning technologies, scholars have increasingly conducted extensive and depth analyses of marine traffic. The Automatic Identification System (AIS) is an automated tracking system that facilitates information exchange between ships and coastal stations. AIS data encompasses both static information, such as vessel identity, dimensions, and status, and dynamic key information, including position, course, and speed. These data are crucial for enhancing the safety and efficiency of marine navigation, providing valuable marine traffic data resources for a wide range of research and applications.

Currently, marine vessel trajectories primarily rely on AIS data. Research areas such as route planning and accident analysis, which are based on vessel trajectories, generally do not demand high-quality trajectory data. However, with the widespread application of artificial intelligence methods, in the marine traffic domain, studies on vessel behavior analysis, navigation trajectory prediction, and intelligent collision avoidance for unmanned vessels are becoming increasingly sophisticated. These studies necessitate high-quality trajectory data to enhance the accuracy of trajectory predictions (Liu *et al.* 2019) and improve the precision of intelligent collision avoidance systems (Gao *et al.* 2024). Nonetheless, AIS data can be corrupted due to transmission equipment issues or interference with radio signals, which can affect the integrity, continuity, and consistency of trajectory data. Therefore, smoothing and filtering trajectory data are essential steps in marine traffic data mining research (Zhang *et al.* 2023).

DOI: 10.1201/9781003624424-9

AIS trajectory data is essentially a type of time-series data, where changes in dynamic features such as position, course, and speed are closely related to time. Therefore, the filtering method proposed in this paper adopts a time window approach, progressively shifting the window to conduct smoothing and filtering operations across the entire trajectory. The primary contributions of this paper are outlined below:

(1) To address issues in the trajectory dataset collected within a specific geographical area, a spatiotemporal segmentation method for trajectories is established. For the segmented trajectories or sub-trajectories, a spline interpolation method is applied to ensure that the changes in position and speed of trajectory points align with the characteristics of physical movement after interpolation.
(2) A trajectory filtering method based on panning time windows is proposed, aimed at reducing errors such as jumps and deviations in the trajectory data. This method ensures that the entire trajectory adheres to the hydrodynamic characteristics of vessels.

2 RELATED WORKS

Many studies that employ ship trajectories as a data application contain a data processing component. There are also dedicated studies addressing issues present in trajectories. Lin *et al.* (2016) proposed a noise filtering technique for processing raw trajectories, which includes segmenting and simplifying trajectories to reduce large volumes of trajectory data. Zhao *et al.* (2018) addressed AIS data errors by pre-processing these errors in terms of physical integrity, spatial logic, and temporal accuracy, thereby enhancing the data quality dimensions of AIS trajectories. Zhang *et al.* (2018) proposed a multi-state vessel trajectory reconstruction model, which reconstructs vessel trajectories under different navigation states through three steps: outlier removal, vessel navigation state estimation, and trajectory fitting. Regmi. and Shah. (2021) introduced a method for smoothing trajectories by learning offsets and predicting estimated trajectory anomalies. Hartawan *et al.* (2021) proposed a preprocessing model comprising three stages: data cleaning, trajectory extraction, and trajectory compression. The results indicate that the proposed model reduces noise and minimizes the effects of trajectories that could impact subsequent clustering processes. Liu. and Yang. (2023) proposed a trajectory smoothing algorithm based on the Kalman filter, evaluating its performance on Gaussian and non-Gaussian noise, with experiments conducted on both simulated and real trajectory datasets. The algorithm effectively smooths Gaussian noise but performs poorly with non-Gaussian noise.

This paper focuses on studying problematic trajectories in vessel positioning data, addressing coarse errors and missing trajectories caused by various interferences. Subsequently, effective segmentation of trajectories based on spatio-temporal gaps features between trajectory sequences is conducted to solve the case of unclear voyages. Finally, it applies effective filtering based on the characteristics of time-series data in trajectories. The purpose of this paper is to establish a high-quality trajectory dataset, providing an effective trajectory data processing procedure for the application of deep learning methods in marine traffic research.

3 METHOD

3.1 *Trajectory processing based on spatio-temporal gaps*

Vessel AIS trajectories, being time-series data, are distinguished by the MMSI identifier, which separates the historical navigation data records of different vessels. Consequently, all other attribute features of each trajectory point are time-referenced. Due to various factors such as signal transmission interference, equipment malfunctions, and human operational errors, AIS trajectory data can exhibit issues in physical integrity, spatial logical integrity, and temporal accuracy (Zhao *et al.* 2018). The loss of communication signals or the brief drifting of vessel signals can lead to local data gaps in AIS trajectory sequences. In practical trajectories, this generally manifests in the following situations. (i) Random errors in the longitude or latitude fields result in position jumps (outliers) in the trajectory data, causing the vessel's trajectory to deviate from the logical spatial position characteristics of vessel movement, as illustrated by the trajectory in Figure 1a. (ii) Vessel AIS trajectories typically exhibit temporal continuity and positional sequence correlation under the same spatio-temporal conditions. However, due to occlusion or interference, continuous data points in a single trajectory might be lost, creating spatiotemporal blanks in the trajectory sequence, as shown in Figure 1b. (iii) When trajectory data are selected over a long time span within a limited geographical area, the same MMSI identifier may repeatedly traverse this area, resulting in unclear voyage demarcations for the same MMSI, as shown in Figure 1c. Therefore, during the processing of trajectories, it is essential to systematically analyze and address these situations.

To address situation (i), defined as coarse errors in the trajectory, these errors are removed during the preprocessing stage. However, this removal creates new issues, resulting in gaps in the trajectory sequence. Situation (ii) involves gaps in the trajectory sequence due to the continuous absence of data points. Situation (iii) focuses on analyzing the time intervals between adjacent trajectory points. To enhance data quality and ensure that each trajectory or sub-trajectory segment for each unique identifier is a spatiotemporally continuous sequence, a splitting algorithm based on the characteristics of spatiotemporal gaps in the trajectory is developed.

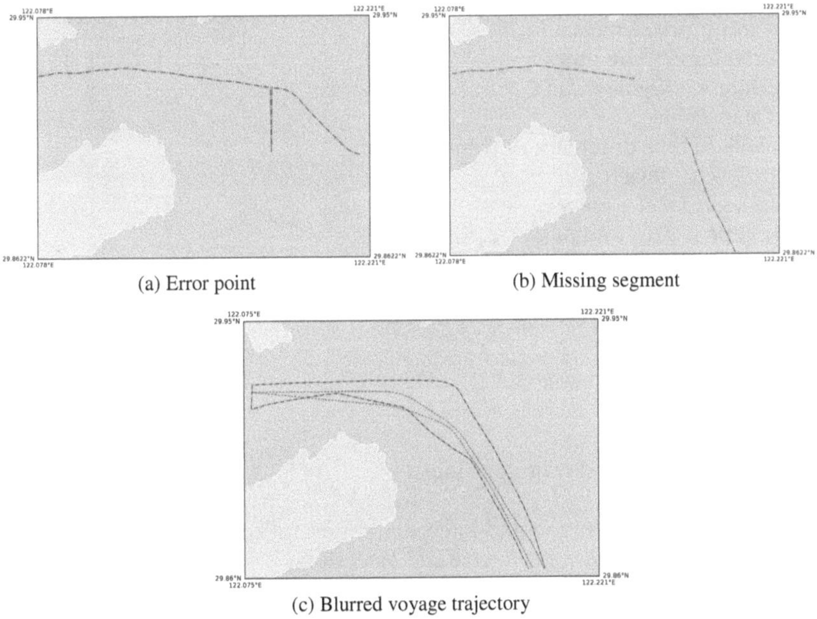

(a) Error point

(b) Missing segment

(c) Blurred voyage trajectory

Figure 1. Several scenarios of issue trajectories.

A trajectory sequence dataset $D = \{P^k; (k = 1, 2\cdots, N) | P \in \Theta\}$ is established based on the identifier MMSI, where Θ represents the area in which the vessel trajectories are located, and P^k represents the trajectory sequences of different MMSI. For any given trajectory segment denoted as $P^k = \{p_1^k p_2^k \cdots p_\tau^k p_{\tau+1}^k \cdots p_T^k; (\tau = 1, \cdots, \mathrm{T})\}$, where p_τ^k represents the spatiotemporal information of the trajectory points, p_τ^k encompasses the spatiotemporal attributes $p = \{t_i, x_i, y_i, v_i, c_i; (i = 1, 2\cdots, n)\}$ of the trajectory points, which represent the longitude, latitude, speed, and heading at time t_i.

3.1.1 *Trajectory splitting*

To address the potential scenarios within the aforementioned trajectories, this section proposes a spatiotemporal discrimination splitting algorithm, as illustrated in Algorithm 1. The decision to segment a trajectory with the same identifier primarily relies on the spatiotemporal correlation between trajectory sequence points. By examining the temporal and spatial intervals between adjacent trajectory points within the trajectory sequence of each MMSI, gaps are identified.

Here, δt denotes the time difference between adjacent trajectory points, and δd denotes the distance difference. T_{th} represents the time threshold, and D_{th} represents the distance threshold. T_{th1} determined based on the maximum AIS device information transmission frequency during the critical transition of a vessel from sailing to anchoring state. The time threshold T_{th2} is determined based on the minimum time required for vessel anchoring or mooring operations, typically derived from the captain's experience.

If the time difference δt or distance difference δd exceeds the corresponding threshold, it is considered a gap and marked as a segmentation label, as indicated in lines 5 and 6 of the pseudocode. If δt and δd do not meet the segmentation thresholds but do not conform to the continuous AIS trajectory time intervals and distance ranges of a navigating vessel, an interpolation operation is performed on the trajectory or trajectory segment.

Algorithm 1 spatio-temporal splitting and reconstruction of sub-trajectories

---

1: **do**: Group by MMSI for D
2: **for** *id* in MMSI $\in D$ **do**
3: Get trajectory sequence $\rightarrow P^{id}$
Sortvalues P^{id} by timestamp $\rightarrow P^k = \{p_1^k \cdots p_\tau^k \cdots p_T^k; (\tau = 1, \cdots, \mathrm{T})\}$
4: Declare var *segment_label* & *interpolate_label*
for p in P^k **do**
Calculate time difference δt from p_τ^k to $p_{\tau+1}^k$
5: **if** $\delta t \geq T_{thr2}$ **then**
6: *segment_label* $\leftarrow$ $index_p$
7: **else if** $\delta t < T_{thr2}$ and $\delta t > T_{thr1}$ **then**
8: Calculate distance p_τ^k to $p_{\tau+1}^k$: *function_geodis* $\rightarrow \delta d$
9: **if** $\delta d < D_{thr}$ **then**
10: *interpolate_label* $\leftarrow$ $index_p$
11: **else**
12: *segment_label* $\leftarrow$ $index_p$
13: **end if**
14: **else**
15: Retain the original trajectory intact.
16: **end if**
Performing segmentation and interpolation on trajectory sequences
17: **for all** $i \in$ *segment_label*, $j \in$ *interpolate_label* **do**
18: Perform trajectory segmentation $\rightarrow$ *function_Seg*(P^k,*i*)
19: Perform interpolation $\rightarrow$ *function_Interp*(P^k,*j*)
20: **end for**
21: **end for**
Input: Trajectory datasets: $D = \{P^k; (k = 1, 2 \cdots, N)\}, T_{thr1}, T_{thr2}, D_{thr}$
Output: The original trajectory sequence or sub-trajectory sequence:
sub_sequence $\Rightarrow P_{sub}^m = \{p_1^m \cdots p_\tau^m; (\tau = 1, \cdots, \mathrm{t})\}$ *or* original sequence $\Rightarrow P^k$

3.1.2 *Trajectory cubic spline interpolation*

Trajectory integrity needs to be repaired to address cases where coarse errors in the trajectory sequence have been eliminated or where trajectory points are missing at short time intervals, despite not reaching the spatiotemporal gap threshold for splitting the trajectory. This section employs the cubic spline interpolation method to repair trajectory sequences. Cubic spline interpolation accurately fits the given discrete data points, providing smoothness, precision, and stability when handling the geographical trajectories of moving objects (Zhao *et al.* 2023). It effectively manages non-uniformly distributed nodes and adapts to complex geographical trajectory shapes.

Given a set of geographical trajectory data points (x_i, y_i), where $i = 0, 1, \ldots, n$, the cubic spline interpolation is performed as follows. Establish the Equations (1) to compute the cubic spline coefficients for each interval $[x_i, x_{i+1}]$. Solve for the coefficients using Equation (2). Construct the cubic spline interpolation function using Equation (3) to repair the missing data points in the trajectory. This method ensures that the reconstructed trajectory maintains the integrity and continuity necessary for accurate analysis.

$$\begin{cases} h_i = x_{i+1} - x_i \\ k = \dfrac{1}{h_i}(y_{i+1} - y_i) - \dfrac{1}{h_{i-1}}(y_i - y_{i-1}) \\ 6k = h_{i-1}S''(x_{i-1}) + 2(h_{i-1} + h_i)S''(x_i) + h_iS''(x_{i+1}) \\ S''(x_0) = 0 \\ S''(x_n) = 0 \end{cases} \tag{1}$$

$$\begin{cases} a_i = \dfrac{S''(x_{i+1}) - S''(x_i)}{6h_i} \\ b_i = \dfrac{S''(x_i)}{2} \\ c_i = \dfrac{y_{i+1} - y_i}{h_i} - \dfrac{h_i}{6}(2S''(x_i) + S''(x_{i+1})) \\ d_i = y_i \end{cases} \tag{2}$$

$$S_i(x) = a_i(x - x_i)^3 + b_i(x - x_i)^2 + c_i(x - x_i) + d_i \tag{3}$$

3.2 *Trajectory smoothing and filtering*

The trajectories are splitted and interpolated as previously described, although the temporal and spatial continuity is guaranteed, but due to potential vessel vibration or other signal interference during the transmission process of AIS data, trajectory data may exhibit irregularities such as jagged, oscillatory, or discontinuous point distributions within a continuous time series, which do not conform to the expected trajectory characteristics of vessel motion. An example of this phenomenon is illustrated in the local trajectory segment indicated by the red arrows within the dashed box in Figure 2, showing jagged, fluctuating, or jumping trajectory point distributions.

To ensure that each trajectory or sub-trajectory accurately represents the true hydrodynamic motion state of the vessel, this paper introduces a time-window trajectory smoothing filter algorithm. This algorithm is designed to mitigate such issues by filtering and smoothing operations applied to the trajectory dataset. Vessel trajectories, characterized as sequential data, offer a unique insight into the temporal evolution of various physical attributes. Parameters such as vessel position, speed, and heading are inherently dependent on temporal indexing. Consequently, following the preprocessing steps including stay segmentation and trajectory reconstruction, the subsequent application of time-window trajectory smoothing filtering necessitates the establishment of a trajectory dataset based on vessel identification numbers (MMSI).

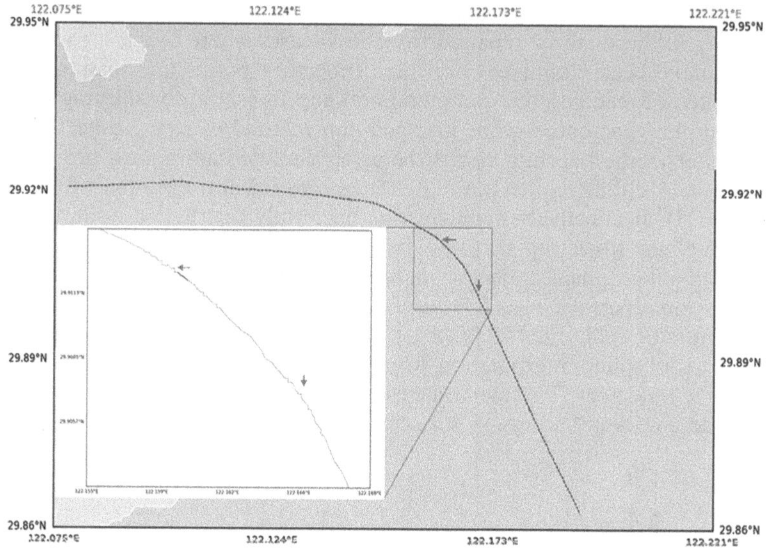

Figure 2. Segments with fluctuations, jumps and jagged sections in the trajectory.

Here, $D = \{P^k; (k = 1, 2, \cdots, N)|P \in \Theta\}$ denotes the trajectory dataset, Θ represents the spatial domain of vessel trajectories, $P^k = \{p_1^k p_2^k \cdots p_\tau^k p_{\tau+1}^k \cdots p_T^k; (\tau = 1, \cdots, \mathrm{T})\}$ signifies the trajectory sequences associated with different MMSI, and $p = \{t_i, x_i, y_i, v_i, c_i; (i = 1, 2 \cdots, n)\}$ encapsulates the temporal attributes of longitude, latitude, speed, and heading at a given time t_i.

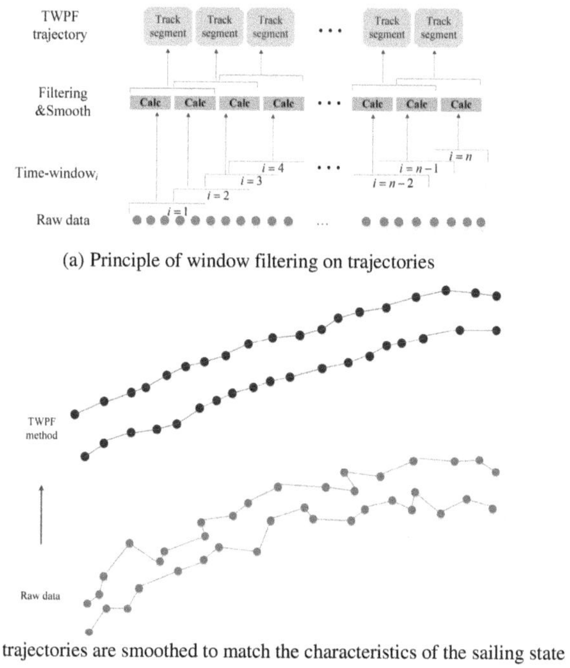

Figure 3. Schematic diagram with time window panning filtering of trajectory.

The smoothing operation on trajectories necessitates the establishment of a model with time as a reference. The first step in conducting trajectory smoothing involves establishing a time window. Within this time window, a time reference point is selected, sequentially corresponding to the order of time points in the window. A kernel function is then constructed by calculating the difference between the times within the window and the time reference point. This kernel function is applied to smooth the trajectory features corresponding to the traversed time. By panning the time window, the entire trajectory point feature data undergoes filtering, gradually achieving the smoothing and filtering of the entire trajectory segment. The principle of window filtering on trajectories is shown in Figure 3a, aiming to smooth problematic trajectories into trajectory points that match the characteristics of the object's water motion, as shown in Figure 3b. When determining the size of the time window, considerations include the AIS transmission frequency and the sampling interval required by the model. In this paper, the time window size is set to 60 seconds. For smoothing continuous adjacent trajectory points, Gaussian kernel functions are utilized to operate on the trajectory feature data within the window, as depicted in Equation (4). The trajectory sequence point attribute features within the time window are computed according to Equation (5) to transform the original trajectory sequence into a smoothed sequence.

$$\begin{cases} \delta T = t_{p_i} - t_r \\ \xi(t) = \dfrac{1}{\sqrt{2\pi} \cdot \Delta t_w} \cdot e^{-\frac{\delta T}{2\Delta t_w}} \end{cases} \tag{4}$$

$$\begin{cases} lon_s^i = \xi_i(t) P_{\Delta t}^k \left[lon_o^i \right] \\ lat_s^i = \xi_i(t) P_{\Delta t}^k \left[lat_o^i \right] \\ sog_s^i = \xi_i(t) P_{\Delta t}^k \left[sog_o^i \right] \end{cases} \tag{5}$$

where, Δt_w represents the size of the time window, t_{p_i} denotes the time of the i point within the window, and t_r represents the reference time within the window. $P_{\Delta t}^k$ represents the trajectory sequence within the time window. lon_o^i, lat_o^i, and sog_o^i represent the longitude, latitude, and speed of the original trajectory sequence, respectively. lon_s^i, lat_s^i, and sog_s^i represent the longitude, latitude, and speed after window smoothing and filtering.

4 TRAJECTORY DATA QUALITY ANALYSIS AND APPLICATION

In trajectory data, the analysis of speed, acceleration, and jerk (the change rate of acceleration) is crucial for assessing data quality. These metrics can reveal anomalies, navigational behaviors, and dynamic characteristics within the trajectory data. For moving vessel trajectories, these indicators are influenced by the constraints of the navigational environment and the various situations encountered by the vessel. Consistent changes in these metrics indirectly reflect the vessel's motion state.

Trajectory speed is the change rate of displacement over time. Analyzing speed variations over time can identify acceleration and deceleration behaviors during navigation. Trajectory acceleration, the change rate of speed over time, can reveal abrupt acceleration or deceleration actions when analyzed over time. Trajectory jerk, the change rate of acceleration over time, is also a critical indicator of data quality. By detecting whether jerk values exceed reasonable limits, potential data errors or abnormal navigational behaviors can be identified.

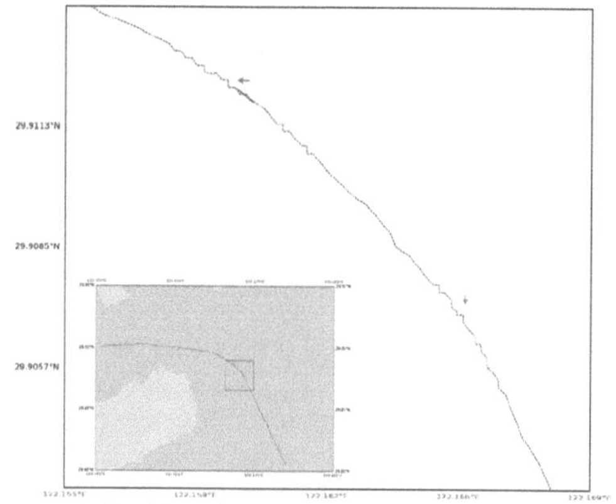

(a) Position change of the issue trajectory

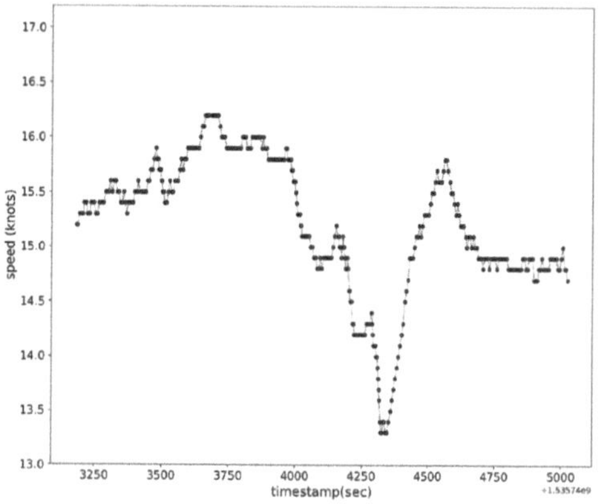

(b) Speed variation of issue trajectory

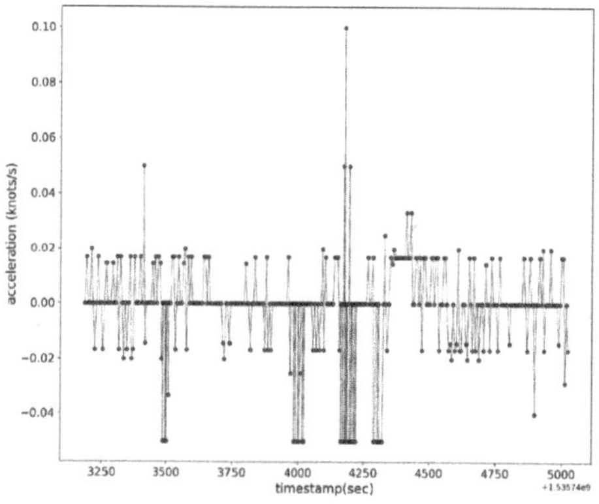

(c) Acceleration curve of issue trajectory

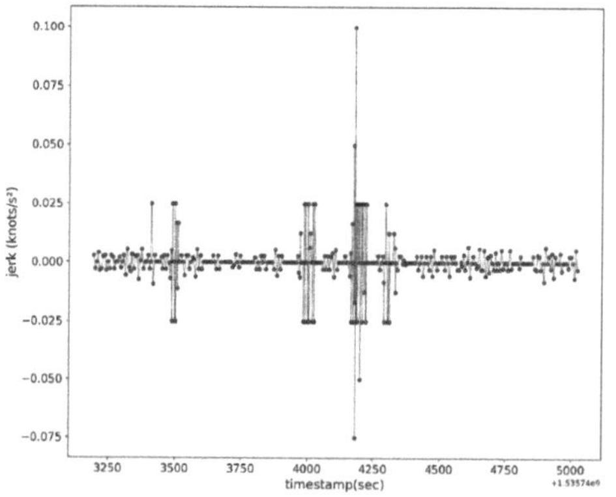

(d) Jerk curve of issue trajectory

Figure 4. Variation of original trajectory indicators.

Figure 4a shows the AIS trajectory data of a container vessel with MMSI 538007938, identified as a problematic trajectory. The figure reveals irregular segments with sawtooth patterns, fluctuations, and jumps. Figure 4b illustrates the vessel's speed variations over time within the channel, predominantly fluctuating between 13 and 16.5 knots, with localized instability in speed changes. Figure 4c and Figure 4d display the changes in acceleration and jerk, respectively. Despite overall stable variations in acceleration and jerk, the figures reveal abrupt acceleration and deceleration anomalies within short time intervals.

To address the issues with the problematic trajectory described above, we implemented the proposed time-window panning filter method for smoothing operations. After smoothing, the distribution of trajectory data points became more uniform and consistent. Compared to the original trajectory, the filtered data more accurately corrects points that deviate from the normal navigation state, ensuring that the trajectory data aligns with the kinematic characteristics of the vessel's movement. This effectiveness in correcting spatial position is demonstrated in Figure 5a.

Figure 5b presents a comparison of the speed values before and after applying the window panning filter method. Evidently, the smoothed trajectory exhibits more gradual speed changes over time, thereby reducing instantaneous fluctuations between adjacent speed values. Figure 5c and Figure 5d illustrate the changes in acceleration and jerk before and

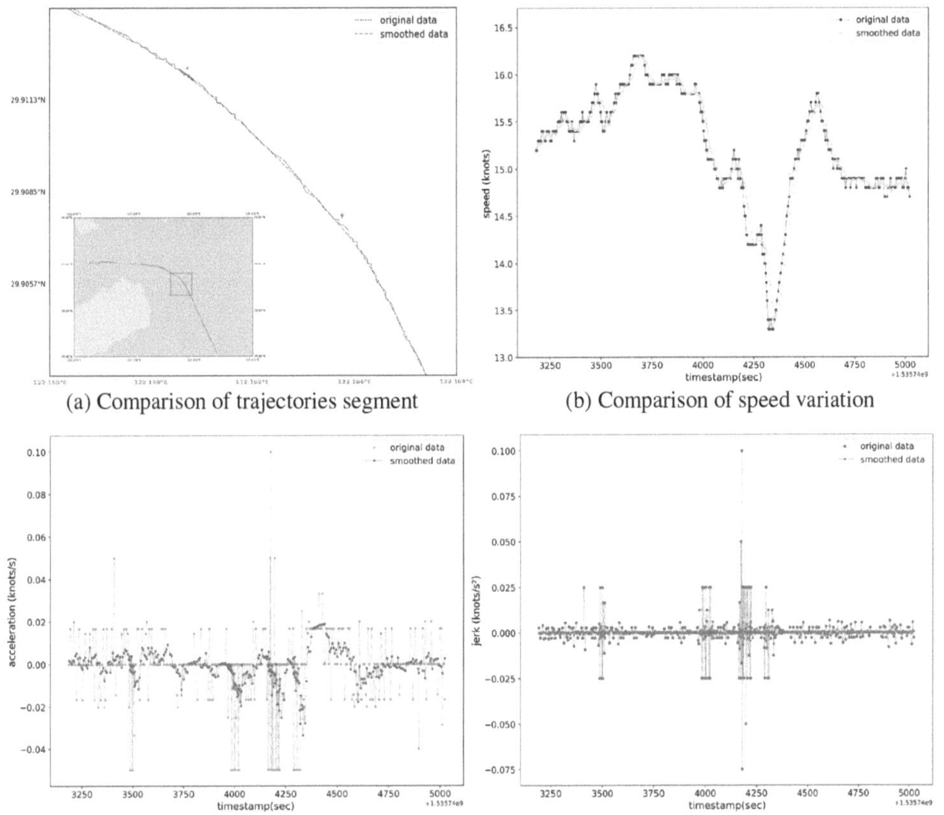

(a) Comparison of trajectories segment

(b) Comparison of speed variation

(c) Comparison of acceleration curve

(d) Comparison of jerk curve

Figure 5. Comparison of indicators for trajectories before and after time window panning filtering.

after filtering, respectively. Clearly, the filtered trajectory exhibits smaller fluctuations in acceleration and jerk, minimizing the impact of anomalous values on the trajectory data quality. This demonstrates the effectiveness of our filtering method in achieving uniform speed characteristics and significantly enhancing data quality.

To validate the effectiveness of our research, we selected one week of vessel trajectory data within the geographical range (122.0780°- 122.2212°E, 29.8624°- 29.9449°N) of Ningbo-Zhoushan Port, China. As shown in Figure 6a, the data within this geographical area primarily consist of inbound and outbound vessel trajectories. The figure demonstrates that, following our data processing and spatiotemporal splitting, errors and missing trajectory points have been effectively corrected. Notably, at the edges of the geographical area, different voyages of the same vessel have been successfully separated. Most importantly, after applying the time window panning filter method, each trajectory forms a continuous and smooth sequence, proving the effectiveness of our method in processing trajectory data. Figure 6b presents a density analysis of the traffic flow within the channel based on the processed navigation trajectories. This analysis clearly identifies high-density navigation areas, with deeper yellow indicating higher frequency of use of specific geographical locations within the channel. This application provides marine management departments with precise navigation data references.

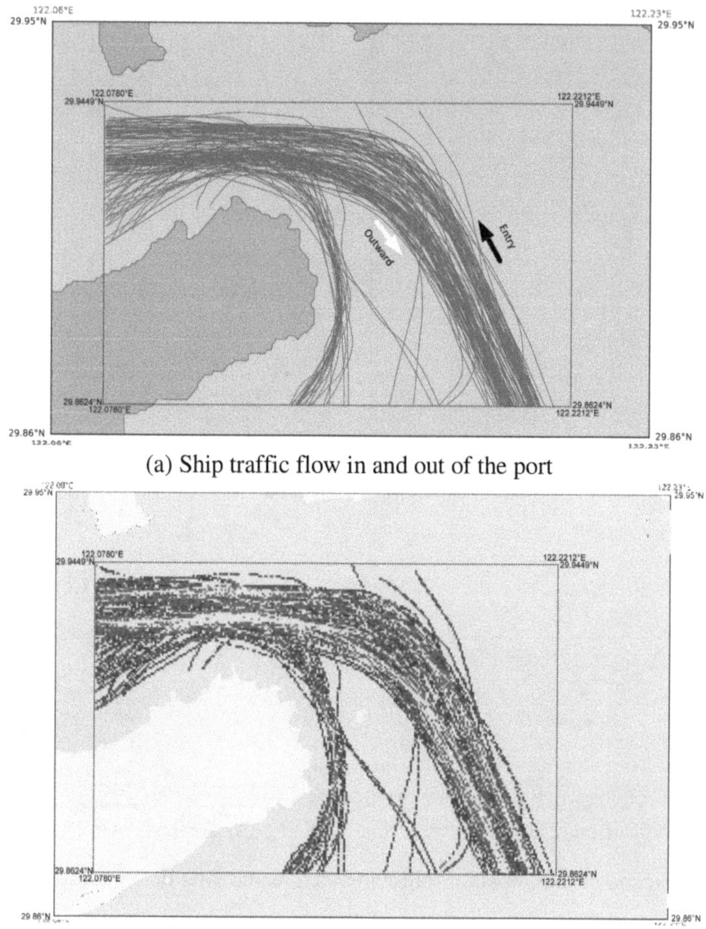

(a) Ship traffic flow in and out of the port

(b) Density distribution of ship traffic

Figure 6. Traffic flow and density distribution of filtered ship trajectory data.

5 CONCLUSIONS

Vessel AIS trajectories, as the primary data source for marine traffic navigation and positioning, are subject to various interferences that can introduce issues within the trajectories. In this paper, we propose a comprehensive framework for processing marine vessel trajectory data. Initially, coarse error removal is applied to eliminate outliers in the trajectory data. Subsequently, we employ a spatiotemporal splitting method to address missing trajectory sequences and unclear voyages within the data. To enhance data quality and ensure that all trajectories and segmented sub-trajectories are smooth and continuous time-series data, we designed a time-window panning filter algorithm. This method progressively filters out unstable point sequences, such as sawtooth patterns, fluctuations, and jumps, by sliding a time window across the data. Our research effectively ensures that each trajectory is a smooth and continuous independent time-series, providing a robust data processing framework. This framework supports the application of machine learning algorithms in the marine traffic domain and facilitates the creation of high-quality datasets.

REFERENCES

Gao, M., Liang, M. H., Zhang, A. M., Hu, Y. J. and Zhu, J. X. (2024), Multi-ship encounter situation graph structure learning for ship collision avoidance based on AIS big data with spatio-temporal edge and node attention graph convolutional networks, *Ocean Engineering*, Vol. 301.

Hartawan, I., Widyantara, I., Karyawati, A., Er, N. I., Artana, K. B., Sastra, N. P. (2021), AIS Data Pre-Processing for Trajectory Clustering Data Preparation, *in 5th IEEE International Conference on Aerospace Electronics and Remote Sensing Technology (ICARES)*, pp. 1–5.

Lin, K., Xu, Z., Qiu, M., Wang, X. and Han, T. (2016), Noise filtering, trajectory compression and trajectory segmentation on GPS data, in *Institute of Electrical and Electronics Engineers Inc.*, Nagoya, Japan, pp. 490–495.

Liu, J., Shi, G. Y. and Zhu, K. G. (2019), Vessel Trajectory Prediction Model Based on AIS Sensor Data and Adaptive Chaos Differential Evolution Support Vector Regression (ACDE-SVR), *Applied Sciences-Basel*, Vol. 9 No. 15, pp. 2983.

Liu, Y. and Yang, Z. (2023), Trajectory Smoothing Algorithm Based on Kalman Filter, *in Institute of Electrical and Electronics Engineers Inc.*, Virtual, Online, China, pp. 52–56.

Regmi, K. and Shah, M. (2021), Video Geo-Localization Employing Geo-Temporal Feature Learning and GPS Trajectory Smoothing, *in Institute of Electrical and Electronics Engineers Inc.*, Virtual, Online, Canada, pp. 12106–12115.

Zhang, L., Meng, Q., Xiao, Z. and Fu, X. (2018), A novel ship trajectory reconstruction approach using AIS data, *Ocean Engineering*, Vol. 159, pp. 165–174.

Zhang, W. B., Jiang, W. Y., Liu, Q. and Wang, W. F. (2023), AIS data repair model based on generative adversarial network, *Reliability Engineering and System Safety*, Vol. 240.

Zhao, L. B., Shi, G. Y. and Yang, J. X. (2018), Ship Trajectories Pre-processing Based on AIS Data, *Journal of Navigation*, Vol. 71 No. 5, pp. 1210–1230.

Zhao, X., Ren, G., Ma, J. F., Wang, S. Y. and Deng, Y. (2023), Reconstruction Method for Multi-Vehicle Trajectories on Arterials Driven by Multi-Source Data, in 23rd COTA International Conference of Transportation Professionals (CICTP) - Innovation-Empowered Technology for Sustainable, *Intelligent, Decarbonized, and Connected Transportation*, pp. 2189–2198.

Proceedings of the International Association of Institutes of Navigation World Congress 2024 – Chen et al. (Eds)

Risk assessment of Vessel Traffic Services (VTS) based on intelligent shipping

Q.Y. Lou
Shanghai Rules & Research Institute of China Classification Society, Shanghai, China

J.P. Liu & K. Liu
Yangshangang Maritime Safety Administration, Shanghai, China

ABSTRACT: With the rapid advancement of digitalization and intelligence, in the context of intelligent shipping, the Vessel Traffic Services (VTS) is undergoing significant transformations. These include the shift from conventional ships to intelligent ships as participants, the improvement allowing intelligent ships to be used as mobile sensors for the VTS, the enhancement of the VTS intelligent decision aid program, the change in communication objects of the VTS from the crew aboard to staff in the Remote Operation Centres (ROCs), and the transition from the manual operation of the VTS to intelligent procedures. While these shifts present opportunities and improvements for the VTS, they also give rise to the related security risks. To address these challenges, this paper conducts a risk assessment of the VTS by using System-Theoretic Process Analysis (STPA), and proposes the corresponding risk control strategies, so as to provide a reference for the safety design of Vessel Traffic Services.

1 INTRODUCTION

With the vigorous development of global intelligent shipping, many companies in the industry are developing intelligent ships such as remote control ships and putting them into commercial operation [1]. At the same time, the functions of the VTS systems are also tending to be highly intelligent [2]. Gradually integrate key technologies such as information perception, data cognition, risk prevention and control, information interaction, service application and regional coordination [3]. The VTS is facing subversive changes from itself to its service objects: VTS participating ships are transformed from conventional ships to intelligent ships, intelligent ships are used as mobile sensors of the VTS, the VTS is assisted by intelligent programs in decision-making, VTS communication contact objects are changed from the crew aboard to staff in ROCs, and VTS manual operation is gradually transformed into intelligent programs. Intelligent shipping not only increases the complexity of VTS system, but also brings security risks. At present, the industry lacks risk assessment of the VTS under the background of intelligent shipping. In order to reduce the risks of the VTS under the background of intelligent shipping, this paper uses STPA method to conduct risk assessment and puts forward risk control strategies.

2 VTS CONCEPTS AND FUNCTIONS

2.1 *VTS definition and main VTS components*

On 15 December 2021, the 32nd Assembly of the International Maritime Organization (IMO) adopted *Guidelines for Vessel Traffic Services* submitted by the Maritime Safety

 DOI: 10.1201/9781003624424-10

Committee (MSC), and recommended that governments encourage captains of ships sailing in the VTS areas to use this service. The guidelines define VTS as follows:

> "Vessel traffic services (VTS) means services implemented by a Government with the capability to interact with vessel traffic and respond to developing situations within a VTS area to improve safety and efficiency of navigation, contribute to the safety of life at sea and support the protection of the environment."

In order to promote the safety of life at sea, enhance navigation safety and efficiency, and support environmental protection in the area covered by the VTS system, the VTS system currently operated by VTS staff mainly consists of five subsystems: Radar system, Very High Frequency (VHF) system, Automatic Identification System (AIS), Closed-circuit Television (CCTV) system, and Management Information System (MIS) [4–9].

2.2 *VTS functions*

According to the working requirements of the VTS, the VTS includes three functions, timely provision of relevant information, monitoring and managing ship traffic, and responding to unsafe situations.

1) Timely provision of relevant information: providing timely and relevant information that may affect ship movements and assist onboard decision-making, including ship position, maritime safety information, allied service support, etc.
2) Monitoring and managing ship traffic: monitoring and managing traffic to ensure the safety and efficiency of ship movement, including planning ship movement in advance, organizing space allocation, providing route advice, etc.
3) Responding to unsafe situations: responding to developing unsafe situations, including ships uncertain about their routes or positions, ships that deviate from the route, ships that require guidance to the anchor position, ships with deficiencies or defects, etc.

3 STPA METHOD

System-Theoretic Process Analysis method is a risk analysis technique based on system theory[10]. In addition to component failures, STPA assumes that hazards may also be caused by unsafe interactions of system components. Compared with traditional risk analysis techniques, STPA can not only analyze complex systems and can be started in early conceptual analysis, but also covers the risks caused by software and human factors that are ignored by traditional risk assessment methods[11]. The method includes four steps: defining the purpose of the analysis, modeling the control structure, identifying unsafe control actions, and identifying loss scenarios, as shown in Figure 1.

1) Define the purpose of the analysis: used to determine the system and system boundaries for this risk assessment, and clarify the losses and system-level hazards that need to be avoided in this risk assessment;
2) Establish a control structure: Establish a control structure to help identify the functional relationships and internal interactions of the system;
3) Identify unsafe control behaviors: Analyze the control behaviors in the control structure to verify how such behaviors lead to the losses mentioned in the first step;
4) Identify loss scenarios: Identify the reasons why unsafe control behaviors might occur in the system. This step can be used to help improve the system architecture and propose mitigation measures.

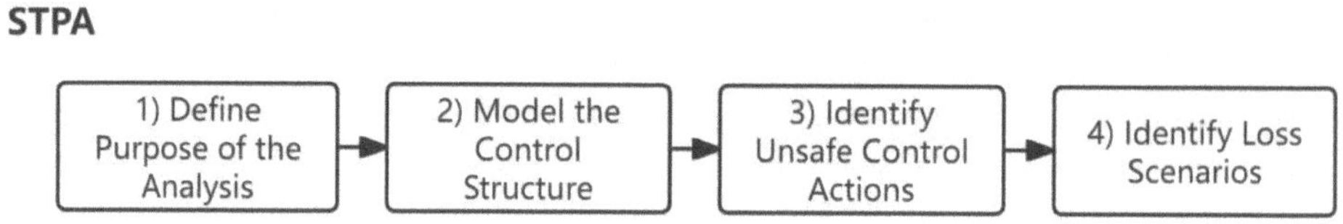

Figure 1. Process of the SPTA method.

4 STPA ANALYSIS OF VTS

4.1 *Overview of STPA analysis of VTS*

In view of the objective phenomenon of the development of traditional ships to intelligent ships under the background of intelligent shipping, two scenarios are discussed. The first step is to analyze the scenarios of the transformation from traditional ships to intelligent ships with the crew onboard (referred to as "primary intelligent scenarios"). The second step is to analyze the scenarios of further intelligence (referred to as "advanced intelligent scenarios"), namely, intelligent ships as mobile sensors of VTS, VTS decision-making assisted by intelligent programs, VTS communication contact objects changed from the crew aboard to staff in ROCs, and VTS manual operation gradually changed to intelligent programs.

4.2 *STPA analysis of VTS in primary intelligent scenarios*

4.2.1 *Overview of STPA analysis of VTS*

Determining the analyzed system and its system boundaries is the basis for identifying system-level hazards. In the primary intelligent scenarios, the VTS is composed of subsystems and VTS personnel, and the boundary is the VTS coverage area. The VTS and its boundary in the primary intelligent scenarios are shown in Figure 2.

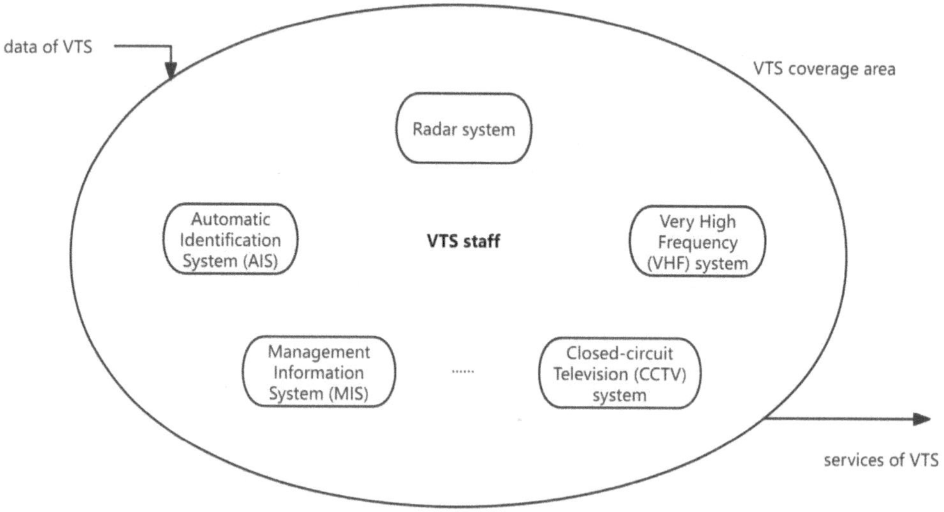

Figure 2. VTS and its boundary in the primary intelligent scenarios.

The stakeholders of the VTS include the contracting government, VTS competent authorities, VTS suppliers, ships in the VTS area, and relevant units of maritime transport. The interests maintained by VTS are the safety and efficiency of navigation, the promotion of safety of life at sea, and support for environmental protection. According to the stakeholders and interests of VTS, six unacceptable losses of the VTS can be deduced, including damage to the ship, loss of life or personal injury to crew or passengers, pollution of navigable waters, service failure, damage to the satisfaction of relevant units of maritime transport, and damage to sensitive information, as shown in Table 1.

Table 1. Loss of VTS in primary intelligent scenarios.

Loss	Loss description
L-1	Damage to the ship
L-2	Loss of life or personal injury to crew or passengers
L-3	Pollution of navigable waters
L-4	Service failure (Information service failed, navigation aid service failed, traffic organization service failed, support joint action service failed)
L-5	Loss of the satisfaction of the relevant units of maritime transport
L-6	Loss of sensitive information of VTS

System-level hazards are defined by the system conditions or states that lead to losses under the worst environmental conditions. The system-level hazards and their associated losses of the VTS in the primary intelligent scenarios are shown in Table 2.

Table 2. System-level hazards and associated losses of VTS in primary intelligent scenarios.

System-level hazards	System-level hazard description	Associated losses
H-1	VTS systems cannot determine the relative positions of vessels.	L-1,L-2,L-4
H-2	VTS staff lose control of the VTS systems.	L-4,L-5
H-3	VTS systems fail to provide broadcast information or service information to the outside world.	L-1,L-2,L-3, L-4,L-5
H-4	VTS systems cannot assist VTS staff in organizing transportation services.	L-1,L-2,L-3, L-4,L-5
H-5	VTS systems cannot collect information.	L-4,L-5
H-6	VTS systems cannot manage information.	L-4,L-5,L-6

The system-level hazards in the table are further sorted out and refined, taking the sub-hazards of H-1 as an example, as shown in Table 3.

Table 3. H-1 sub-hazards and their descriptions of VTS systems in primary intelligent scenarios.

System-level hazards	System-level hazards description	Associated losses	Sub-hazards	Sub-hazards description
H-1	VTS systems cannot determine the relative positions of vessels	L-1,L-2, L-4	H-1.1	The radar system does not receive the command.
			H-1.2	The radar system sensing device fails or could not sense normally.
			H-1.3	The sensed data is wrong or has a large deviation from the actual data.
			H-1.4	Failure of radar system to output information.

(*continued*)

Table 3. Continued

System-level hazards	System-level hazards description	Associated losses	Sub-hazards	Sub-hazards description
			H-1.5	AIS system equipment fails or is unable to operate normally.
			H-1.6	AIS system signal is interfered with, ship position is invisible or inaccurate.
			H-1.7	AIS system usage error or configuration problem.
			H-1.8	There is human damage or malicious interference, thus the AIS system receives changed ship information or false AIS signals.
			H-1.9	VHF system equipment fails or is unable to operate normally.
			H-1.10	VHF system communication signal is weak.
			H-1.11	VHF system has no signal, thus is unable to communicate.
			H-1.12	VHF system cross-frequency or frequency hopping.
			H-1.13	VTS staff fail to listen to the VHF system channel as required.
			H-1.14	The MIS cannot display information.
			H-1.15	VTS staff make operational errors.

4.2.2 *Modeling the control structure*

The second step of STPA is to establish the control structure model. Combined with the system boundary of the VTS in the primary intelligent scenarios proposed in the previous article, the control structure is modeled as shown in Figure 3 below. The VTS control structure consists of three parts, from top to bottom: personnel control module, traffic display and control module, and information management module. Personnel input instructions or information to the traffic display and control module, and the traffic display and control module arranges the information management module to retrieve or process information, or arranges broadcasting to release water safety information. At the same time, considering the complexity of information sources, such as information from the VTS subsystem, hydrological and meteorological information, external requests, etc., internal and external information will be identified and recorded, and feedback or alarm will be given to the upper-level information management module, traffic display and control module, and personnel control module.

4.2.3 *Identifying unsafe control actions*

According to the STPA method, unsafe control actions can be classified into four categories: "not providing causes hazard", "providing causes hazard", "too early, too late, out of order", and "stopped too soon, applied too long". Unsafe control actions are identified for the main control actions of the VTS, and the unsafe control actions are sorted out as shown in Table 4.

4.2.4 *Identifying loss scenarios*

Loss scenarios refer to the factors that may lead to unsafe control actions and to hazards. By identifying the loss scenarios of unsafe control actions of VTS in primary intelligent

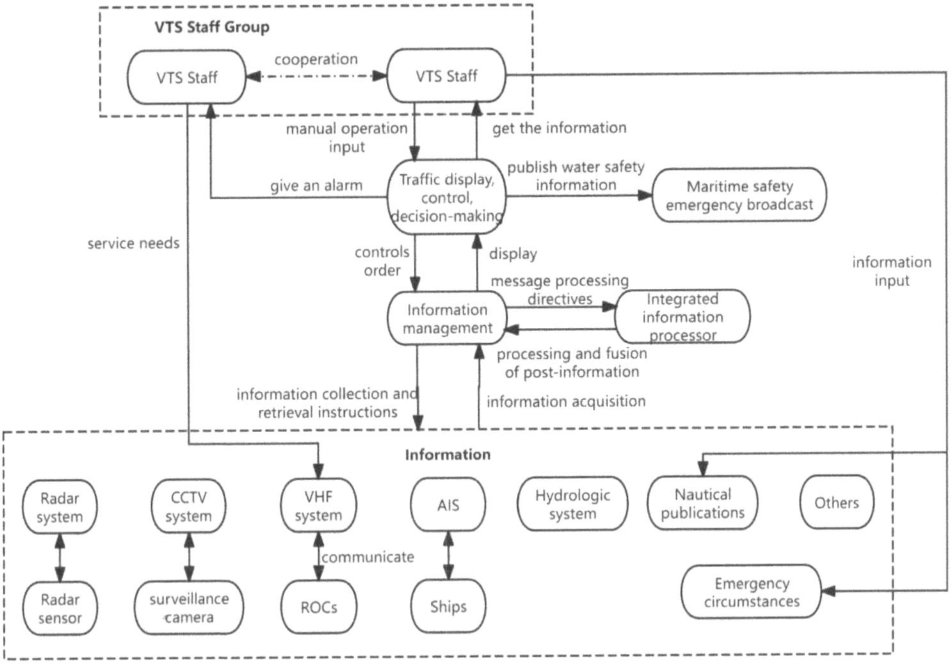

Figure 3. The control structure of VTS in primary intelligent scenarios.

Table 4. Unsafe control actions of VTS in primary intelligent scenarios.

Control Actions	Not providing causes hazard	Providing causes hazard	Too early, too late, out of order	Stopped too soon, applied too long
Access to information	UCA-1: Information management module does not obtain the required information when it is requested.	UCA-2: The information obtained is wrong or deviates greatly from the actual situation	N/A	N/A
Information processing	UCA-3: Failure to process required information when information management module requires it.	UCA-4: Failure to process information or failure to meet requirements.	N/A	N/A
Information retrieval	UCA-5: Information management module does not receive the requested information when it requested it.	UCA-6: Search information is incorrect or missing.	N/A	N/A
Information broadcast	UCA-7: Information broadcasts does not issue water safety information when requested.	UCA-8: Incorrect water safety information is published.	N/A	UCA-9: Safety information broadcast lasts too long.

(continued)

Table 4. Continued

Control Actions	Not providing causes hazard	Providing causes hazard	Too early, too late, out of order	Stopped too soon, applied too long
				UCA-10: Safety information broadcast stops too early.
Information services	UCA-11: Information service is not provided at fixed times or at any other time deemed necessary by the VTS or at the request of the vessel.	UCA-12: Wrong information service is provided.	N/A	N/A
Navigational assistance services	UCA-13: Failure to provide additional information services or to assist shipboard navigational decision making by providing timely essential navigational information.	UCA-14: Providing incorrect navigation information and failing to assist shipboard navigational decision making.	N/A	N/A
Traffic organization	UCA-15: Unorganized transportation to ensure safe and efficient transportation.	UCA-16: Providing incorrect transportation organization plan.	UCA-17: Too late to organize transportation.	UCA-18: Organizing traffic for too long.
Supporting joint action	UCA-19: No support for joint services or other services.	UCA-20: Providing incorrect support services.	N/A	N/A
Emergency	UCA-21: When a system failure occurs, personnel	UCA-22: When a system fails, the wrong	UCA-23: When a system fails,	N/A

(*continued*)

scenarios, it is possible to propose risk mitigation measures. An analysis is conducted on the unsafe control actions listed above, and the loss scenarios are obtained by analysis, taking UCA-1 and UCA-2 as examples, as shown in Table 5.

Table 5. Unsafe control actions of VTS and their loss scenarios in preliminary intelligent scenarios.

Unsafe Control Actions	Loss scenarios
UCA-1	The MIS fails or loses power and cannot operate normally. The control algorithm of the MIS is defective, resulting in control not meeting expectations. Radar system, AIS, VHF system, CCTV system or other systems are faulty or power outage occurs and cannot operate normally. Signals of Radar system, AIS, VHF system, CCTV system or other systems are interfered.

(*continued*)

Table 5. Continued

Unsafe Control Actions	Loss scenarios
UCA-2	Weather data, tide information, nautical publications or other information are not updated in a timely manner. VTS staff fail to listen to the VHF system channel as required. The information has been collected, but the storage device is faulty or the storage space is insufficient, so it cannot be stored normally. The MIS is not responding. The equipment of Radar systems, CCTV systems or other systems lack accuracy and stability. Severe weather conditions exceed the normal operating range of the equipment, causing information distortion. There is human sabotage or malicious interference, and the AIS receives altered ship information or false AIS signals. The MIS responds but receives information from other components.

4.3 *STPA analysis of VTS in advanced intelligent scenarios*

4.3.1 *Determination of system-level hazards*

Four further intelligent scenarios are analyzed, namely, intelligent ships as mobile sensors of VTS, VTS decision-making assisted by intelligent programs, VTS communication contact objects changed from the crew aboard to staff in ROCs, and VTS manual operation gradually changed to intelligent programs. The VTS and its boundary are re-identified. The VTS and its boundary in advanced intelligent scenarios are shown in Figure 4.

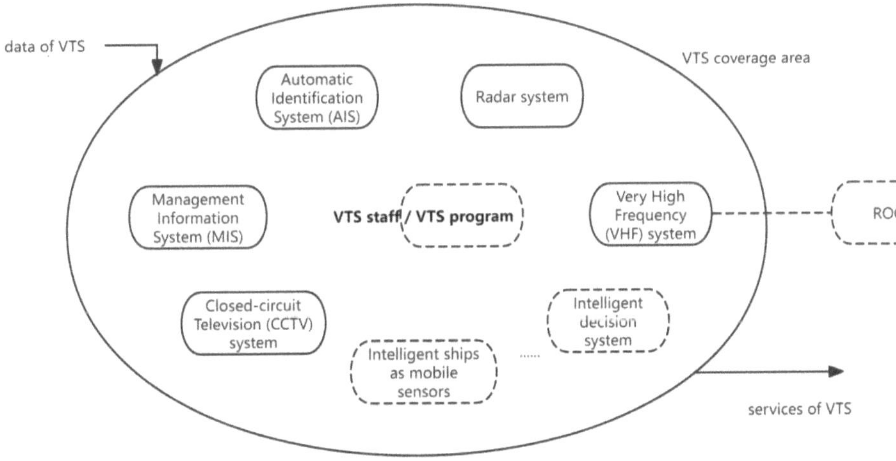

Figure 4. VTS and its boundaries in advanced intelligent scenarios

According to the delineated boundary analysis of losses and system-level hazards, it is determined that the losses of the VTS in advanced intelligent scenarios are consistent with those in primary intelligent scenarios. Since the functions of the VTS have not changed, the system-level hazards of the VTS in advanced intelligent scenarios are also consistent with those in primary intelligent scenarios. Therefore, only the differences at the sub-hazard level are analyzed. According to the analysis of the sub-hazards of the VTS in advanced intelligent scenarios, the H-1 sub-hazards are still taken as an example, as shown in Table 6 below.

Table 6. H-1 sub-hazards, sub-hazards descriptions and revision factors of VTS in advanced intelligent scenarios.

System-level hazards	System-level hazards description	Associated losses	Sub-hazards	Sub-hazards description	Revision factors
H-1	VTS systems cannot determine the relative positions of vessels	L-1,L-2, L-4	H-1.16	The VTS cannot identify error information and deviation information of acquisition equipment.	Manual operation of VTS is gradually changing to intelligent programs.
			H-1.17	The VTS cannot clearly recognize the information input via the VHF system.	Manual operation of VTS is gradually changing to intelligent programs.
			H-1.18	The VTS staff do not take over the VTS system in an emergency.	Manual operation of VTS is gradually changing to intelligent programs.
			H-1.19	The VTS cannot communicate with the remote control station.	The communication contact objects changed from the crew aboard to staff in ROCs.
			H-1.20	The intelligent ships do not receive instructions.	Intelligent ships as mobile sensors of VTS
			H-1.21	The intelligent ships' sensing equipment is faulty and cannot sense normally.	Intelligent ships as mobile sensors of VTS
			H-1.22	The intelligent ships have insufficient perception accuracy or unstable performance, thus the perceived data is wrong or deviates greatly from the actual data.	Intelligent ships as mobile sensors of VTS
			H-1.23	The VTS cannot receive the ship position transmitted by the intelligent ships.	Intelligent ships as mobile sensors of VTS

4.3.2 *Modeling the control structure*

The control structure of VTS in advanced intelligent scenarios has a similar basic structure to that in primary intelligent scenarios. The difference lies in the changes brought about by the development direction of intelligent shipping. The control structure model of the VTS in advanced intelligent scenarios is shown in the below Figure 5.

4.3.3 *Identifying unsafe control actions*

Since the VTS functions in advanced intelligent scenarios are the same as those in primary intelligent scenarios, the control actions and their corresponding unsafe control actions are also the same, as shown in Table 4 in the previous 4.2.3.

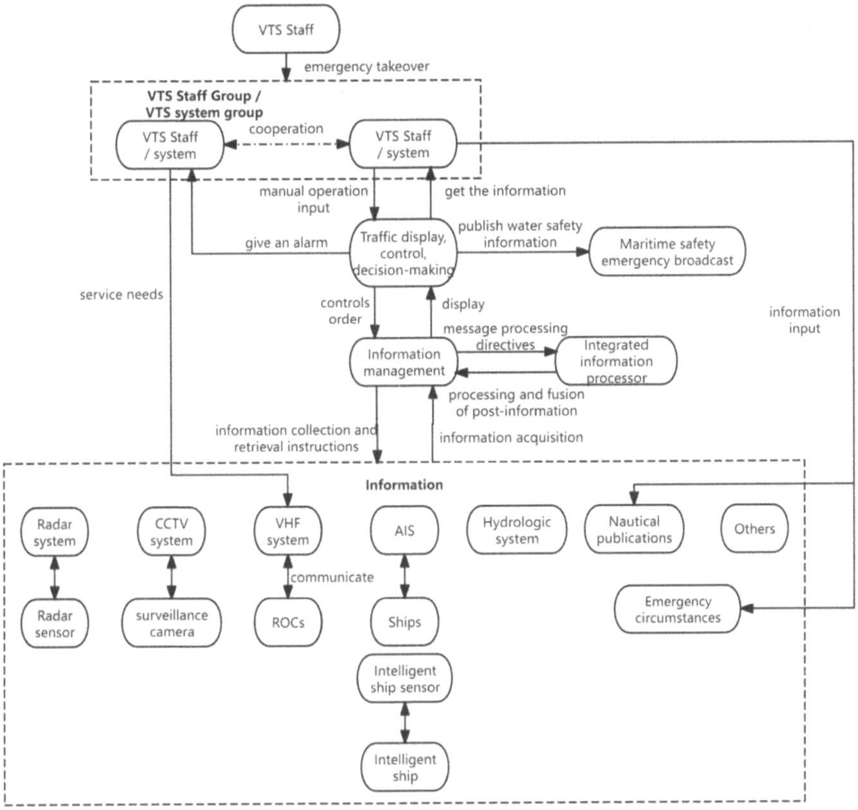

Figure 5. The control structure of VTS in advanced intelligent scenarios.

4.3.4 *Identifying loss scenarios*

In order to identify loss scenarios, the unsafe control actions of VTS in advanced intelligent scenarios were analyzed. In addition to the loss scenarios analyzed in 4.3.4, 31 new loss scenarios are added.

Table 7. Unsafe control actions of VTS and their loss scenarios in advanced intelligent scenarios.

Controlling Actions	UCA	Loss scenarios
Access to information	UCA-1	When an intelligent ship is used as a mobile sensor for VTS, the intelligent ship's perception system fails or loses power, and cannot operate normally. When an intelligent ship is used as a mobile sensor for VTS, the equipment signal of the intelligent ship perception system is interfered. When manual VTS operations shift to intelligent programs, the information transmitted by the VHF system cannot be identified by programs. When the VTS communication contact object changes from the crew aboard to the staff in ROCs, ROCs is out of the VHF signal range or the network fails.

(continued)

Table 7. Continued

Controlling Actions	UCA	Loss scenarios
	UCA-2	When an intelligent ship is used as a mobile sensor for VTS, the accuracy and stability of intelligent ship perception equipment are insufficient. When an intelligent ship is used as a mobile sensor for VTS, severe weather conditions exceed the normal operating range of the equipment, causing information distortion.
Information broadcast	UCA-7	When manual VTS operations shift to intelligent programs, information broadcasting does not receive the broadcast instructions of the VTS system. When the VTS communication contact object changes from the crew aboard to the staff in ROCs, ROCs is out of the VHF signal range or the network fails.
	UCA-8	When manual VTS operations shift to intelligent programs, information broadcasts receive broadcast instructions from the VTS system but the response does not match expectations.
	UCA-10	When manual VTS operations shift to intelligent programs, information broadcasts receive broadcast instructions from the VTS system but the response does not match expectations.
Information services	UCA-11	When VTS is assisted by intelligent programs in decision-making, the decision-making module of the VTS system provides too many or too few information service solutions. When manual VTS operations shift to intelligent programs, traffic display and control systems do not respond to VTS system information service instructions.
	UCA-12	When VTS is assisted by intelligent programs, VTS staff fail to select the correct information service plan based on the decision of the VTS intelligent programs. When manual VTS operations shift to intelligent programs, traffic display and control systems do not respond appropriately to VTS system information service instructions.
Navigational assistance services	UCA-13	When manual VTS operations shift to intelligent programs, the traffic display and control system does not respond to VTS system navigation service instructions. When VTS is assisted by intelligent programs, the decision-making module of the VTS system provides too many or too few solutions for navigation services.
	UCA-14	When manual VTS operations shift to intelligent programs, traffic display and control systems do not respond appropriately to VTS navigation service instructions. When VTS is assisted by intelligent programs, the VTS staff fail to select the correct navigation aid according to the decision of the VTS programs.
Traffic organization	UCA-15	When manual VTS operations shift to intelligent programs, the traffic display and control system does not respond to VTS system navigation service instructions. When VTS is assisted by intelligent programs, the decision-making module of the VTS system provides too many or too few solutions for traffic organization. When manual VTS operations shift to intelligent programs, traffic display and control systems fail or lose power, and thus cannot operate normally.

(*continued*)

Table 7. Continued

Controlling Actions	UCA	Loss scenarios
	UCA-16	When manual VTS operations shift to intelligent programs, the traffic organization plan reserve is insufficient. When VTS is assisted by intelligent programs, the VTS staff fail to select the correct traffic organization plan based on the decision of the VTS programs.
	UCA-17	When manual VTS operations shift to intelligent programs, traffic display and control systems do not respond appropriately to VTS system traffic organization instructions.
	UCA-18	When manual VTS operations shift to intelligent programs, traffic display and control systems do not respond appropriately to VTS system traffic organization instructions.
Supporting joint action	UCA-19	When manual VTS operations shift to intelligent programs, the traffic display and control system did not respond to VTS system support joint action service instructions. When VTS is assisted in decision-making by intelligent programs, the decision-making module of the VTS system provides too many or too few options to support joint actions.
	UCA-20	When manual VTS operations shift to intelligent programs, traffic display and control systems respond inappropriately to VTS system support joint action instructions. When VTS is assisted by intelligent programs, the VTS staff fail to select the correct supporting joint action plan based on the decision of the VTS programs.
Emergency	UCA-21	When manual VTS operations shift to intelligent programs, the VTS staff do not take over the control of VTS system.

5 STRATEGIES FOR CONTROLLING RISKS

Based on the SPTA risk analysis in primary and advanced scenarios, the following strategies are proposed to control risks:

1) Risk constraints : Targeted constraints are imposed based on the dangerous and unsafe control actions of VTS to ensure that overall risks are controllable.
2) Classification of remote-controlled ships and other ships: For intelligent ships whose VTS communication targets have changed from the crew onboard to the staff in ROCs, it is necessary to provide differentiated services from other ships because there are no crew members on board. This approach can not only solve the impact of the change in communication targets on subsystems such as the VHF system, but also provide stable services to the other category when one category encounters hazards.
3) Personnel monitoring and takeover: In view of the gradual transition from manual operation of VTS to intelligent programs, for safety reasons, VTS personnel are required to take over quickly in emergency situations or when the machine cannot provide stable services. Relevant VTS staff should receive training and regular emergency drills to ensure that the VTS functions can be completed after rapid takeover. In addition, operation manuals and emergency response plans for the VTS with intelligent programs are necessary.
4) Common standards for communication terms: As manual operation of VTS gradually shifts to intelligent programs, it is necessary to establish mutually recognized, common terminology standards for communication between VTS systems and ships to ensure that

the information transmitted and received by VTS intelligent programs is concise and reliable.

5) Screening and accumulating solutions: Since VTS is assisted by intelligent programs and requires sufficient historical data and service solutions as a basis, the solutions used as models need to be typical but comprehensive, otherwise too many or too few solutions will affect the quality of service provided. In addition, service solutions need to be regularly reviewed and updated to ensure availability.
6) Backup plans: Intelligent ships as mobile sensors of VTS broaden the field of view of the VTS system. However, factors such as the accuracy of the sensors of mobile ships, the difficulty of converting information, and the stability of signals may cause the field of view to disappear at any time. Therefore, it is necessary to prepare backup plans to ensure the acquisition of information. In addition, it is necessary to pay attention to and evaluate the reliability of the cooperating ships.

6 CONCLUSION

This paper uses the STPA method to conduct risk assessment of VTS from different levels of intelligent shipping scenarios, carries out a qualitative safety analysis of each part of the VTS control structure, and proposes six targeted risk control strategies: risk constraints, classification of remote-controlled ships and other ships, personnel monitoring and takeover, common standards for communication terms, screening and accumulation solutions, and backup plans, which provide a reference for the safety design of VTS.

REFERENCES

[1] Sawyer, B., Teitenborn, A., Leloudas, G., *et al.* (2022). Remote control and autonomous Shipping: a case study based on the UK. *Haida Law Review* (00): 396–425.

[2] Zhou, Z. S. and Li L. (2023). Development of Ship traffic management System under the background of intelligent shipping. *China Ship Inspection* (11):44–48.

[3] Zhao, X. F. and Tian, C. (2021). Research on Key Capabilities of New generation VTS System Adapted to Intelligent shipping. *Water Transport in China* (4): 67–68.

[4] Xia, M. C. and Liang, J. S. (2017). The role, existing problems and reflections of VTS in China's maritime development. *China Water Transport (Second Half)* 17(07):49–51.

[5] Czaplewski, K., Wąż, M. and Zienkiewicz, M. H. (2019). A novel approach of using selected unconventional geodesic methods of estimation on VTS areas. *Marine Geodesy* 42(5): 447–468.

[6] Brödje, A., Lützhöft, M., and Dahlman, J. (2010, June). The whats, whens, whys and hows of VTS operator use of sensor information. In International conference on human performance at sea, Glasgow. *University of Strathclyde, Glasgow* pp. 161–172.

[7] Kos, S., Vukić, M., and Barić, M. (2016). AIS AS VHF COMMUNICATION IMPROVEMENT. *In ISEP Symposium, Ljubljana, Slovenia.*

[8] Zhao, H., Li, D., Yuan, J. H., *et al.* (2016). A China central television (CCTV) in VTS system on ship tracking monitoring method. *Journal of Radar and Confrontation* 4(02): 12–14 + 18.

[9] Chen, J., Wang, J., and Lu, H. (2022). Information fusion structure of VTS sensors for multi-target tracking of vessels. In *ITM Web of Conferences* Vol. 45, p. 01008. EDP Sciences.

[10] Ishimatsu, T., Leveson, N. G., Thomas, J., Katahira, M., Miyamoto, Y., and Nakao, H. (2010). Modeling and hazard analysis using STPA.

[11] Sulaman, S. M., Beer, A., Felderer, M., and Höst, M. (2019). Comparison of the FMEA and STPA safety analysis methods–a case study. *Software Quality Journal* 27, 349–387.

*Proceedings of the International Association of Institutes of Navigation World Congress 2024 –
Chen et al. (Eds)*

Research on collision risk evaluation method considering ship avoidance ability

Zhitao Yuan
Wuhan University of Technology, Wuhan, China
Hubei Key Laboratory of Inland Shipping Technology, Wuhan, China

Yurou Jiao, Jiamin Gan, Xiaolie Wu, Mozi Chen & Xiyang Zhang
Wuhan University of Technology, Wuhan, China

ABSTRACT: Ship maneuverability plays an important role in ship risk avoidance and directly affects the evaluation of collision risk. However, the current risk evaluation methods often ignore the important factor of ship maneuverability. Therefore, a collision risk evaluation method considering ship collision avoidance ability is proposed in this paper. The method combines the Automatic Identification System (AIS) data, the NOMOTO maneuvering motion equation, the Convention on the International Regulations for Preventing Collisions at Sea (COLREG rules) and the ship domain intrusion measurement model to establish two risk evaluation indicators related to the latest rudder application time and the safety avoidance Angle set, i.e., Avoidance Difficulty (AD) and Avoidance Time Urgency (ATU). Then, the criterion Importance Through Intercriteria Correlation (CRITIC) method is used to determine the weight of the two indicators. Finally, the method proposed in this paper is used to identify the risk of ships in the offshore waters of Ningbo Zhoushan port area. Risk comparison experiments are conducted by comparing with Collision Risk Index (CRI) method and domain-based method. The results show that the method not only takes into account the differences in maneuverability of different ships, but also focuses on the relative movement trend between ships in the risk evaluation, which can realize the risk evaluation more reasonably.

Keywords: collision risk, avoidance ability, ship domain, latest rudder application time, safety avoidance angle set

1 INTRODUCTION

1.1 *Background*

Due to the frequent occurrence of maritime accidents as well as the serious consequences and poor social impact of ship collision avoidance, the current research on ship collision avoidance has become one of the popular topics in the maritime community. Technically speaking, ship collision detection is to issue early warning to encountering ships by assessing the collision risk. Therefore, the premise of ship collision avoidance is to conduct effective ship collision risk assessment (Chen, P. *et al.* 2019).

There are three main types of methods for ship collision risk assessment. The first method is based on indicators, and the most widely used indicators are Distance to Closest Point of Approach (DCPA) and Time to Closest Point of Approach (TCPA). In order to make collision risk more intuitive for collision avoidance decisions, DCPA and TCPA have been merged into Collision Risk Index (CRI), which was initially synthesized based on techniques

DOI: 10.1201/9781003624424-11

such as linear regression (Chin, H. C., and Debnath, A. K. (2009). Modeling perceived collision risk in port water navigation. *Safety Science*, *47*(10), 1410–1416., weighted coefficients (Zhen, R. *et al.* 2017), and fuzzy theory (Lee, H. J., and Rhee, K. P. (2001). Development of collision avoidance system by using expert system and search algorithm. *International shipbuilding progress*, *48*(3), 197–212.), and has been widely used for collision avoidance of ships in open waters.

Another approach is based on the concept of ship domain. Ship's domain is the area around the ship that avoids the entry of other obstacles to ensure navigation safety. When another ship enters the ship's field or the fields of two ships overlap, there is a danger of collision between the two ships. Most of the differences between different ship domain models are the differences in the consideration of key factors, e.g., based on the ship's own attributes and the navigator's knowledge (Liu, J. *et al.* 2016), based on the definition of the domain shape (Szlapczynski, R., and Szlapczynska, J. (2016). An analysis of domain-based ship collision risk parameters. *Ocean Engineering*, *126*, 47–56.), and based on the specification of the model training method (Szlapczynski, R., and Szlapczynska, J. 2017).

There is also a velocity-based approach, this idea was introduced by Degré and Lefèvre in their 1981 study (Chen, P. *et al.* 2019, Lenart, A. S. 2015) formulated this idea by assuming that the ship maintains its speed and heading constant during the voyage as Collision Threat Parameters Area (CTPA), and the idea subsequently became known as the Velocity Obstacle (VO) method. Researchers have also relaxed the linear motion assumption to nonlinear, (Huang, Y. 2017; Huang, Y., Van Gelder, P. H. A. J. M., and Wen, Y. 2018) proved that CTPA is identical to linear VO and that they are another form of TCPA and DCPA in velocity space. In addition, (Huang Y., and Van Gelder, P. H. A. J. M. 2020) utilized the set of VOs to measure the collision risk, which is formulated as the proportion of reachable velocities that lead to a collision.

1.2 *Motivition*

Based on the above research background, current collision hazard studies usually ignore ship maneuverability, which is characterized by large inertia and long maneuvering time delay. Therefore, if the ship maneuverability is ignored, the collision hazard may be underestimated. (Wang X. 2017) considered the ship maneuverability when using DCPA and TCPA, and found that the real DCPA is usually smaller than the DCPA without considering the maneuverability. However, the method requires too many ship parameters (e.g., ship's squareness coefficient, propeller thrust coefficient, propeller damping coefficient, etc.), and it is not possible to know his ship maneuverability during the real ship It is impossible to know the excessive parameters of other ships or even this ship during the voyage.

Moreover, when measuring the danger, the officer of the watch will subconsciously consider COLREG rules and good seamanship. For example, when two ships meet, the pilot usually chooses a collision avoidance action that conforms to good seamanship. Therefore, collision avoidance measures that ignore COLREG rules or good seamanship will not be in line with the good habits of drivers. (Yang, T. *et al.* 2023) proposed a new ship collision hazard parameter Current State of Domain (CSD) using ship domain combined with ship maneuvering, although the parameter considers different encounter situations, the selection of its ship domain is based on the decision of the larger ship between two ships, and it does not reflect the difference in maneuverability of different ships in the simulation.

In brief, there is currently no method for assessing the risk of ship collision from the ship's avoidance ability. And it does not take into account the variability in the maneuverability of different ships, as well as the limitations of the COLREG rules and good seamanship.

1.3 *Contributions*

This paper proposes a collision hazard assessment method that takes into account a ship's avoidance ability, taking into account the variability of a ship's maneuverability performance and the limitations imposed on a ship by COLREG rules and good seamanship during avoidance.

2 TECHNOLOGY

2.1 *Ship motion model*

In 1957, Kensaku Nomoto analyzed the influence of the ship's heading angle on the response of the steering using the method of automatic adjustment, and proposed the NOMOTO maneuvering equation of motion in the case of infrequent steering, which was applied to the calculation of ship motion due to its simplicity and fast calculation speed, as Eq. (1).

$$T\ddot{\varphi} + \dot{\varphi} = K\delta \tag{1}$$

where K and T represent the turning ability index and the turning lag index; $\dot{\varphi}$, $\ddot{\varphi}$ represent the angular velocity and angular acceleration of the ship; δ indicates the rudder angle of the ship.

When the rudder direct ship is rotated with the rudder angle δ, in the case of the initial conditions $t = 0, \dot{\varphi} = 0$, the general solution for the first-order linear differential equation is as Eq. (2).

$$\dot{\varphi} = K\delta\left(1 - e^{-\frac{t}{T}}\right) \tag{2}$$

There is a certain time delay between the steering of the ship and the transfer to the specified rudder angle during the actual steering of the ship, which is recorded as the rudder turning time, t_1, according to Article 29 of Chapter II-1 of the International Convention for the Safety of Life at Sea: "The main steering device shall enable the ship to turn the rudder from 35° on one side to 35° on the other when moving at maximum operating speed. The rudder should be able to turn from 35° to 30° from the other within 28 seconds at maximum operating speed. "Because there are too many factors that can affect the rudder rotation time of the ship, and when the ship turns, the selection of the rudder angle mainly depends on the preference of the navigator and the danger degree of the current situation, and the selection of the rudder angle of the ship is different in different situations. The proposed method pays more attention to the ship's latest avoidance ability, so the rudder angle is taken as the maximum, that is, the ± 35°, and the ship's rudder rotation time t_1 is set to a fixed value of 14s. In this case, the cumulative value of the ship's heading from the start of steering to the steering process over time is:

$$\Delta\varphi_1 = \begin{cases} \dfrac{K\delta}{t_1}\left[\dfrac{t^2}{2} - Tt + T^2\left(1 - e^{\frac{-t}{T}}\right)\right], 0 \le t < t_1 \\ K\delta\left[t - \left(T + \dfrac{t_1}{2}\right) + \dfrac{T^2}{t_1}\left(e^{\frac{t_1}{T}} - 1\right)e^{\frac{-t}{T}}\right], t \ge t_1 \end{cases} \tag{3}$$

where $\Delta\varphi_1$ represent the change in course over time after rudder turning; The Figure 1 is a simple schematic diagram, in which "o" indicates the position of the ship in the process of steering to the specified rudder angle, and "+" indicates the process of the ship starting to turn.

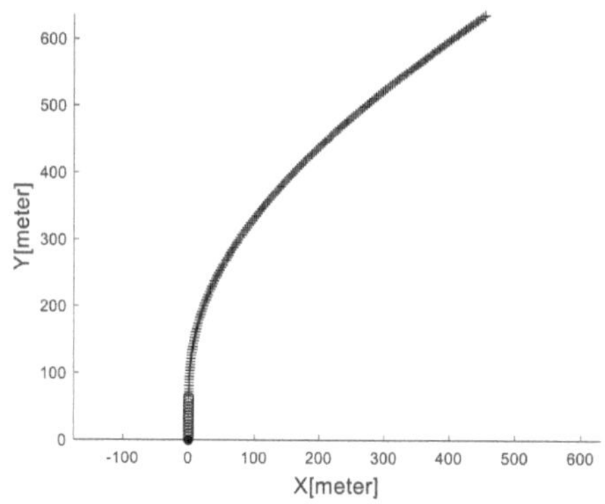

Figure 1. Schematic diagram of ship steering.

The cumulative value of the ship's heading at the midship:

$$\triangle\varphi_2 = \dot{\varphi}_0\left(T + \frac{t_1}{2}\right)\left(1 - e^{\frac{-t}{T}}\right) \tag{4}$$

where $\Delta\varphi_2$ represents the change in heading over time after rudder return; $\dot{\varphi}_0$ represents the velocity's angular velocity at rudder return; t represents the change in time after the rudder return operation is the initial point.

Therefore, the change in the course of a ship under a complete ship steering operation is:

$$\triangle\varphi = \triangle\varphi_1 + \triangle\varphi_2 \tag{5}$$

In general, K and T are calculated using the Z-shaped experimental method and the regression analysis estimation formula method. The Z-shaped test method of the ship needs to be carried out in specific stable sea conditions, and this method requires precise operation and complex calculations, so it is very difficult for the crew to obtain K and T values in practice. In order to obtain the maneuverability index of the ship in the whole water area, the ship data in the ship AIS data can be used to calculate the ship maneuverability index K and T by combining the regression analysis estimation formula method. The fitting equation (Hong, B.K., and Yu, Y. 2000) is:

$$K * \frac{L}{V} = K' = 0.458 + 0.191\frac{L}{B} \tag{6}$$

$$T * \frac{V}{L} = T' = 0.756 + 0.208\frac{L}{B} \tag{7}$$

where K', T' represent indicates the maneuverability index of the unfactorized treatment; L, V, B the length, speed, and width of the ship.

By substituting the calculated operability indices K and T into the ship's first-order NOMOTO equations, the steering trajectory of the ship can be understood. Defines that the speed of the ship does not change during the steering process, and the rudder return operation is carried out after only one rudder turn.

2.2 *Domain intrusion metrics model*

Generally speaking, the domain intrusion of a ship means that another ship has entered the domain of this ship or the domains of the two ships overlap. The domain intrusion metric model used in the paper takes into account the relative orientation of ships on the basis of domain overlap, which is measured as follows: at the current moment, the distance between two ships is greater than the sum of the distances from the intersection points of the two ships' lines and the domain boundary to their respective ship positions, i.e., the Boundary Distance (BD) of the domain is greater than 0, and there is no conflict between the two ships. (Xin, X. *et al.* 2023)

$$BD_{AB} = d_{AB}(t) - [d_A(t) + d_B(t)] \tag{8}$$

where $d_{AB}(t)$ represents the distance between the two ships at the current moment; $d_A(t)$, $d_B(t)$ represent the distance from the intersection of the boundary of the two ships and the domain boundary of ship A (or ship B) to ship A (or ship B) at the current moment.

$d_A(t)$, $d_B(t)$ is calculated as follows:

$$d_A(t) = \left(\frac{1 + tan^2(\beta_{AB}(t) - \varphi_A(t))}{\frac{1}{R_{LA}^2} + \frac{tan^2(\beta_{AB}(t) - \varphi_A(t))}{R_{SA}^2}} \right)^{\frac{1}{2}} \tag{9}$$

where $\beta_{AB}(t)$ represents that ship B is located in the opposite direction of ship A at the current moment; $\varphi_A(t)$ represents the course of ship A; R_{LA}, R_{SA} represent the major and minor semi-axes of the elliptical domain of ship A; In this study, the classical Fujii model (Wang, S. *et al.* 2021) is used, in which the values of the major semi-axis is 4 times the length of the ship and the minor semi-axis is 1.6 times the length of the ship.

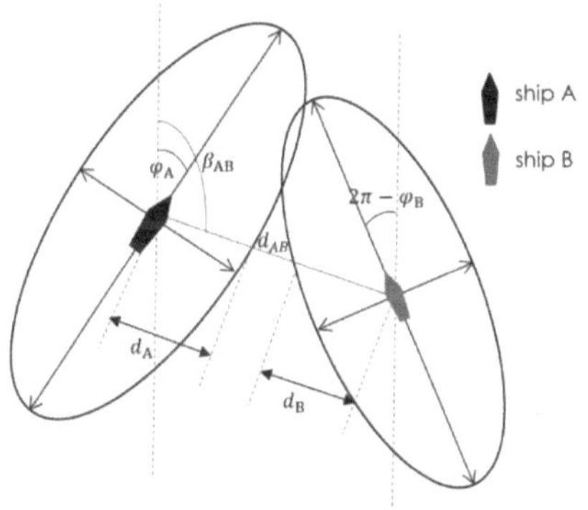

Figure 2. Schematic diagram of ship domain intrusion measurement model.

2.3 *Ship collision risk indicators*

When there is a danger of collision between two ships, first clarify the encounter situation and identify the giving way ship and the stand-on ship before proceeding with the avoidance maneuver. In principle, if COLREG rules and good seamanship are ignored, the giving way vessel can change any course. However, depending on the navigator's habits and rudder limitations, the range of course change is usually not greater than 90° (Li M. *et al.* 2021). In addition, the course changes available to the give way vessel in different encounter situations should vary.

According to COLREG rules, there are three types of encounter situations between two vessels, namely overtaking, head-on and crossing. The encounter situations can be categorized according to Table 1. The encounter situations are categorized into four types: head-on, small-angle crossing, large-angle crossing, and overtaking situation. A positive value in the heading change interval R indicates a change of heading to starboard and a negative value indicates a change of heading to port.

Table 1. Criteria for dividing the situation.

	encounter situation		relative bearing conditions	R
There will be domain intrusion in the future	head-on		$Q_1 \in [0°, 5°] \cup [355°, 360°]$, $Q_2 \in [0°, 5°] \cup [355°, 360°]$	$\{1, 2, 3, \ldots, 90\}$
			$Q_1 \in [0°, 112.5°]$, $Q_2 \in [0°, 112.5°]$	$\{1, 2, 3, \ldots, 90\}$
			$Q_1 \in [247.5°, 360°]$, $Q_2 \in [247.5°, 360°]$	$\{1, 2, 3, \ldots, 90\}$
	overtaking	Giving way ship	$Q_1 \in [0°, 90°] \cup [270°, 360°]$, $Q_2 \in [112.5°, 247.5°]$	$\{-90, -89, \ldots, -1, 1, 2, 3, \ldots, 90\}$
		Stand-on ship	$Q_1 \in [112.5°, 247.5°]$, $Q_2 \in [0°, 90°] \cup [270°, 360°]$	NULL
	small-angle crossing	Giving way ship	$Q_1 \in [0°, 67.5°]$, and it's not crossing or overtaking	$\{1, 2, 3, \ldots, 90\}$
		Stand-on ship	$Q_1 \in [292.5°, 360°]$, and not crossing or overtaking	NULL
	large-angle crossing	Giving way ship	$Q_1 \in [67.5°, 112.5°]$, and not crossing or overtaking	$\{-90, -89, \ldots, -1\}$
		Stand-on ship	$Q_1 \in [247.5°, 292.5°]$, and not crossing or overtaking	NULL

According to the situation division standard, it can be seen that the relative bearing between the two ships are the key elements of the situation division, and the relative bearings between the ships are shown in Figure 3.

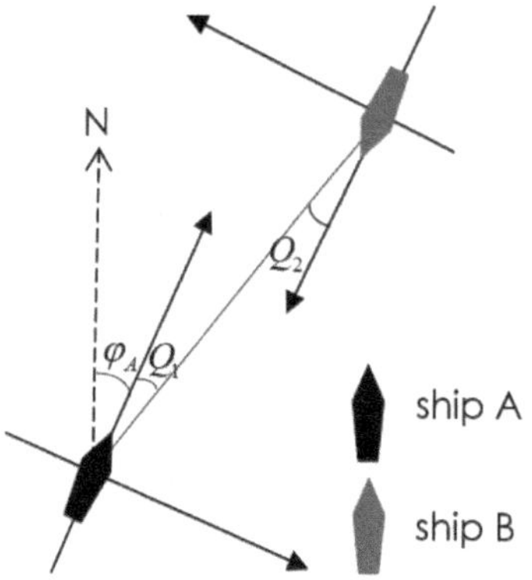

Figure 3. Schematic diagram of the relative bearing.

The bearing of ship B relative to ship A is Q_1, and the bearing of ship A relative to ship B is Q_2, which is calculated as shown in Eq. (10).

$$Q_1 = \begin{cases} \arcsin\left(\dfrac{X_B - X_A}{d_{AB}}\right) - \varphi_A, Y_A \leq Y_B \\ \pi - \arcsin\left(\dfrac{X_B - X_A}{d_{AB}}\right) - \varphi_A, Y_A > Y_B \end{cases} \tag{10}$$

where (X_A, Y_A) represents coordinates of the ship A's location; (X_B, Y_B) represents coordinates of the ship B's location; d_{AB} is the distance between the two ships.

When there is a danger of collision between two ships, the giving way ship needs to determine the steering angle of the ship's safe avoidance according to the heading change interval, that is, as shown in Figure 4. After a certain period of time, two ships will invade each other's ship domain, as shown by the ship domain drawn by the dotted line in the Figure 4.

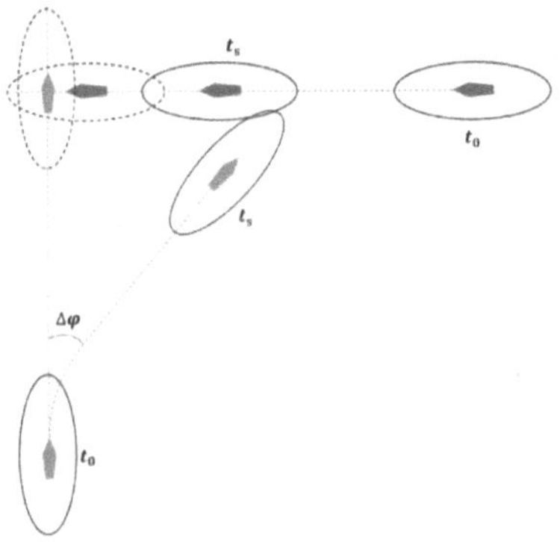

Figure 4. Schematic diagram of ship safety avoidance angle.

The ship needs to start steering at the time of t_0., that is, to start steering, and the course change $\Delta\varphi$ so that the two ships can pass safely, t_s. is not a fixed value, it only indicates that there is no domain intrusion between the two ships within 20 minutes from the start of the rudder turn, and defines the set of all the heading change angles $\Delta\varphi$ of the ship to achieve safety avoidance within the set course change interval R, r_{safe}, that is, the safety avoidance angle set. Through the set of safety avoidance angles and the course change interval of the ship, it can obtain one of the collision risk indexes of the ship, the difficulty of avoiding (AD).

$$AD = 1 - \frac{|r_{safe}|}{|R|} \tag{11}$$

where $|r_{safe}|$, $|R|$ represent the cardinality of two sets, that is, the number of elements in the set.

At the current moment, t_0., when there is a danger of collision between the two ships, when the two ships keep the current speed and course for a long time, t_1., the safe avoidance angle set of the giving way ship is not empty, and at the next moment, the safe avoidance angles set of the giving way ship is empty, then the latest rudder application time of the giving way ship is obtained, $T_{\text{In}} = t_1 - t_0$.

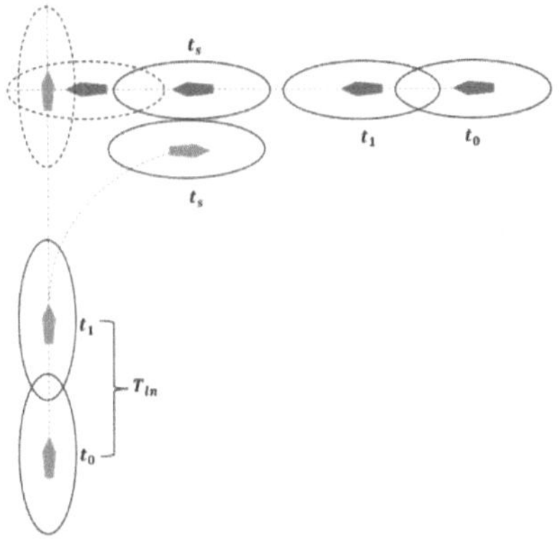

Figure 5. Schematic diagram of the latest rudder application time.

During the course of a ship's voyage, there may be non-avoidance steering behavior on the part of the ship due to the unknown voyage plan, and it is not meaningful to calculate the risk of ship collision for too long a time in the future. (Ha J. *et al.* 2021) used standard TCPA in the improved CRI using DCPA and TCPA and set it to 20 minutes, so the maximum latest rudder application time is set to 20 minutes, i.e., T_m=1200s. That is, through the latest rudder application time of the ship and the maximum latest rudder application time, another ship collision risk index avoidance time urgency (ATU) can be obtained.

$$ATU = 1 - \frac{T_{\ln}}{T_m} \tag{12}$$

2.4 *Indicators weight calculation*

According to the two indexes obtained in the previous section, it can be understood that the increase of the value of the two indicators will lead to the increase of the collision risk of ships, and in order to obtain the accurate collision risk, the CRITIC method is used to calculate the weights of the two indicators. The CRITIC method is an objective weighting method that determines the objective weight of evaluation indicators based on the contrast intensity of evaluation indicators and the conflict between evaluation indicators (Zhang, L., and Zhang, X. 2015). This approach takes into account both the size of the indicator variability and the correlation between the indicators. For the CRT1C method, when the standard deviation is constant, the less the conflict between the indicators, the smaller the weight. The greater the conflict, the greater the weight; In addition, when the positive correlation between the two indicators is greater, the less conflicting it is, which indicates that the information reflected by the two indicators is more similar in the evaluation of the merits of the program. The obtained data of the two indicators are standardized, and the standardized formula is as follows:

$$x'_{ij} = \frac{\max(x_j) - x_{ij}}{\max(x_j) - \min(x_j)} \tag{13}$$

where x_{ij} represents the value in row i, column j, in the initial data; x'_{ij} represents the value in row i, column j, in the standardized data; $\max(x_j)$, $\min(x_j)$ represent the maximum and minimum values of the j-th indicator.

Calculate the volatility of the data:

$$S_j = \sqrt{\sum_{i=1}^{m} \left(x'_{ij} - \overline{x_j}'\right)^2} \tag{14}$$

where S_j represents the volatility of the j-th indicator; m represents the number of rows for the initial data; $\overline{x_j}'$ represents the average of the j-th indicator after normalization.

Before calculating the conflicting of the data, it is necessary to calculate the correlation matrix between the indicators:

$$r_{kj} = \frac{\sum_{j,k=1}^{n} \left(x'_{ij} - \overline{x_j}'\right)\left(x'_{ik} - \overline{x_k}'\right)}{\sqrt{\sum_{j=1}^{n} \left(x'_{ij} - \overline{x_j}'\right)^2 \sum_{k=1}^{n} \left(x'_{ik} - \overline{x_k}'\right)^2}} \tag{15}$$

where r_{kj} represents the correlation matrix between the k-th indicator and the j-th indicator; n represents the number of columns for the initial data, that is, the number of indicators. The conflict is:

$$A_j = \sum_{i=1}^{n} (1 - r_{ij}) \tag{16}$$

where A_j represents the conflict of the j-th indicator; The amount of information in the data is:

$$C_j = S_j \times A_j \tag{17}$$

where C_j represents the amount of information carried by the first indicator. The weights of the indicators are:

$$W_j = \frac{C_j}{\sum_{j=1}^{n} C_j} \tag{18}$$

where W_j represents the weight of the j-th indicator.

3 EXPERIMENT

3.1 *Overview of the waters*

Ningbo Zhoushan Port is located in the middle of China's coastline, the southern wing of Hangzhou Bay, the northeast coast of Zhejiang, and the south side of the intersection of China's coastline and the Yangtze River. In the sea off Ningbo Zhoushan Port, the inbound and outbound ship routes intersect with China's coastal merchant shipping routes, and high-density ships meet in this water, and the sea waters off Ningbo Zhoushan Port are selected as the research scope, and the selected waters are about 28 nautical miles long in the east-west direction and about 17 nautical miles wide in the north-south direction.

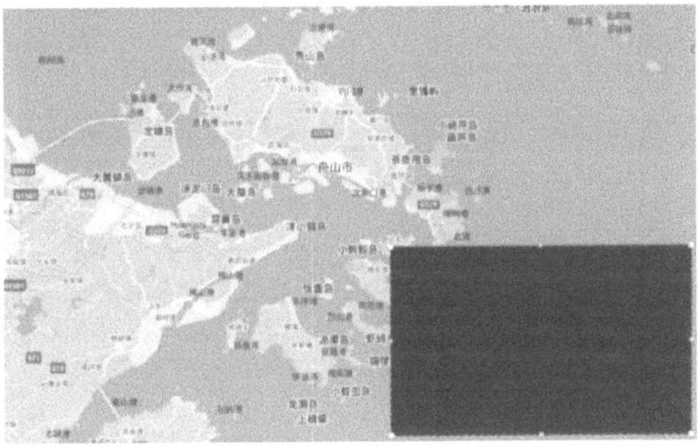

Figure 6. Research scope of waters.

3.2 *Comparison in different scenarios*

The AIS data in the research scope in March 2024 is cleaned and interpolated, and then slice every 30 minutes from 0:00 on March 1, 2024, and the values of the two indicators of the ship with collision danger at that moment are calculated, the avoidance difficulty AD and the avoidance time urgency ATU of the ship with collision risk are obtained, and the values of all fragments are summarized to construct a matrix of 9790×2, and the weight calculation is carried out by substituting the data obtained in the study water area into the CRITIC method. It can be obtained that the volatility of the indicators AD and ATU is 0.30 and 0.43, respectively, and their conflicts are 0.24. Finally, the weights of AD and ATU are 0.59 and 0.41, respectively, that is, the risk calculation formula is as follows:

$$risk = 0.59 \times AD + 0.41 \times ATU \tag{19}$$

In order to reflect the rationality of the proposed risk method, different risk methods are used for comparative analysis, and the CRI method is used firstly, and its formula (Zhen R. *et al.* 2022) is as follows:

$$RI = \omega_1 CRI_{DCPA} + \omega_2 CRI_{TCPA} + \omega_3 ADRI \tag{20}$$

where CRI_{DCPA} represents the risk of spatial collision caused by DCPA; CRI_{TCPA} represents the risk of time collision caused by TCPA; $ADRI$ represents the risk of collision based on the polymer coefficient; ω_1,ω_2,ω_3 represent the weights of CRI_{DCPA},CRI_{TCPA},$ADRI$.

In addition, a domain-based risk method is adopted, and the risk equation (Liu K. *et al.* 2021) is:

$$R_{ij} = \begin{cases} \left(\dfrac{d_i + d_j - \|\vec{D}_{ij}\|}{d_i + d_j} \right)^{3.03}, \|\vec{D}_{ij}\| \leq d_i + d_j \\ 0, \|\vec{D}_{ij}\| > d_i + d_j \end{cases} \tag{21}$$

where R_{ij} represents the severity of the ship i-to-ship j conflict; d_i, d_j represent the distance from the intersection of the two ship lines and the ship i (or ship j) domain boundaries to the ship j (or ship i); $\|\vec{D}_{ij}\|$ represents the distance from ship to ship.

3.2.1 *Comparison in the two ships keeping speed and direction scenarios*

Taking the trajectory point of the real ship as the research object, the relative position of the current ship and the ship in the next 6 minutes is shown in Figure 7. The solid line ellipse represents the ship domain at the current moment, the dotted line ellipse represents the ship domain of the ship after 6 minutes, and Figure 7. shows that the two ships encounter each other, and if the two ships keep the course and speed, the ship domain intrusion will occur.

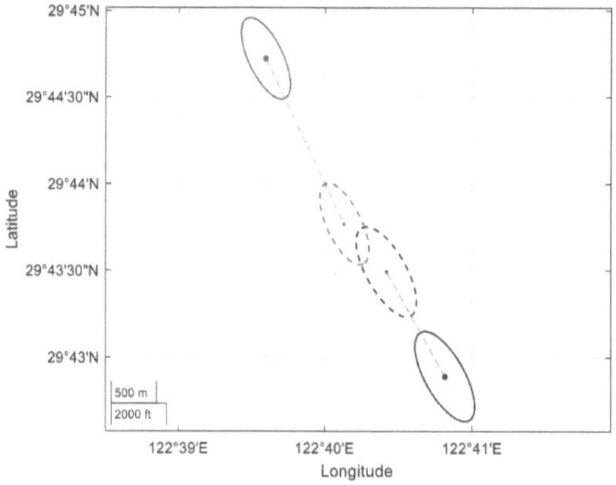

Figure 7. Schematic diagram of head-on situation.

Figure 8 shows the changes of multi-risk and distance between the two ships in the process of maintaining speed and direction. According to Figure 8, it can be seen that the risk calculated by the domain-based method is steadily increasing and at a high level, and the method can only reflect the distance between the two ships by the degree of domain invasion between the two ships, but cannot reflect the movement trend between the ships. The risk calculated by the CRI method mainly takes into account DCPA, TCPA and the density of ships in the vicinity of the water area where the ship is located, and does not take into account the operational performance of the ship and the movement trend of the ship. In the same scenario, different ships have different tolerances for DCPA and TCPA, and the CRI method cannot reflect their differences, nor does it consider the differences in the difficulty of avoiding different situations encountered by ships. The proposed risk method takes into account the different difficulty of ship avoidance under different encounter situations, and the risk of the two ships at the initial moment is small in the encounter scenario, which also leads to the low risk at the initial moment.

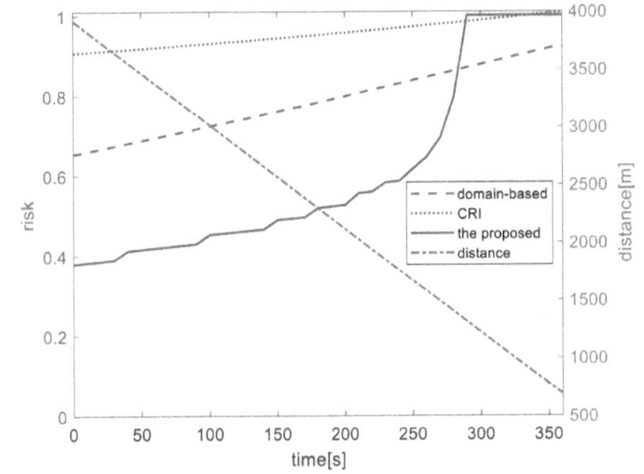

Figure 8. Diagram of change of multi-risk and distance in head-on situation.

3.2.2 *Comparison of the actual trajectory of the two ships and the multi-risk*

Two ship trajectories with collision risk are selected, and their trajectories are shown in Figure 9, in which "★" represents the starting point of the ship, "●" represents the end point, and triangle represents the closest distance point of the two ships. Ship1 size is 145m×21m, and ship2 size is 100m×19m. According to the trajectory of the two ships, it can be concluded that at the initial position, the two ships form a crossing situation, and according to rules 15 of COLREG rules, ship1 needs to turn to the right to avoid collision between the two ships. Looking at the trajectory of the subsequent ship, it can be seen that ship1 did turn to the right.

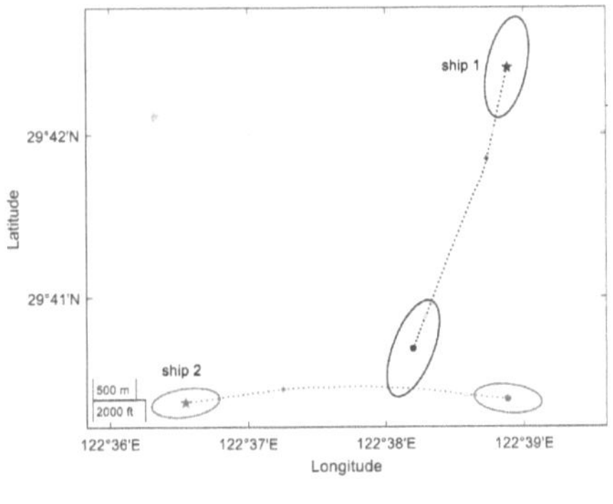

Figure 9. True trajectory diagram of ship.

By analyzing the actual trajectories of the two ships, the change of the multi-risk and the distance between the two ships over time can be obtained, as shown in Figure 10. According to the multi-risk, it can be seen that the risk calculated by the domain-based method increases with the passage of time from the initial position, and finally remains at a high level, because the distance between the two ships is decreasing, and the degree of intrusion in the ship domain is increasing. There is a certain fluctuation in the CRI method, but it remains at about 0.8, because although the distance between the two ships is far away at the initial moment, the DCPA is small, and the subsequent distance between the two ships decreases as the ship approaches, but the steering of the ship causes the DCPA to increase.

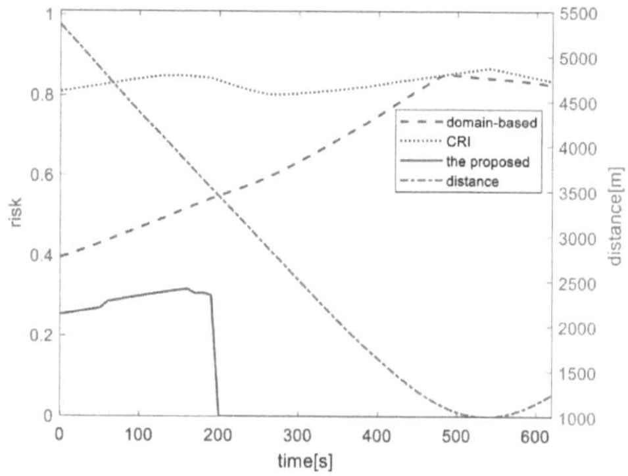

Figure 10. Diagram of change of multi-risk and distance during the voyage of a ship.

The risk of the proposed method remained at 0.25~0.35 from the initial moment to 190s, and after 190s, the risk of the two ships based on the proposed method is in a downward trend until the risk reached 0. The reason why the danger degree of the 20th trajectory point becomes 0 is as follows: according to the observation of the real trajectory map of the ship, it can be found that the two ships have steering behavior after the 19th trajectory point (Figure 9 "*"), taking the 20th trajectory point as the original data, the change of the domain boundary distance BD with time under the direction and speed maintenance is shown in Figure 11, it can be found that there is no domain intrusion between the two ships in the future with the track point at this time, and in the subsequent time, the two ships still turn to the right, It means that there will be no domain intrusion between the two ships in the future, that is, the risk is zero.

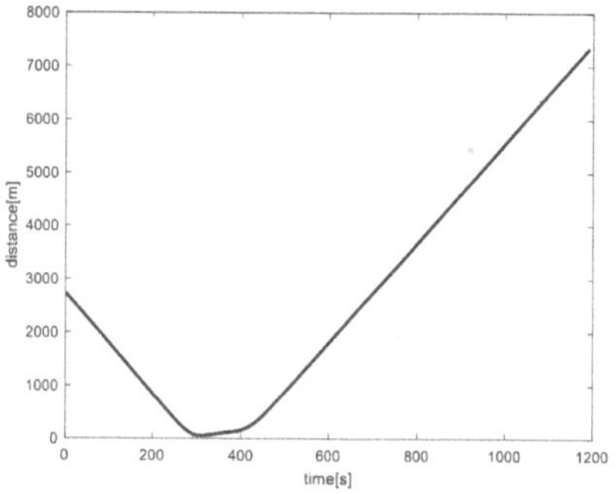

Figure 11. Boundary distance variation diagram under the direction and velocity maintenance.

Therefore, the proposed method can effectively understand the steering situation of the ship, better understand the current situation of the ship, and avoid the collision between the two ships. According to the distance curve between the two ships, it can be understood that the distance between the two ships is the lowest at 550s (Figure 9 "triangle"), which is 1000m, and the risk of the CRI method and domain-based method is still high, but the risk of the proposed method is 0, mainly because the two ships are in a state of departure and the subsequent TCPA is negative, and there is no ship domain intrusion situation between the two ships at the closest distance, which indicates that there will be no probability of domain intrusion under the normal driving state of the subsequent two ships. This reflects that the proposed method pays more attention to the relative direction of motion of the two ships than to the distance between the two ships.

4 RESULTS AND DISCUSSIONS

In order to solve the problem that the current collision risk ignores the maneuverability of the ship, a collision risk method considering the avoidance ability of the ship is proposed. The steering trajectory of the ship is calculated by combining the ship AIS data with the first-order NOMOTO manipulation equation, and then the indicators related to the ship's latest rudder application time and safety avoidance angle set are calculated by combining the ship domain and COLREG rules, and then the weights of the two indicators are calculated by using the CRITIC method, and finally the ship collision risk calculation formula is obtained by taking the offshore waters of Ningbo Zhoushan Port as the experimental object. Compared with the CRI method and the domain-based method, the results show that the proposed method can not only easily consider the ship's maneuverability, but also join the ship field and consider the influence of different ship types on the collision risk. Through experimental comparison, it is found that the proposed risk model can effectively understand the steering behavior of the ship, and the calculation of the risk pays more attention to the relative direction of motion of the two ships rather than the distance between the two ships.

Due to the inaccuracy of the calculation of K and T values, the assumption that the ship only performs one rudder turning and rudder return operation during the avoidance operation, the threshold of the latest rudder application time is set to a fixed value, and the selection of the ship domain in the model is slightly unreasonable, which leads to some shortcomings in the method. In the future, the accurate method is used to obtain the ship avoidance trajectory that is more in line with the actual navigation, and the selection of the latest rudder application time threshold considers the ship type, ship speed, water area and other factors, and in the future research, AIS data analysis and mining will be used to design the ship domain model, so as to obtain the risk more accurately.

REFERENCES

Chen, P., Huang, Y., Mou, J., and Van Gelder, P. H. A. J. M. (2019). Probabilistic risk analysis for ship-ship collision: State-of-the-art. *Safety science*, *117*, 108–122.

Chin, H. C., and Debnath, A. K. (2009). Modeling perceived collision risk in port water navigation. *Safety Science*, *47*(10), 1410–1416.

Ha, J., Myung-Il, R., and Hye-Won, L. (2021). Quantitative calculation method of the collision risk for collision avoidance in ship navigation using the CPA and ship domain [J]. *Journal of Computational Design and Engineering*, *8*(3), 894–909.

Hong, B.K., and Yu, Y. (2000). Statistical analysis of ship maneuverability indices K, T. *Journal of Dalian Maritime University*, 26(4), 29–33.

Huang, Y., and Gelder, P. H. A. J. M. V. (2017). Non-linear velocity obstacles with applications to the maritime domain. *Maritime transportation and harvesting of sea resources*, 999–1007.

Huang, Y., and Van Gelder, P. H. A. J. M. (2020). Time-varying risk measurement for ship collision prevention. *Risk Analysis*, *40*(1), 24–42.

Huang, Y., Van Gelder, P. H. A. J. M., and Wen, Y. (2018). Velocity obstacle algorithms for collision prevention at sea. *Ocean Engineering*, *151*, 308–321.

Lee, H. J., and Rhee, K. P. (2001). Development of collision avoidance system by using expert system and search algorithm. *International shipbuilding progress*, *48*(3), 197–212.

Lenart, A. S. (2015). Analysis of collision threat parameters and criteria. *The Journal of Navigation*, *68*(5), 887–896.

Li, M., Mou, J., Chen, L., He, Y., and Huang, Y. (2021). A rule-aware time-varying conflict risk measure for MASS considering maritime practice. *Reliability Engineering and System Safety*, *215*, 107816.

Liu, J., Zhou, F., Li, Z., Wang, M., and Liu, R. W. (2016). Dynamic ship domain models for capacity analysis of restricted water channels. *The Journal of Navigation*, *69*(3), 481–503.

Liu, K., Yuan, Z., Xin, X., Zhang, J., and Wang, W. (2021). Conflict detection method based on dynamic ship domain model for visualization of collision risk Hot-Spots. *Ocean engineering*, *242*, 110143.

Szlapczynski, R., and Szlapczynska, J. (2016). An analysis of domain-based ship collision risk parameters. *Ocean Engineering*, *126*, 47–56.

Szlapczynski, R., and Szlapczynska, J. (2017). Review of ship safety domains: Models and applications. *Ocean Engineering*, *145*, 277–289.

Wang, S., Zhang Y, and Hu X. (2021). Ship collision risk calculation method considering uncertainty of ship position prediction. *Chinese Journal of Ship Research*, *16*(1), 114–120.

Wang, X., Liu, Z., and Cai, Y. (2017). The ship maneuverability based collision avoidance dynamic support system in close-quarters situation. *Ocean Engineering*, *146*, 486–497.

Xin, X., Liu, K., Loughney, S., Wang, J., Li, H., and Yang, Z. (2023). Graph-based ship traffic partitioning for intelligent maritime surveillance in complex port waters. *Expert Systems with Applications*, *231*, 120825.

Yang, T., Wang, X., and Liu, Z. (2023). A Novel Ship Domain-oriented Parameter of Ship Collision Risk Considering the Ship Maneuverability and Encounter Situation. *Journal of Marine Science and Application*, *22*(2), 181–198.

Zhang, L., and Zhang, X. (2015). A weighted clustering method based on improved CRITIC method. *Statistics and Decision Making*, (22), 65–68.

Zhen, R., Riveiro, M., and Jin, Y. (2017). A novel analytic framework of real-time multi-vessel collision risk assessment for maritime traffic surveillance. *Ocean Engineering*, *145*, 492–501.

Zhen, R., Shi, Z., Shao, Z., and Liu, J. (2022). A novel regional collision risk assessment method considering aggregation density under multi-ship encounter situations. *The Journal of Navigation*, *75*(1), 76–94.

Proceedings of the International Association of Institutes of Navigation World Congress 2024 – Chen et al. (Eds)

Parametric modeling method of effective sound speed error and its application in underwater navigation

J. Zhou, S.Q. Xue, J.C. Bian & S. Zhao
Chinese Academy of Surveying and Mapping, Beijing, China

N.L. Han
China University of Geosciences, Beijing, China

ABSTRACT: In dynamic ocean environments, ranging errors caused by the temporal and spatial variations of sound speed are the main challenges faced by underwater navigation. High-precision acoustic navigation technology usually relies on complex sound ray-tracing algorithms and requires in-situ sound speed profile observations, which makes it difficult to meet real-time navigation and positioning needs. For this, we propose a parametric modeling approach for effective sound speed errors. By constructing a function model that correlates effective sound speed errors with the elevation angle and depth of sound rays, combined with observed time delay information, corrections for ranging errors are achieved. It shows that the proposed method can accurately model effective sound speed errors and improve computational efficiency by more than 15 times while maintaining an accuracy loss of less than decimeters in positioning and less than centimeters per second in measuring speed, which provides favorable technical support for real-time navigation services.

1 INTRODUCTION

First paragraph: Seafloor acoustic navigation positioning technology plays a vital role in ensuring the safety and efficiency of marine operations in modern oceanographic research and marine engineering. It is extensively utilized in various applications, including ocean resource exploration, seafloor topography mapping, and military operations (Poutanen et al, 2020; Yang *et al*, 2020; Xu *et al*, 2023). Acoustic navigation positioning technology utilizes the propagation characteristics of sound waves in water, which have the advantages of strong penetration, long propagation distance and wide application range (Maki *et al*, 2013; Chen et al, 2011). However, the complex and variable marine environment has resulted in severe sound ray bend. This means that the distances calculated from fixed reference signal speeds are not always the same as the actual signal propagation distances (Wang *et al*, 2020; Xue *et al*, 2024). This variation can greatly inhibit the accuracy and reliability of underwater navigation positioning, commonly referred to as acoustic ranging errors. At present, identifying and correcting this error has become a critical issue for achieving high-accuracy underwater navigation positioning.

For this, numerous scholars have conducted comprehensive studies. The current study focuses on eliminating ranging errors by using correction methods for sound speed or by directly correcting the ranging error values, the former of which is currently the research of scholars. The correction methods can be mainly divided into two types, the first type includes the Taylor series expansion method (RK 1978), empirical formula method (Christopher et al, 1997), weighted average sound speed method, and effective sound speed method (Sun *et al*, 2019). These methods are computationally efficient and straightforward, but they are less accurate and prone to calculation errors. The second group of methods, which are

DOI: 10.1201/9781003624424-12

extensively employed and have higher precision, primarily consist of sound ray-tracing (Chadwell et al, 2010; Sakic *et al*, 2018), equivalent sound speed profile method (He *et al*, 2019), sound ray bending correction method (Zhao *et al*, 2018), and similar approaches. The most popular and effective way to get rid of acoustic ranging errors right now is the sound ray-tracing method, which is based on Snell's law and uses the layer addition technique to caclulate parameters like sound ray propagation time and displacement.

However, there are two problems with sound ray-tracing methods. The complexity of the model requires solving numerous inverse problems to calculate initial incidence angles, resulting in a significant reduction in computational efficiency, especially when dealing with a large number of control points or while deploying navigation beacons in deep-sea circumstances. In order to address this issue, relevant scholars have proposed a variety of improved methods based on the Newton method and the secant method for the iterative algorithm of sound ray incident angle, which can significantly improve the computational efficiency (Liu *et al*, 2023; Yang *et al*, 2021). Second, sound ray-tracing is highly dependent on sound speed profile data in practical applications. At the same time, field measurements of sound speed profiles are time-consuming and obtaining real-time data is challenging (Zhao *et al*, 2023; Xue *et al*, 2023), which limits the real-time capability and applicability of the sound ray-tracing method. Therefore, it is essential to develop an efficient method to rapidly correct sound ray ranging errors in order to better real-time navigation and positioning services.

Aiming at the above problems, this paper proposes a parametric modeling method for effective sound speed errors. Initially, sound speed errors are defined and calculated using a constant gradient sound ray-tracing model. Subsequently, a functional model relating sound speed errors to the sound ray elevation angle and depth is constructed, incorporating observed time delay information to indirectly correct ranging errors. The organization of the paper is as follows: Section 2, introduces the window estimation approach for an active acoustic navigation model and presents the definition and parametric model of effective sound speed errors. Section 3, tests the performance of the proposed method using both simulation and actual sea trial data from Japan. Finally, conclusions are summarized in the last section.

2 METHODOLOGY

2.1 *Window estimate active acoustic navigation model*

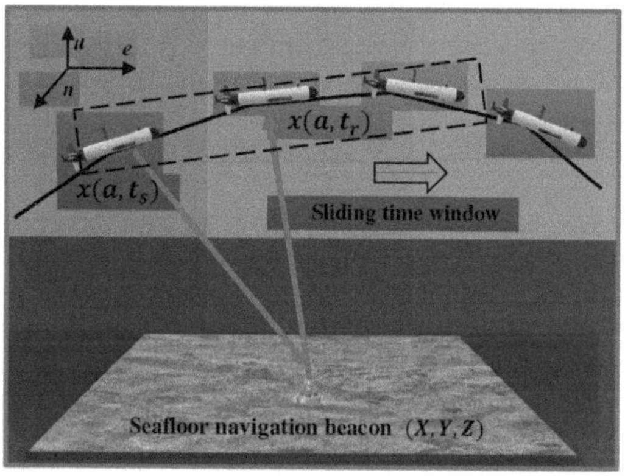

Figure 1. Underwater vehicle navigation.

As shown in Figure 1, the spatial coordinates of the seafloor navigation beacon are denoted by(X, Y, Z). Within a given time window, the carrier trajectory can be described as a function

$\boldsymbol{x}(\boldsymbol{a}, t)$that varies with time t, then:

$$\boldsymbol{x}(\boldsymbol{a}, t) = \begin{bmatrix} x(\boldsymbol{a}_x, t) \\ y(\boldsymbol{a}_y, t) \\ z(\boldsymbol{a}_z, t) \end{bmatrix} \tag{1}$$

Where $\boldsymbol{a}_k$ represents the carrier trajectory parameters for the $k \in \{x, y, z\}$ coordinate component. Assuming the undetermined coordinate for the carrier acoustic signal transmission time t_s is $\boldsymbol{x}(\boldsymbol{a}, t_s)$, and the undetermined coordinate of the signal reception time t_r is $\boldsymbol{x}(\boldsymbol{a}, t_r)$, the active acoustic observation equation can be expressed as:

$$c_0 \cdot T_{\text{obs}} = f[\boldsymbol{x}(\boldsymbol{a}, t_s)] + f[\boldsymbol{x}(\boldsymbol{a}, t_r)] + \varepsilon \tag{2}$$

Where T_{obs} is the two-way propagation time of the acoustic signal, $f[\boldsymbol{x}(\boldsymbol{a}, t)] = \sqrt{[x(\boldsymbol{a}_x, t), -X]^2 + [x(\boldsymbol{a}_y, t) - Y]^2 + [x(\boldsymbol{a}_z, t) - Z]^2}$ is the distance function between the carrier acoustic position and the seafloor beacon, ε represents the measurement random error, and c_0 is the effective sound speed of the signal propagation.

In the practical navigation process, the effective sound speed can be derived through constant gradient sound ray-tracing. However, due to its computational complexity and high cost, real-time continuous navigation work is challenging. Therefore, this paper proposes a parametric modeling approach to address this issue, with detailed methodology provided in Section 2.2.

For vertical positioning information, the underwater vehicle can utilize a pressure sensor. So the model only navigates the underwater vehicle in the plane dimension. Let $F[\boldsymbol{x}(\boldsymbol{a}, t_s, t_r)] = f[\boldsymbol{x}(\boldsymbol{a}, t_s)] + f[\boldsymbol{x}(\boldsymbol{a}, t_r)]$, assume that the time of transmission and reception of the i th ranging period is t_{si}, t_{ri} and the time of signal round-trip propagation is $T_{obs,i}$, then the observation equation for the i th ranging period can be expressed as follows :

$$c_0 \cdot T_{obs,i} = F_i[\boldsymbol{x}(\boldsymbol{a}, t_{si}, t_{ri})] + \varepsilon_i \tag{3}$$

By linearizing (3) and discarding the residual term, we can obtain:

$$dL_i = B_i d\boldsymbol{a} + \varepsilon_i \tag{4}$$

Where $dL_i = c_0 \cdot T_{obs,i} - (F_i[\boldsymbol{x}(\boldsymbol{a}^0, t_{si}, t_{ri})])$, $\mathbf{a}^0$ is the initial value of the carrier trajectory parameters, $\mathrm{d}\boldsymbol{a}$ is the correction of the carrier trajectory parameters, and B_i is the acoustic navigation design matrix:

$$B_i = \frac{\partial F_i}{\partial \boldsymbol{x}} \cdot \frac{\partial \boldsymbol{x}}{\partial \boldsymbol{a}} \tag{5}$$

Due to the local support characteristics of B-splines, it has gradually become one of the preferred methods for curve fitting. Within the time window $[t_1 t_m]$, the carrier's trajectory is estimated using a cubic B-spline model, thus the carrier's position at time t within the time window can be expressed as:

$$\begin{bmatrix} x(\boldsymbol{a}_x, t) \\ y(\boldsymbol{a}_y, t) \end{bmatrix} = \begin{bmatrix} \sum_{s=0}^{K} b_{x,s} \Phi_{x,s}(t) \\ \sum_{s=0}^{K} b_{y,s} \Phi_{y,s}(t) \end{bmatrix} \tag{6}$$

Where $\Phi_{I,k}, I \epsilon \{x, y\}$ represents the B-spline basis function describing the coordinate component, which can be obtained using the de-Boor Cox recursive formula (Boor 1978),

$b_{I,s}$ is the B-spline basis function coefficient for the coordinate component, and K is the total number of B-spline coefficients, which indicates the time window $[t_1 t_m]$ is divided into K-2 equally spaced node intervals. The design matrix can thus be expressed as:

$$B_i = \left[\frac{\partial F_i}{\partial \boldsymbol{x}} \cdot \frac{\partial \boldsymbol{x}}{\partial \boldsymbol{a}}\right] = [\, e_{x,s}\Psi_{x,s} + e_{x,r}\Psi_{x,r} \quad e_{y,s}\Psi_{y,s} + e_{y,r}\Psi_{y,r} \,] \tag{7}$$

Where

$$\begin{cases} e_{x,s} = \dfrac{x(\boldsymbol{a}_x, t_s) - X}{f[\boldsymbol{x}(\boldsymbol{a}, t_s)]} \\ e_{y,s} = \dfrac{y(\boldsymbol{a}_y, t_s) - Y}{f[\boldsymbol{x}(\boldsymbol{a}, t_s)]} \end{cases}; \begin{cases} e_{x,r} = \dfrac{x(\boldsymbol{a}_x, t_r) - X}{f[\boldsymbol{x}(\boldsymbol{a}, t_s)]} \\ e_{y,r} = \dfrac{y(\boldsymbol{a}_y, t_r) - Y}{f[\boldsymbol{x}(\boldsymbol{a}, t_s)]} \end{cases} \tag{8}$$

$$\begin{cases} \Psi_{I,s} = [\, \Phi_{I,0}(t_s) \quad \Phi_{I,1}(t_s) \quad \cdots \quad \Phi_{I,K}(t_s) \,] \\ \Psi_{I,r} = [\, \Phi_{I,0}(t_r) \quad \Phi_{I,1}(t_r) \quad \cdots \quad \Phi_{I,K}(t_r) \,] \end{cases} \tag{9}$$

By combining the observation equations of each epoch within the time window, the least squares solution can be derived as follows:

$$d\widehat{X} = (\boldsymbol{B}^T \boldsymbol{P}\boldsymbol{B})^{-1}\boldsymbol{B}^T\boldsymbol{P}\boldsymbol{L} \tag{10}$$

where $\boldsymbol{P}$ is the observation weight matrix. When the initial value accuracy is not high, the Gauss-Newton iteration can be used to obtain the least squares estimation of the nonlinear model. After determining the trajectory parameters within the time window, the carrier's position at any time in the window can be calculated according to Equation (1).

2.2 *Effective sound speed error modeling method*

Since the propagation path of sound rays in marine environments is not a straight line but rather a curved path due to continuous refraction, calculating distances using a fixed sound speed, instead of the actual signal propagation distance, can introduce significant model errors. Effective sound speed is an important method to correct these distance errors. Therefore, an effective sound speed v related to the observation direction can be introduced as follows:

$$v_\theta = \frac{\sqrt{Y^2 + Z^2}}{\tau} \tag{11}$$

Where θ is the sound ray height angle, Y is the horizontal distance of sound ray propagation, Z is the depth of sound ray propagation, and τ is the sound ray propagation time. These parameters are determined using a layered constant gradient sound ray-tracing algorithm. It is evident that the effective sound speed is related to θ and Z. In order to simplify the model, a constant sound speed C_0 is introduced to obtain the effective sound speed error at any position, as follows:

$$\Delta C_{\theta,Z} = C_0 - v_{\theta,Z} \tag{12}$$

Where ΔC is the effective sound speed error at any elevation angle and depth. Through the above process, the functional relationship between θ and ΔC at a fixed depth is first established, which can be expressed as:

$$\Delta C_\theta = f(\sin\,\theta, \mathbf{p}) \tag{13}$$

Where $\sin\,\theta$ is the sine of the height angle , and $\mathbf{p} = [p_1, p_2, \ldots, p_m]$ represents the function model parameter, which can be solved by the nonlinear least squares criterion, as shown in the following formula:

$$\min g(\sin\,\theta, \mathbf{p}) = \mathbf{V}^T(\sin\,\theta, \mathbf{p})\mathbf{V}(\sin\,\theta, \mathbf{p}) \tag{14}$$

Where g is the sum of squared residuals , $\mathbf{V}(\sin\ \theta, \mathbf{p}) = \Delta C_\theta - f(\sin\ \theta, \mathbf{p})$ is the residual vector, and the least squares solution can be finally expressed as:

$$\mathbf{p} = \mathbf{p}_0 + \mathbf{dp} = \mathbf{p}_0 + \left(\mathbf{J}^T\mathbf{J}\right)^{-1}\mathbf{J}^T\mathbf{V} \tag{15}$$

Where p is the set initial parameter value, **d***a*is the model parameter correction, and **J**is the Jacobian matrix:

$$\mathbf{J} = \begin{bmatrix} \frac{\partial f(\sin\ \theta_1, p)}{\partial p_1} & \frac{\partial f(\sin\ \theta_1, p)}{\partial p_2} & \cdots & \frac{\partial f(\sin\ \theta_1, p)}{\partial p_m} \\ \frac{\partial f(\sin\ \theta_2, p)}{\partial p_1} & \frac{\partial f(\sin\ \theta_2, p)}{\partial p_2} & \cdots & \frac{\partial f(\sin\ \theta_2, p)}{\partial p_m} \\ \vdots & \vdots & & \vdots \\ \frac{\partial f(\sin\ \theta_n, p)}{\partial p_1} & \frac{\partial f(\sin\ \theta_n, p)}{\partial p_2} & \cdots & \frac{\partial f(\sin\ \theta_n, p)}{\partial p_m} \end{bmatrix} \tag{16}$$

By examining the relationship between ΔC and $\sin\ \theta$, this paper will parametrize the effective sound speed error using the following three function models:

M1: Exponential model

$$\Delta C = p_1 \cdot e^{p_2 \cdot\ \sin\ \theta} + p_3 \tag{17}$$

M2: Composite model (linear + exponential)

$$\Delta C = p_1 + p_2 \cdot\ \sin\ \theta + p_3 \cdot e^{p_4 \cdot\ \sin\ \theta} \tag{18}$$

M3: 4th-polynomial model

$$\Delta C = p_1 \cdot\ \sin\ \theta^4 + p_2 \cdot\ \sin\ \theta^3 + p_3 \cdot\ \sin\ \theta^2 + p_4 \cdot\ \sin\ \theta + p_5 \tag{19}$$

Where $[p_1, p_2, \ldots, p_m]$ are the function coefficients for each function model at depth Z. On this basis, a cubic polynomial model (referred to as S1) is then used to further fit the model coefficients for all depths twice, as shown in the following formula:

$$p_i = b_1 \cdot Z^3 + b_2 \cdot Z^2 + b_3 \cdot Z + b_4 i = 1, 2, 3, 4 \tag{20}$$

Where $[b_1, b_2, b_3, b_4]$ are the model parameters obtained by fitting the coefficients at each depth. Through this process, effective sound speed error modeling is completed. In the navigation process, the effective sound speed at any elevation angle θ and depth Z can be obtained by adding the correction value on the basis of C0. Combined with the observation delay data, the correction for ranging errors can be further achieved. At this point, the observation model (2) can be rewritten as:

$$\left(C_0 - \frac{\Delta C_{std,i} + \Delta C_{res,i}}{2}\right) \cdot T_{obs,i} = F_i[\mathbf{x}(\boldsymbol{a}, t_{si}, t_{ri})] + \varepsilon_i \tag{21}$$

Where ΔC_{std} is the effective sound speed error of the forward sound ray, and ΔC_{res} is the effective sound speed error of the return sound ray. It is important to note that the magnitude of each depth error is closely related to the actual sound speed profile data and C_0 of the layer. Therefore, the rationality of the constant sound speed selection in the practical application process also determines the pattern and fitting effectiveness of the effective sound speed error.

3 RESULTS AND DISCUSSION

In order to evaluate the effectiveness of the proposed model, this paper conducts tests through both simulation and measured experiments. Navigation results based on sound ray-tracing are used as reference values to compare the accuracy of different models after ranging error correction. The measured data are derived from GNSS-A acoustic observation data at the CHOS station in Japan in August 2015 (Watanabe *et al.* 2020). Simulation data are generated using the GNSS-A simulation program (Xiao *et al.* 2023), with sound speed profile data determined using the Munk function. Refer to the measured trajectory of the carrier and the spatial dimensions of the seafloor navigation beacons, as shown in Figure 2. The setup includes four navigation beacons deployed at a depth of 1500 meters, with the carrier moving at a speed of 2.5 m/s and sampling intervals of 10 seconds. Additionally, GNSS antenna errors with standard deviations of 3 cm, 3 cm, and 5 cm in the E, N, and U directions are introduced, along with a round-trip propagation time error with a standard deviation of 2.5×10^-5 seconds.

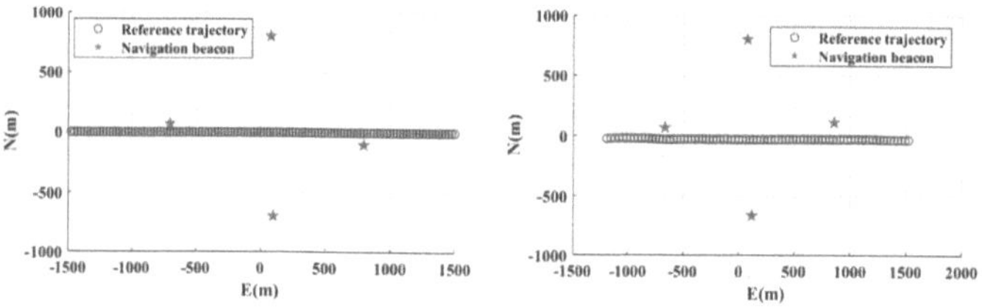

Figure 2. Carrier trajectory and seafloor beacon position: (a) simulation data; (b) measured data.

Since the cubic B-spline model is comprised of at least four basis functions, the B-spline time window navigation and positioning model contains at least eight trajectory parameters under two-dimensional conditions. In this experiment, the time window is set to encompass 16 epochs of observations. For each window, the position and velocity at the middle time (MW) and at the current time (EW) are calculated. Finally, the root mean square error (RMSE) between the navigation positions corrected by the ranging model and the true positions is calculated to serve as the accuracy evaluation index.

3.1 *Result of simulation test*

According to the simulation settings described, the experiment was repeated 100 times to compare the navigation accuracy loss associated with different ranging correction models. The solution results are shown in Figures 3, 4. It is evident that the navigation results obtained by the three correction models proposed in this paper are in good agreement with the results obtained by the ray-tracing method. Where the 'M2+S1' and 'M3+S1' models demonstrate superior accuracy, with positional accuracy loss at intermediate times maintained within 0.1 m, and at current times generally controlled within 0.2 m. The speed measurement accuracy loss for all three models is contained within centimeters per second.

Figure 5 displays the differences in computed distance residuals between the various correction models and the ray-tracing method, and the high-accuracy models 'M2+S1' and 'M3+S1' are able to keep the residuals within 0.15 m.

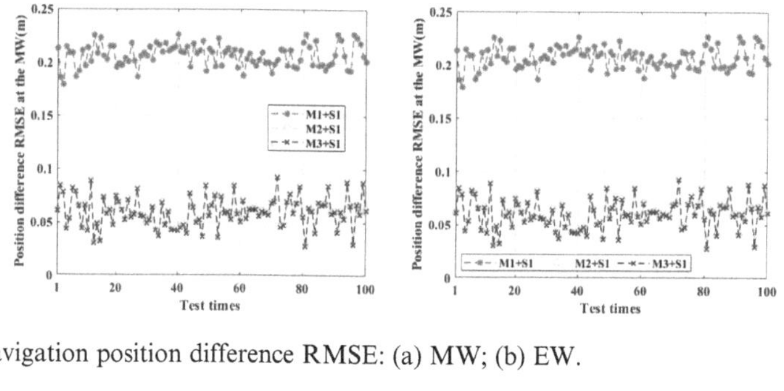

Figure 3. Navigation position difference RMSE: (a) MW; (b) EW.

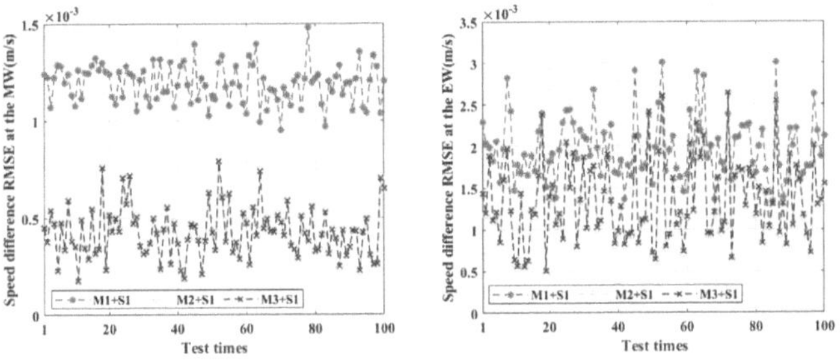

Figure 4. Navigation speed difference RMSE: (a) MW; (b) EW.

Table 1. Calculation accuracy statistics.

	Position difference RMSE (m)		Speed difference RMSE (m)		distance residual
Model	MW	EW	MW	EW	difference RMSE(m)
M1+S1	0.2065	0.2801	0.0011	0.0019	0.1844
M2+S1	0.0596	0.1080	0.0004	0.0013	0.1057
M3+S1	0.0603	0.1089	0.0004	0.0013	0.1070

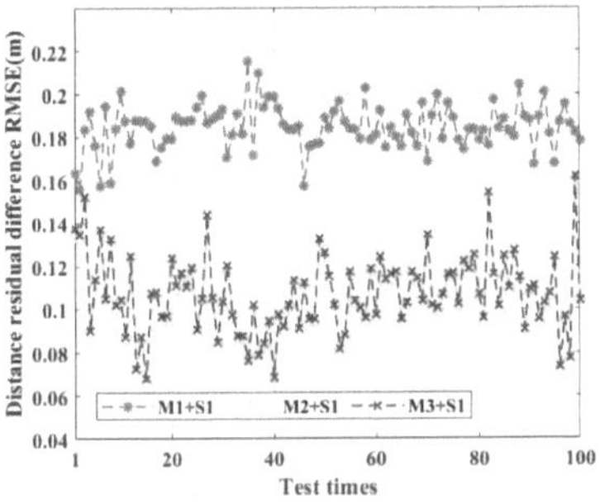

Figure 5. Different model distance residual difference sequence.

To precisely describe the accuracy of the statistical information of the above errors, the RMSE from the 100 experimental results is averaged, as shown in Table 1. It is clear that the 'M2+S1' model exhibits the highest solution stability, with RMSE values of 0.0596 m and 0.1080 m for the position differences at the current and intermediate moments, respectively, and an RMSE of 0.1057 m for the distance residual difference. Additionally, as shown in Figure 6, the proposed method significantly improves computational efficiency by over 15 times compared to the traditional ray-tracing methods.

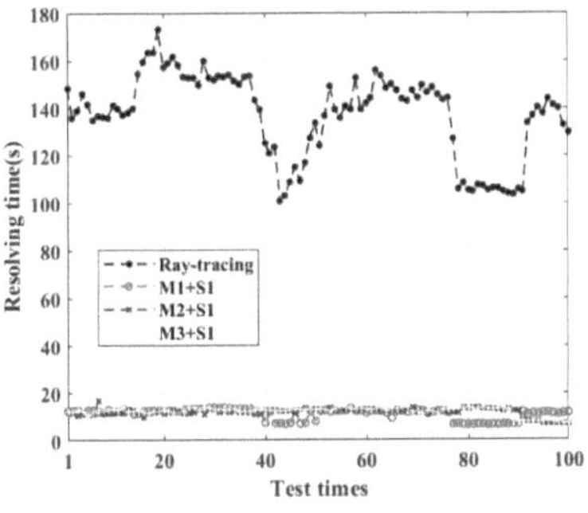

Figure 6. Comparison of solution time of different model.

3.2 *Result of measured test*

The real-time position of the carrier transducer is calculated based on the GNSS antenna coordinates, arm length parameters, and attitude angles, which serve as the reference position for the carrier. The calculation results based on the sound ray-tracing method are shown in Figure 7. The RMSE for position at the current moment and middle moment is 2.4220 m and 2.8893 m, respectively, while the RMSE for speed is 0.0706 m/s and 0.0850 m/s. These ray-tracing method-derived trajectories at the current and intermediate moments are used as reference trajectories to compare the navigation accuracy loss of different correction models.

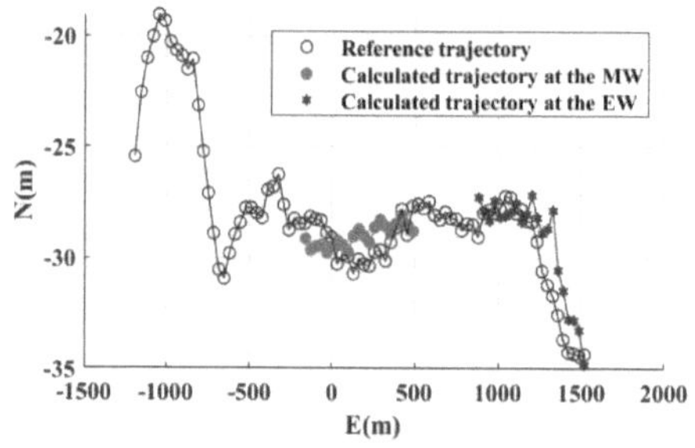

Figure 7. Navigation results based on sound ray-tracing method.

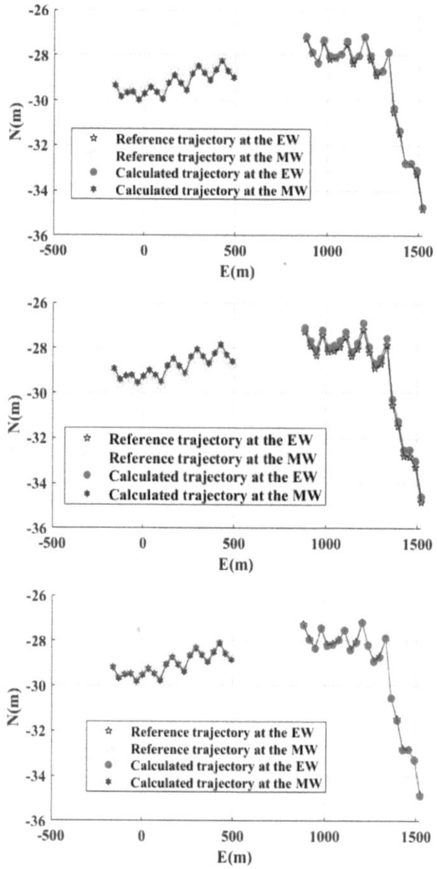

Figure 8. Navigation results of different models: (a) M1+S1; (b) M2+S1; (c) M3+S1.

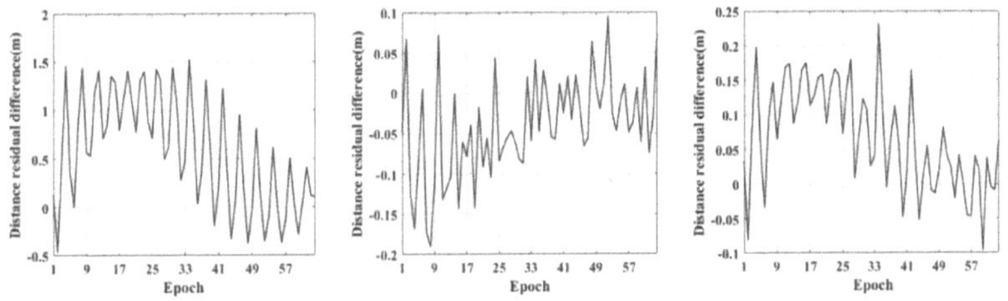

Figure 9. Different model distance residual difference sequence: (a) M1+S1; (b) M2+S1; (c) M3+S1.

The differences in navigation solutions are illustrated in Figure 8, with the residual difference sequence shown in Figure 9. It is evident that 'M2+S1' exhibits the highest accuracy, whereas 'M1+S1' demonstrates the lowest accuracy, consistent with the conclusions from the simulation results. As shown in Table 2, the RMSE for position difference at the current moment using 'M2+S1' is 0.1804 m, whereas 'M1+S1' shows a navigation accuracy loss reaching the meter level. At the middle moment, the navigation accuracy is notably improved, with the RMSE for the position difference of 'M2+S1' being 0.0844 m, and the ranging accuracy loss maintained at the centimeter level. Furthermore, as shown in Figure 10, the proposed method enhances computational efficiency by over 15 times compared to the ray-tracing method.

Table 2. Navigation accuracy statistics of different correction models.

	Position difference RMSE (m)		Speed difference RMSE (m)		distance residual difference RMSE(m)
Model	MW	EW	MW	EW	
M1+S1	0.4755	1.0979	0.0207	0.0057	0.8190
M2+S1	0.0844	0.1804	0.0010	0.0007	0.0750
M3+S1	0.2377	0.2732	0.0005	0.0010	0.1009

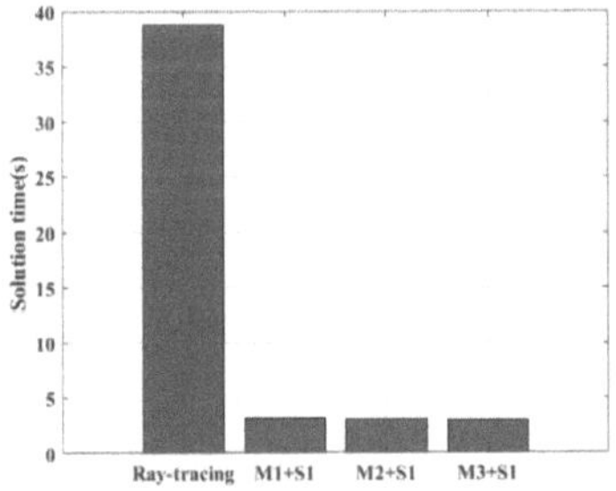

Figure 10. Result of navigation solution.

4 CONCLUSION

Although traditional ray-tracing methods are accurate in eliminating environmental range errors in ocean acoustic navigation and positioning, they are computationally intensive and heavily dependent on sound speed profile data, which presents challenges for real-time applications. To address these issues, this paper introduces a functional model that correlates sound speed errors with ray elevation angles and depths, and utilizes observed time delay information to correct range errors. Validated through simulations and actual sea trial data from the CHOS station in Japan, the results demonstrate that this method effectively models sound speed errors. Compared to the traditional sound ray-tracing method, this method improves the computational efficiency by over 15 times, and maintains centimeter-level navigation accuracy with minimal loss in speed accuracy.

This parametric modeling approach provides a feasible solution for real-time underwater navigation and positioning, significantly improving operational efficiency without compromising range accuracy. It holds substantial practical value, particularly in deep-sea environments where rapid and precise navigation is crucial.

REFERENCES

Amiet R.K. (1978). Refraction of sound by a shear layer. *Journal of Sound and Vibration*, 58(4), 467–482.

Boor, C.D. (1978). A practical guide to spline. *Mathematics of Computation*, 27(149), 157–157(1).

Chen, H. (2011). Accuracy assessment of GPS/Acoustic positioning using a Seafloor Acoustic Tran sponder System. *Ocean Engineering*, 38, 1472–1479.

Christopher, S.M. and Watts, D.R. (1997). Further evidence that the sound-speed algorithm of Del Grosso is more accurate than that of Chen and Millo. *Journal of the Acoustical Society of America*, 102: 2058–2062.

Chadwell, C.D. and Sweeney, A.D. (2010). Acoustic Ray-Trace Equations for Seafloor Geodesy. *Marine Geodesy*, 33(2–3), 164–186.

He, C., Wang, Y., Yu, W., and Song, L. (2019). Underwater Target Localization and Synchronization for a Distributed SIMO Sonar with an Isogradient SSP and Uncertainties in Receiver Locations Sensors 19, 9: 1976.

Liu, Y.F, Xu, T. and Wang, J. (2023). An improved method for computing acoustic ray inci dent angle based on secant method *Applied Acoustics*, 205.

Maki, T. Matsuda, T. Sakamaki, T. Ura and J. Kojima (2013). Navigation method for underwater vehicles based on mutual acoustical positioning with a single seafloor station. *Oceanic Engineering*, 38(1), 167–177.

Poutanen, M., and Rózsa, S. (2020). The Geodesist's handbook 2020. *Journal of Geodesy*, 94(11), 1–343.

Sun, D., Li, H., Zheng, C., and Li, X. (2019). Sound speed correction based on effective sound speed for underwater acoustic positioning systems. *Applied Acoustics*, 151, 55–62.

Sakic, P., Ballu, V., Crawford, W. C. and Wöppelmann, G. (2018). Acoustic Ray Tracing Comparisons in the Context of Geodetic Precise off-shore Positioning Experiments. *Marine Geodesy*, 41(4), 315–330.

Wang, H., Yang W., Xin Y., and Zhang S. (2020). The construction of sound speed field based on back propagation neural network in the global ocean. *Marine Geodesy*, 43(1), 1–22.

Watanabe, S., lshikawa, T., Yokota, Y. and Nakamura, Y. (2020). GARPOS: Analysis Software for the GNSS-A seafloor positioning with simultaneous estimation of sound speed structure. *Frontiers in Earth Science*, 8.597532.

Xu, P., Liu, J. and Shi, Y. (2023). Unbiased weighted least squares location estimation. *Journal of Geodesy*, 97 (68): 1–22.

Xue, S., Yang, Y., Yang, W., and Li, J. (2024). GNSS-A network solution with zenith acoustic delay estimation. *Marine Geodesy*, 47(3), 237–268.

Xue, S., Li, B., Xiao, Z., Sun, Y., and Li, J. (2023). Centimeter-level-precision seafloor geodetic positioning model with self-structured empirical sound speed profile. *Satelliate Navigation*, 4(30).

Xiao, Z., Xue, S., Han, B. (2023). Simulation analysis on reference sound speed profile error influence on active acoustic positioning. *ChineseJ.Geophys*, 66(12), 4889–4899

Yang, Y., Liu, Y., Sun D, Xu, T, Xue, S, Han, Y and Zeng, A. (2020). Seafloor geodetic network establishment and key technologies. *Science China Earth Sciences*, 63(8), 1188–1198.

Yang, W., Xue, S., and Liu, Y. (2021). P-order secant method for rapidly solving the ray inverse problem of underwater acoustic positioning. *Marine Geodesy*, 46(1), 3–15.

Yang, Y., Liu, Y. and Sun, D. (2020). Seafloor geodetic network establishment and key technologies. *Science China Earth Sciences*, 63, 1188–1198.

Zhao, S., Wang, Z., He, K. and Ding, N. (2018). Investigation on underwater positioning stochastic model based on acoustic ray incidence angle. *Applied Ocean Research*, 77, 69–77.

Zhao, S., Liu, H., Xue, S., Wang, Z. and Xiao, Z. (2023). Two-step correction based on in-situ sound speed measurements for USBL precise real-time positioning. *Remote Sensing*, 15(20), 5046.

Proceedings of the International Association of Institutes of Navigation World Congress 2024 – Chen et al. (Eds)

Augmented acoustic positioning model for submarine vehicle navigation in a seafloor geodetic array

S.Q. Xue, Z. Xiao, J.C. Bian & S. Zhao
Chinese Academy of Surveying and Mapping, Beijing, China

W.L. Yang
Shandong University, Weihai, China

B.J. Li & J.X. Zhu
Chinese Academy of Surveying and Mapping, Beijing, China

ABSTRACT: Acoustic positioning systems, like the long baseline positioning system based on a sea-surface or seabed acoustic array, are often used for submarine vehicle navigation. The active-sonar positioning mode is however ill-posed and suffer from the sound speed error effect. We first propose an augmented acoustic navigation service based on a sea-surface and seafloor network by monitoring the acoustic delay errors in the acoustic observations, which is parameterized by a B-spline model. Then, we recommend a vehicle trajectory estimation model by accumulating a piece of acoustic observations with the acoustic delay error corrections within a specific time window. In the time window, the B-spline model is used for fitting the trajectory to conquer the ill-posed problem. The estimated B-spline trajectory can then be utilized to determine the state of the vehicle at any given time, including its location and velocity. It shows that the proposed augmented acoustic positioning model can achieve a navigation accuracy better than two meters.

1 INTRODUCTION

First paragraph: Global Navigation Satellite System (GNSS)-Acoustic (GNSS-A) positioning technique has become a vital tool for seafloor geodesy and crustal deformation applications of submarine offshore regions (Roland Bürgmann and Chadwell 2014; Iinuma *et al.* 2021). GNSS-A positioning technique was pioneered by Fred Spiess (Spiess 1985; Spiess *et al.* 1998) and further developed for more than two decades (Chadwell *et al.* 1997; Chadwell and Sweeney 2010). Regional seafloor geodetic networks were established in the past (Yang *et al.* 2020; Poutanen and Rózsa 2020) and had provided key observations for constraining tectonic motion, crustal deformation and the earthquake cycle (Gagnon *et al.* 2005; Gagnon and Chadwell 2007; Iinuma *et al.* 2021; Kido *et al.* 2006; Sato *et al.* 2011). GNSS-A technique achievements made in the past are on the one hand from observation scheme advances (Kido *et al.* 2008; Sato *et al.* 2013; Spiess *et al.* 1998) and on the other hand from acoustic positioning model improvements (Spiess 1980; Asada and Yabuki 2001; Fujita *et al.* 2006; Yokota *et al.* 2019; Watanabe *et al.* 2020a; YANG and Qin 2020).The acoustic positioning is seriously affected by the sound speed variations (Kido *et al.* 2008), which is very familiar to the atmosphere variation affecting GNSS positioning (Chen and Herring 1997). Inspired by this, the Nadir Total Delay (NTD) model which is analogous to the zenith total delay in GNSS was proposed recently (Honsho *et al.* 2019; Honsho and Kido 2017). This idea developed in the GNSS-A positioning can be also found in analyzing the sound speed variation for GNSS-A measurements (Kido *et al.* 2008). Besides NTD model, a

DOI: 10.1201/9781003624424-13

generalized GNSS-A positioning model was also developed to correct the effect of the sound speed variation (Watanabe *et al.* 2020a; Yokota *et al.* 2022).

Submarine acoustic navigation usually adopts an active-sonar positioning system to overcome the clock synchronization challenge in water (Bingham and Seering 2006; Zhang *et al.* 2016; Yang 2018). The active-sonar positioning pattern has an advantage in the timekeeping capability of the underwater clocks installed at the seafloor navigation beacons; on the other hand, the positioning system has a fatal disadvantage due to the objective condition, which requires estimating the vehicle states at transmission and reception epochs during each signal sending-receiving cycle. In addition, this pattern greatly limits the user capacity for its request-response mode. In this pattern, the submarine vehicle emits a positioning request signal, and then an underwater navigation beacon transmits the corresponding response signals when receiving the request signal, by which the round-trip travel time is measured to determine the location of the vehicle (Gao *et al.* 2007; Webster *et al.* 2012). Apparently, as each round-trip ranging observation corresponds to two vehicle's states at transmission and reception times, it is challenging to obtain the two vehicle's states by the one ranging observation, meaning that this issue is typically ill-posed (Xue and Yang 2017; Qi *et al.* 2019).

Positioning for submarine vehicles can be realized based on the traditional ranging intersection principle, in which half of the round-trip time of the ranging signal is used for ignoring the vehicle motion during the request-response cycle (Xue *et al.* 2024). However, this inevitably introduces a model error under the above assumption, which is proportional to the vehicle speed and the signal propagation time. This modelling error is more significant in deep-sea scenarios or high-speed submarine vehicle positioning applications (Sun *et al.* 2019; Wang *et al.* 2023). From the aspect of the seafloor geodetic positioning based on a sea-surface vehicle equipped with Global Navigation Satellite System (GNSS), as the vehicle's positions at the transmission and reception times are known (Xiao *et al.* 2023), the impact of the aforementioned modelling error can be reduced by the round-trip intersection principle (Xue *et al.* 2024), but not vice versa.

To address the ill-posed issue of the aforementioned active sonar navigation pattern, an alternative feasible approach is to incorporate an Inertial Navigation System (INS) (A. Bucci *et al.* 2023). By integrating the time derivatives of the vehicle's acceleration and angular velocity variation measured by the Inertial Measurement Unit (IMU) (Liu *et al.* 2022), the velocity and attitude change information of the submarine vehicle is obtained and then fused with active acoustic observations to overcome the ill-posed issue in active-sonar navigation, which is similar to the application in GNSS/INS navigation applications (Li *et al.* 2017) (J. Song *et al.* 2022; C. Hu *et al.* 2022). The acoustic navigation system and INS operate independently. However, the accumulation error of INS cannot be ignored and needs to be calibrated by the acoustic navigation system. This indicates that the acoustic navigation system should be operated independently of INS navigation. On the other hand, the high-precision INS sensors come with elevated costs and increased volume, thereby augmenting the size and expenses of the acoustic navigation terminals.

This contribution is to propose an augmented acoustic navigation service based on a sea-surface and seafloor network by monitoring the environmental error in the acoustic observations, and then recommend a vehicle trajectory estimation model by accumulating a piece of acoustic observations within a specific time window in which a B-spline model is used for fitting the trajectory to conquer the ill-posed problem. Section 2 is to introduce the augmented positioning service based on a seafloor geodetic array with resolving the sound speed effect on the acoustic observations. Section 3 is to discussed the trajectory estimation model with the sound speed variation effect corrections. Section 4 is to validate the proposed results by the actual seafloor geodetic array observations.

2 AUGMENTED POSITIONING SERVICE OF A SEAFLOOR GEODETIC ARRAY

As shown in Figure 1, there are acoustic observation platforms located at different positions on the sea-surface and seabed, corresponding to a sea-surface GNSS network or trajectory and a seafloor geodetic acoustic array or network. The above-mentioned sea-surface network is connected to the seafloor geodetic network by acoustic observations, forming an integrated communication and navigation service network for submarine vehicles. Each observation platform collects, processes, and applies acoustic observation data by transmitting and receiving acoustic signals. The real-time and long-term observation data obtained by the network can be used to monitor the effect of the ocean sound speed variation on the acoustic ranging to provide a network augmentation navigation service.

As shown in Figure 2, there is a seafloor geodetic array composed of M seafloor geodetic points $\mathbf{x}_j$ (j=1,2, ... ,M), the acoustic transducer installed on the sea-surface vehicle with known position transmits an acoustic ranging signal to the array, and then it is received and retransmitted to the transducer to measure the round-trip travel time to perform the array positioning (Asada and Yabuki 2001). Then, we can establish an observation model as

$$\mathbf{L} = \mathbf{F}(\mathbf{X}) + \delta\mathbf{c} + \boldsymbol{\varepsilon} \tag{1}$$

where **L** is the round-trip travel time, **F(X)** is the calculated travel time based on a reference SSP, **X** is the array coordinate vector; $\delta\mathbf{c} = \mathbf{BY}$ is the sound seed variation effect on the acoustic observation, **Y** is the nuisance parameter vector to compensate for the sound speed variation effect and **B** is the corresponding design matrix, ε denotes the measurement noise. As there are M seafloor geodetic points, the model (1) essentially provides an array solution regarding to the effect of the sound speed variation on the acoustic observations.

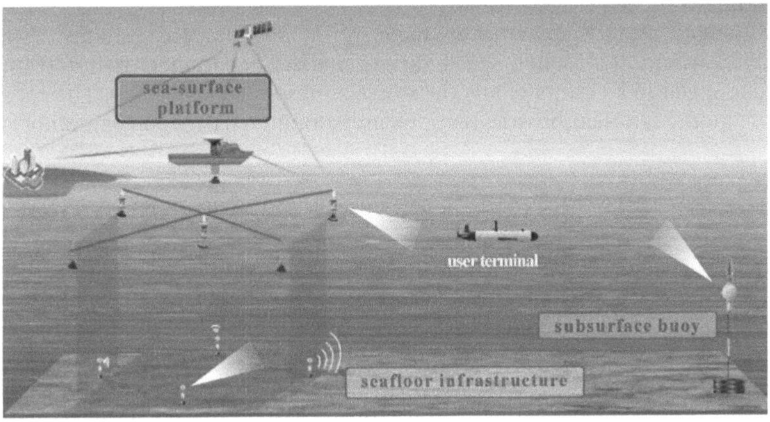

Figure 1. Sketch map of an augmented positioning service network with sea-surface and seafloor infrastructure for underwater navigation users.

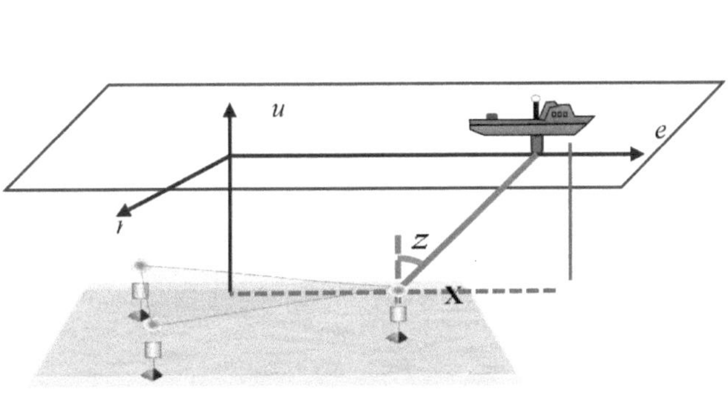

Figure 2. Network solution of the environmental error in the acoustic observation.

The sound speed variation effect parameterization **BY** is vitally important for centimeter-precision seafloor geodetic positioning, which is nonunique, such as the models used in the GNSS-Acoustic Ranging combined Positioning Solver (GARPOS) (Watanabe *et al.* 2020b), the resilient model (Yang and Qin 2021) developed recently and the zenith acoustic delay (ZAD) model from (Xue *et al.* 2023). The ZAD model contains the three ZAD components and for the *i*th observation of the *j*th seafloor geodetic point, the entries of the matrix **B** reads (Xue *et al.* 2023).

$$b_{k,j}=\left[b_{\mathrm{u},k,j} b_{\mathrm{n},k,j} b_{\mathrm{e},k,j}\right] \quad k=1,2,\ldots,S; \quad j=1,2,\ldots,M \tag{2}$$

where

$$b_{J,k,j}=\left[m_{J,k,j}\Phi_{J,0}(t_k) \quad m_{J,k,j}\Phi_{J,2}(t_k) \quad \cdots \quad m_{J,k,j}\Phi_{J,Q_J}(t_k)\right] \quad J\in\{\mathrm{u},\mathrm{n},\mathrm{e}\} \tag{3}$$

and $\Phi_{J,s}(t)$ ($s = 0, 1, \ldots, Q_J$) is the kth base function about the time t, Q_J is the total number of base functions,t_k is the observation epoch,

$$\begin{cases} m_{\mathrm{u},k,j}=\left(\cos z_{k,j}\right)^{-1} \\ m_{\mathrm{n},k,j}=\left(\cos z_{k,j}\right)^{-1}\tan z_{k,j}\cos\alpha_{k,j} \\ m_{\mathrm{e},k,j}=\left(\cos z_{k,j}\right)^{-1}\tan z_{k,j}\sin\alpha_{k,j} \end{cases} \tag{4}$$

is the function mapping the line-of-sight (LOS) delay to the zenith direction, z and α are the zenith angle and azimuth angle of the sea-surface trajectory observed at the i th seafloor geodetic point.

We recommend using B-spline to characterize the sound speed variation effect on the acoustic observations. The sound speed variation effect is parameterized and treated as a common parameter to be estimated by the whole seafloor geodetic array. The coefficients of the B-spline model can then broadcast to the users to improve the navigation precision.

3 POSITIONING MODELS FOR SUBMARINE VEHICLE NAVIGATION

For submarine vehicles, the acoustic positioning and navigation based on the seafloor geodetic array have advantages such as high accuracy and nondivergence compared to the inertial navigation. Actually, underwater acoustic positioning is typically achieved by processing sound speed observations using empirical values or historical sound speed profiles, which form the basis of acoustic navigation services provided by the seafloor acoustic array. However, due to the dynamic and complex nature of the marine environment, sound speed variation errors can be significant. Relying on traditional fixed sound speed values or profiles often severely limits navigation and positioning accuracy, necessitating corrections for acoustic static and dynamic delays. The integrated communication and navigation network enhances real-time high-precision navigation for underwater vehicles by calculating and broadcasting sound speed correction information, thereby providing augmented navigation services.

Let $[x(t_s), y(t_s), z(t_s)]$ and $[x(t_r), y(t_r), z(t_r)]$ be the vehicle's positions at the signal transmission time and reception time, respectively. It is true that the navigation model is always ill-posed, as there are at least six parameters to be estimated for each round-trip time observation. However, this problem may be solved by simulating a piece of round-trip time observation within a specific time window. Assuming there are n acoustic beacons forming a long baseline array and their positions are $[X_j, Y_j, Z_j], j = 1, 2, \ldots, n$, we can obtain an active acoustic navigation model as

$$L_j = T_j \times c_0 = F_j[x(t_{s,j}), y(t_{s,j}), z(t_{s,j}); x(t_{r,j}), y(t_{r,j}), z(t_{r,j})] + \delta c_j + \varepsilon_j \quad j = 1, 2, \ldots, n \tag{5}$$

where T_j is the signal round-trip time during requesting the jth acoustic beacon, c_0 is the effective sound speed based on a reference sound speed profile (SSP) (Huang and Yan 2018), n is the total number of acoustic beacons

$$F_j = f_{s,j}[x(t_{s,j}), y(t_{s,j}), z(t_{s,j})] + f_{r,j}[x(t_{r,j}), y(t_{r,j}), z(t_{r,j})] \tag{6}$$

is a nonlinear round-trip distance function of the following form

$$\begin{cases} f_{s,j} = \sqrt{[x(t_{s,j}), -X_j]^2 + [y(t_{s,j}) - Y_j]^2 + [z(t_{s,j}) - Z_j]^2} \\ f_{r,j} = \sqrt{[x(t_{r,j}) - X_j]^2 + [y(t_{r,j}) - Y_j]^2 + [z(t_{r,j}) - Z_j]^2} \end{cases} \tag{7}$$

are the forward trip distance and backward distance, respectively. Note that the positioning model can adopt a ray-tracing positioning algorithm instead of the above ranging positioning model based on the effective sound speed, δ_c is the effect of the sound speed variation on the navigation observation.

As pointed out at the very beginning of this section, it is useless for resolving the vehicle position by blindly increasing the number of acoustic beacons, i.e., the unknown parameters of this model are always three times the number of the acoustic beacons. The simplest way to solve the ill-posed problem (1) is to assume the vehicle is nearly stationary while requesting the n acoustic beacons. At this time, we can assume that $[x(t_{s,j}), y(t_{s,j}), z(t_{s,j})] = [x(t_{r,j}), y(t_{r,j}), z(t_{r,j})]$ denoted as (x, y, z), and then the model (1) can degenerate into the traditional intersection positioning model as

$$l_j = f_j(x, y, z) + \delta c_j + \varepsilon'_j \; j = 1, 2, \ldots, n \tag{8}$$

Where $l_j = L_j/2, \varepsilon'_j = \varepsilon_j/2$, $f_j(x, y, z) = \sqrt{[x, -X_j]^2 + [y - Y_j]^2 + [z - Z_j]^2}$. Obviously, this model is validated under a strong assumption that is hardly suitable for a wide range of navigation scenarios. If the displacement of the vehicle during the ranging requests is considerable, it will be inevitable to damage the navigation precision, so the model error caused by the stationary approximation cannot be ignored. An alternative way is to divide one round-trip observation into two single-pass distances by an initial state about the vehicle, but this way might be invalidated for lacking an initial state precisely enough.

To solve the above-mentioned ill-posed problem, we recommend to use an active acoustic navigation model as follow

$$L_j = F_j[\boldsymbol{x}(\boldsymbol{P}, t_{s,j}); \boldsymbol{x}(\boldsymbol{P}, t_{r,j})] + \delta c_j + \varepsilon_j \; j = 1, 2, \ldots, n \tag{9}$$

Where

$$\boldsymbol{x}(\boldsymbol{P}, t) = \begin{bmatrix} x(\boldsymbol{p}_x, t) \\ y(\boldsymbol{p}_y, t) \\ z(\boldsymbol{p}_z, t) \end{bmatrix} \tag{10}$$

is a trajectory description model with an unknown parameter vector $\boldsymbol{P} = [\boldsymbol{p}_x; \boldsymbol{p}_y; \boldsymbol{p}_z]$, the parameter $\boldsymbol{p}_K$ for $K \in \{x, y, z\}$ is the trajectory parameter used to describe the vehicle's trajectory curve along the K direction. $\delta c_i$is the provided by the network solution,ε is random noise.

It needs to be pointed out that, the trajectory contains vehicle state information including the position, velocity, and acceleration, e.g., the vehicle's velocity can be extracted by the derivative $d\boldsymbol{x}(t)/dt$. For a given initial value $\boldsymbol{P}^0$ about the trajectory parameter vector, linearizing the model (5) and ignoring the linearization remainder, we can obtain a linearization model as follow

$$dL_j = \boldsymbol{a}_j d\boldsymbol{P} + \varepsilon_j \tag{11}$$

where $dL_j = L_j - F_j(\boldsymbol{P})$ is the observation residual, $d\boldsymbol{P}$ is the correction vector of the trajectory parameter vector,

$$\boldsymbol{a}_j = \frac{\partial F}{\partial \boldsymbol{x}}\frac{\partial \boldsymbol{x}}{\partial \boldsymbol{P}} = \left[\boldsymbol{n}_j \boldsymbol{e}_j \boldsymbol{u}_j\right] \tag{12}$$

is the design matrix of the active acoustic navigation model, which can also be called navigation geometry. Rewrite (8) into matrix form $\boldsymbol{Y} = \boldsymbol{A}d\boldsymbol{X} + \boldsymbol{\varepsilon}$, then we can obtain the Least Squares (LS) solution as

$$d\widehat{\boldsymbol{X}} = (\boldsymbol{A}^T \boldsymbol{W}\boldsymbol{A})^{-1}\boldsymbol{A}^T \boldsymbol{W}\boldsymbol{Y} \tag{13}$$

where $\boldsymbol{W}$ is the weight matrix. For low-precision t initial values, we recommend adopting Gauss-Newton method to iteratively solve the LS solution of the nonlinear model (5).

B-spline model has strong flexibility and local detail depiction capability, gradually becoming an effective model for fitting a nonstationary signal. Let $[t_{st} t_{ed}]$ be the time window, when using the B-spline model to describe the trajectory, at an arbitrary time t, the vehicle trajectory reads

$$\begin{bmatrix} x(t) \\ y(t) \\ z(t) \end{bmatrix} = \begin{bmatrix} \sum_{s=0}^{K} a_{x,s}\Phi_{x,s}(t) \\ \sum_{s=0}^{K} a_{y,s}\Phi_{y,s}(t) \\ \sum_{s=0}^{K} a_{z,s}\Phi_{z,s}(t) \end{bmatrix} \tag{14}$$

where $\Phi_{I,k}$ represents the B-spline base function for the coordinate component $I \in \{x, y, z\}$, which can be obtained by using the Cox-de Boor recursion formula (De Boor 1978); $a_{I,s}$ represents the corresponding coefficient of the B-spline base function; K+1 is the total number of B-spline base functions used, while the time window $[t_{st} t_{ed}]$ is equally split into K-p+1 intervals; p is the degree of B-spline base function. Therefore, Eq. (7) can be expressed as:

$$\begin{cases} \boldsymbol{n}_j = e_{x,s,j}\Psi_{x,s,j} + e_{x,r,j}\Psi_{x,r,j} \\ \boldsymbol{e}_j = e_{y,s,j}\Psi_{y,s,j} + e_{y,r,j}\Psi_{y,r,j} \\ \boldsymbol{u}_j = e_{z,s,j}\Psi_{z,s,j} + e_{z,r,j}\Psi_{z,r,j} \end{cases} \tag{15}$$

Where

$$\begin{matrix} \Psi_{I,s,j} = \left[\begin{matrix} \Phi_{I,0}(t_{s,j}) & \Phi_{I,1}(t_{s,j}) & \cdots & \Phi_{I,K}(t_{s,j}) \end{matrix}\right] \\ \Psi_{I,r,j} = \left[\begin{matrix} \Phi_{I,0}(t_{r,j}) & \Phi_{I,1}(t_{r,j}) & \cdots & \Phi_{I,K}(t_{r,j}) \end{matrix}\right] \end{matrix} \tag{16}$$

And

$$\begin{cases} e_{x,s,j} = \dfrac{x(t_{s,j}) - X_j}{f_{s,j}} \\ e_{y,s,j} = \dfrac{y(t_{s,j}) - Y_j}{f_{s,j}} \\ e_{z,s,j} = \dfrac{z(t_{s,j}) - Z_j}{f_{s,j}} \end{cases} ; \begin{cases} e_{x,r,j} = \dfrac{x(t_{r,j}) - X_j}{f_{s,j}} \\ e_{y,r,j} = \dfrac{y(t_{r,j}) - Y_j}{f_{s,j}} \\ e_{z,r,j} = \dfrac{z(t_{r,j}) - Z_j}{f_{s,j}} \end{cases} \tag{17}$$

The cubic B-spline is recommended, i.e., the degree of the above base function is 3.

4 RESULTS AND DISCUSSION

To evaluate the precision of the motion state estimation, the actual seafloor geodetic data which is publicly available was employed (Watanabe *et al.* 2020c). The actual data, including the GNSS position of the sea-surface vehicle and the position of the seafloor beacon, was collected at the KAMN station in June 2011. The trajectory of the vehicle and the position of the beacon are shown in Figure 3.

As depicted in Figure 3, there are four beacons deployed on the seabed. which are denoted by red pentagrams. The Global Navigation Satellite System (GNSS) position of the vehicle is represented by blue dots. A straight trajectory represented by red dots has been chosen as the experimental trajectory.

The acoustic observations are corrected by the sound speed variation effect estimated by the network. As shown in Table 1, the observational precision with the correction is improved significantly, the mean error is reduced from 3.76m to -0.03m while the STD is reduced from 0.93m to 0.1m, and the RMS is reduced from 3.87m to 0.1m. This indicates that the sound speed variation effect mainly causes a systematic error and magnifies the STD of the observations.

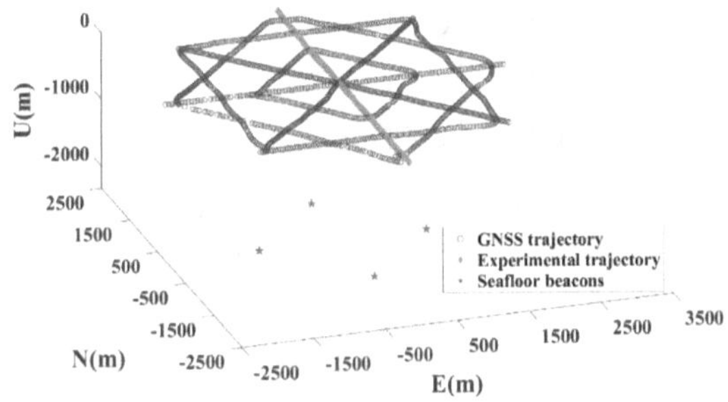

Figure 3. Experimental network with four seafloor beacons.

Table 1. Observation precision improvement by the corrections.

Solutions	MEAN (m)	STD (m)	RMS (m)
without the corrections	3.76	0.93	3.87
With the corrections	−0.03	0.10	0.10

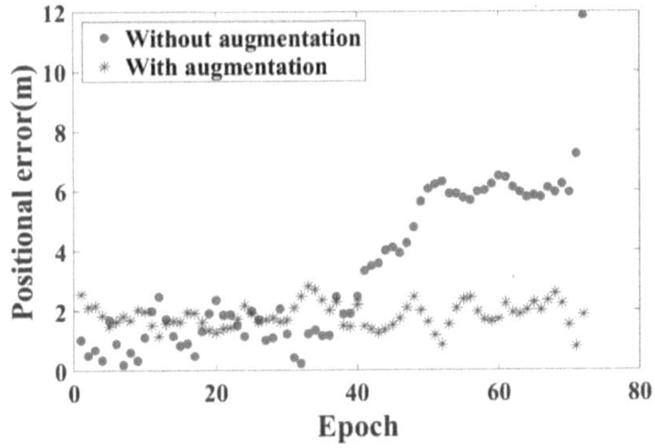

Figure 4. Navigation results with and without the augmentation.

The number N of the B-spline-based functions is set to 24 and therefore there are a total of 48 parameters ito be estimated within window. Calculating the trajectory parameters, the vehicle's position was estimated, see Figure 4. Relative to the GNSS trajectory, the RMSE of the positioning precision was calculated to evaluate the state estimation accuracy in Table 2.

Table 2. Navigation precision improvement by the corrections.

Solutions	Positioning error RMS (m)
Without the correction	4.31
With the correction	1.89

As shown in Table 2, the navigation precision with the acoustic delay correction can be significantly improved. The positioning accuracy is evaluated by the GNSS trajectory and the positioning error RMS is reduced from 4.31m to 1.89m by the correction, indicating that the improvement is over two times.

REFERENCES

Asada A and Yabuki T. (2001). Centimeter-level positioning on the seafloor. *Proceedings of the Japan Academy* 77:7–12

Bingham B. and Seering W. (2006). Hypothesis grids: improving long baseline navigation for autonomous un derwater vehicles. *IEEE Journal of Oceanic Engineering* 31:209–218

Bucci A., Franchi M., Ridolfi A., *et al.* (2023). Evaluation of UKF-Based Fusion Strategies for Autonomous Underwater Vehicles Multisensor Navigation. *IEEE Journal of Oceanic Engineering* 48:1–26. https://doi.org/10.1109/JOE.2022.3168934

Chadwell CD, Spiess FN, Hildebrand JA, *et al.* (1997). *Sea Floor Strain Measurement Using GPS and Acous tics.* In: Segawa J, Fujimoto H, Okubo S (eds) *Gravity, Geoid and Marine Geodesy.* Springer Berlin Heidelberg, Berlin, Heidelberg, pp 682–689

Chadwell C.D. and Sweeney A.D. (2010). Acoustic Ray-Trace Equations for Seafloor Geodesy. *Marine Geodesy* 33:164–186

Chen G and Herring T.A. (1997). Effects of atmospheric azimuthal asymmetry on the analysis of space geodetic data. *Journal of Geophysical Research: Solid Earth* 102:20489–20502

De Boor C. (1978). *A Practical Guide to Splines.* springer-verlag New York

Fujita M, Ishikawa T, Mochizuki M, *et al.* (2006). GPS/Acoustic seafloor geodetic observation: method of data analysis and its application. *Earth, Planets and Space* 58:265–275

Gagnon K, Chadwell C and Norabuena E. (2005). Measuring the onset of locking in the Peru-Chile trench with GPS and acoustic measurements. *Nature* 434:205

Gagnon KL and Chadwell CD. (2007). Relocation of a seafloor transponder—Sustaining the GPS-Acoustic technique. *Earth Planets and Space* 59:327–336

Gao W, Yang Y, Cui X and Zhang S. (2007). Application of adaptive Kalman filtering algorithm in IMU/GPS integrated navigation system. *Geo-spatial Information Science* 10:22–26

Honsho C and Kido M. (2017). Comprehensive Analysis of Traveltime Data Collected Through GPS-Acoustic Observation of Seafloor Crustal Movements. *Journal of Geophysical Research: Solid Earth* 122:8583–8599. https://doi.org/10.1002/2017JB014733

Honsho C, Kido M, Tomita F and Uchida N. (2019). Offshore postseismic deformation of the 2011 tohoku earthquake revisited: Application of an improved GPS-acoustic positioning method considering horizontal gradient of sound speed structure. *Journal of Geophysical Research: Solid Earth* 124:5990–6009

Hu C., Zhu S., Liang Y. and Song W. (2022). Tightly-Coupled Visual-Inertial-Pressure Fusion Using Forward and Backward IMU Preintegration. *IEEE Robotics and Automation Letters* 7:6790–6797. https://doi.org/10.1109/LRA.2022.3177847

Huang J and Yan S. (2018). An Improvement of long baseline system using particle swarm optimization to optimize effective sound speed. *Marine Geodesy* 41:439–456

Iinuma T, Kido M, Ohta Y, *et al.* (2021). GNSS-acoustic observations of seafloor crustal deformation using a wave glider. *Frontiers in Earth Science* 9:

Song J., Li W., Zhu X., *et al.* (2022). Underwater Adaptive Height-Constraint Algorithm Based on SINS/ LBL Tightly Coupled. *IEEE Transactions on Instrumentation and Measurement* 71:1–9. https://doi.org/10.1109/TIM.2022.3160527

Kido M, Fujimoto H, Miura S, Yukihito Osada, *et al.* (2006). Seafloor displacement at Kumano-nada caused by the 2004 off Kii Peninsula earthquakes, detected through repeated GPS/Acoustic surveys. *Earth Planets and Space* 58:

Kido M, Osada Y and Fujimoto H. (2008). Temporal variation of sound speed in ocean: a comparison between GPS/acoustic and in situ measurements. *Earth Planets and Space* 60:229–234

Li Z, Gao J, Wang J and Yao Y. (2017). PPP/INS tightly coupled navigation using adaptive federated filter. *Gps Solutions* 21:137–148. https://doi.org/10.1007/s10291-015-0511-z

Liu Y, Wang L, Hu L, *et al.* (2022). Analysis of the influence of attitude error on underwater positioning and its high-precision realization algorithm. *Remote Sensing* 14:3878. https://doi.org/10.3390/rs14163878

Poutanen M and Rózsa S. (2020). The Geodesist's Handbook 2020. *Journal of Geodesy* 94:1–343. https://doi.org/10.1007/s00190-020-01434-z

Qi K, Qu G, Xue S, *et al.* (2019). Analytical optimization on GNSS buoy array for underwater positioning. *Acta Oceanologica Sinica* 38:137–143. https://doi.org/10.1007/s13131-019-1465-1

Roland Bürgmann and Chadwell D. (2014). Seafloor Geodesy. *Annual Review of Earth and Planetaryences* 42:509–534

Sato M, Fujita M, Matsumoto Y, *et al.* (2013). Improvement of GPS/acoustic seafloor positioning precision through controlling the ship's track line. *Journal of Geodesy* 87:825–842

Sato M, Ishikawa T, Ujihara N, *et al.* (2011). Displacement Above the Hypocenter of the 2011 *Tohoku-Oki Earthquake. ence* 332:1395-

Spiess F. (1985). Analysis of a possible sea floor strain measurement system. *Marine Geodesy* 9:385–398

Spiess F. (1980). Acoustic techniques for marine geodesy. *Marine Geodesy* 4:13–27

Spiess FN, Chadwell CD, Hildebrand JA, *et al.* (1998). Precise GPS/Acoustic positioning of seafloor refer ence points for tectonic studies. *Physics of the Earth and Planetary Interiors* 108:101–112

Sun D, Ding J, Zheng C and Huang W. (2019). Array geometry calibration for underwater compact arrays. *Applied Acoustics* 145:374–384. https://doi.org/10.1016/j.apacoust.2018.10.004

Wang J, Xu T, Liu Y, *et al.* (2023). Augmented underwater acoustic navigation with systematic error modeling based on seafloor datum network. *Marine Geodesy* 46:129–148. https://doi.org/10.1080/01490419.2022.2162646

Watanabe S, Ishikawa T, Yokota Y and Nakamura Y. (2020). GARPOS: Analysis Software for the GNSS-A Seafloor Positioning With Simultaneous Estimation of Sound Speed Structure. *Frontiers in Earth Science* 8:508. https://doi.org/10.3389/feart.2020.597532

Webster SE, Eustice M, Singh H and Whitcomb L.L. (2012). Advances in single-beacon one-way-travel-time acoustic navigation for underwater vehicles. *The International Journal of Robotics Research* 31:935–950. https://doi.org/10.1177/0278364912446166

Xiao Z, Xue S, Han B, *et al.* (2023). Simulation analysis on reference sound velocity profile error influence on active acoustic positioning. *Chinese Journal of Geophysics* 66:4889–4899

Xue S, Li B, Xiao Z, *et al.* (2023). Centimeter-level-precision seafloor geodetic positioning model with self-structured empirical sound speed profile. *Satell Navig* 4:30. https://doi.org/10.1186/s43020-023-00120-7

Xue S, Yang Y. (2017). Understanding GDOP minimization in GNSS positioning: Infinite solutions, finite solutions and no solution. *Advances in Space Research* 59:775–785

Xue S, Yang Y, Yang W and Li J. (2024). GNSS-A Network Solution with Zenith Acoustic Delay Estimation. *Marine Geodesy* 47:237–268. http://dx.doi.org/10.1080/01490419.2024.2307932

Yang Y. (2018). Resilient PNT ConceptFrame. *Acta Geodaetica et Cartographica Sinica* 47:893–898

Yang Y, Liu Y, Sun D, *et al.* (2020). Seafloor geodetic network establishment and key technologies. *Science China Earth Science* 8:1188–1198

Yang Y and Qin X. (2020). Resilient Observation Models for Seafloor Geodetic Positioning. *Journal of Geodesy* 1–13

Yang Y and Qin X. (2021). Resilient observation models for seafloor geodetic positioning. *Journal of Geodesy* 95:79. https://doi.org/10.1007/s00190-021-01531-7

Yokota Y, Ishikawa T and Watanabe S. (2019). Gradient field of undersea sound speed structure extracted from the GNSS-A oceanography: GNSS-A as a sensor for detecting sound speed gradient. *Marine Geophysical Research* 40:493–504. https://doi.org/10.1007/s11001-018-9362-7

Yokota Y, Watanabe S, Ishikawa T and Nakamura Y. (2022). Temporal change of km-scale underwater sound speed structure and GNSS-A positioning accuracy. *Earth and Space Science Open Archive* 27

Zhang J, Han Y, Zheng C and Sun D. (2016). Underwater target localization using long baseline positioning system. *Applied Acoustics* 111:129–134

*Proceedings of the International Association of Institutes of Navigation World Congress 2024 –
Chen et al. (Eds)*

Utilizing radar landmarks for self-localization in GNSS-restricted marine environments

Shuai Sun, Hongguang Lyu & Xiao Yang
Dalian Maritime University, Dalian, Liaoning, China

Yang Yang
University of New South Wales, Sydney, Australia

ABSTRACT: Enabling self-localization for autonomous ships using marine radar data is of great importance, particularly when satellite signals are unavailable. This requires radar detectable objects with known locations (landmarks) to assist in inference of own-ship location. For unmanned ships, it is often difficult to search and select appropriate radar landmarks (RLs) to calibrate the position of own-ship. Properly taking account of the issues associated with radar echoes from these selected RLs for own-ship self-localization is also non-trivial. In this paper, we consider a harborage area where the global navigation satellite system (GNSS) based localization approach may suffer from reliability issues arising from multipath interference. Instead of using GNSS, location data (latitude and longitude) from anchored ships that are received from automatic identification system (AIS) is chosen to provide the RLs for self-localization. A recursive self-localization algorithm is proposed to estimate the position of own-ship using AIS provided location data of anchored ships. Simulation and real-data experiments demonstrate that the proposed solution can facilitate self-localization in GNSS-restricted environments.

Keywords: Self-localization, Radar landmark, GNSS-restricted environments, Autonomous ship

1 INTRODUCTION

In marine environments, most vessels rely on global navigation satellite system (GNSS) to obtain their own position. However, in GNSS-restricted marine environments where the satellite signal is severely affected or unavailable, it becomes vital to conduct autonomous self-localization [1]. In this aspect, marine radar may be a good choice. As a ship-borne equipment, marine radar has a long detection range and can still function well in adverse weather [2]. Through properly selecting and making use of existing landmarks whose locations are known in priori, such as static landmarks obtained from electronic navigation chart (ENC), anchored ships received from their reported AIS data, achieving self-localization is possible. For simplicity, these landmarks are all termed as radar landmarks (RLs).

In the literature, a well-known solution for self-localization is to use dead-reckoning methods [3]. However, it may suffer from drifting issues due to error accumulation. Most of existing works on self-localization focus on applications in underwater vehicles [4], where satellite signal is not available, or land vehicles [5][6], where the GNSS localization accuracy cannot meet the application requirement. Research on ship self-localization in a GNSS-restricted environment is relatively scarce. The importance of autonomous ship localization has raised due to the application of unmanned surface vehicles (USVs) and maritime autonomous surface ship (MASS). For instance, in [7], an image registration based solution is proposed using

 DOI: 10.1201/9781003624424-14

marine radar image and satellite image for USV self-positioning in GNSS-restricted environments. The main drawback lies in the aspect that it requires additional satellite image and the localization accuracy cannot be guaranteed when some images suffer from severe distortion. A digital map can also be pre-loaded to assist sensor data for self-localization [8]. In [9], the idea of simultaneous localization and mapping (SLAM) has been introduced for USV self-localization in unknown costal environments without GNSS signal. However, since the accuracy and resolution of marine radar is not high enough, it appears difficult to conduct fine-grained calibration with respect to the constructed environment map during localization. Recent works [10][11] using LIDAR data for self-positioning have also been applied in GNSS-restricted environments. Compared with marine radar, LIDAR can achieve higher accuracy. Nevertheless, its limited detection range (usually less than 200 meters) may constrain its application only in certain environments. In other words, there must be near-range landmarks available to facilitate properly use of LIDAR data for self-localization.

In this paper, we propose a self-localization method using marine radar measurements obtained from anchored ships whose positions are known by their reported AIS data to conduct own-ship positioning in GNSS-restricted environments. It is a light-weight self-localization solution because only marine radar data and AIS data are used, without employing additional third-party equipment. A Bayesian recursive scheme is applied for own-ship location estimation to avoid degraded localization performance when the number of RLs is insufficient or a missed radar detection of RLs occurs.

2 PROBLEM FORMULATION

At each time k, the problem is formulated as how to estimate the position of own-ship $\boldsymbol{x}_k = [x_k, y_k]^T$, defined in the global frame as shown in Figure 1, given marine radar data z_k^r and AIS data z_k^a. We use radar data update interval as the reference time interval and define Δk being the time interval between time $k-1$ to time k. For instance, if a 24r/min radar antenna is used, the corresponding time interval is 2.5 seconds. Note that though these two types of data do not arrive at the same time, we can assume AIS data is weakly synchronized with radar data because only reports from anchored ships are considered. Assume the speed over ground (SOG) v_k and course over ground (COG) θ_k of own-ship are available, which can be measured by the corresponding speed and distance measurement equipment (SDME) and Gyrocompass from own-ship. Strictly speaking, Gyrocompass measures the heading of own-ship, which may have a big difference with COG if own-ship is at low speed and in an environment with strong wind or current. Under calm situation, the motion of own-ship from $k-1$ to k is modelled as,

$$\boldsymbol{x}_k = \begin{bmatrix} x_k \\ y_k \end{bmatrix} = f(\boldsymbol{x}_{k-1}, v_k, \theta_k) = \begin{bmatrix} x_{k-1} \\ y_{k-1} \end{bmatrix} + \begin{bmatrix} v_k \cos(\theta_k) \\ v_k \sin(\theta_k) \end{bmatrix} \Delta k. \tag{1}$$

Radar data can be considered as periodically updated whereas AIS data update rate is subject to several factors including ship speed and maneuvering status [12]. Since only anchored ships are used, it is assumed that their positions remain the same as the previous reported data, if no new AIS data is received. The original radar data obtained from an anchored ship is relative range r_k and bearing φ_k. Let $\boldsymbol{z}_k^r = \left[z_k^r(x), z_k^r(y)\right]^T$ denote the corresponding converted radar measurement component defined in the global frame. According to [13],

$$\boldsymbol{z}_k^r = \begin{bmatrix} z_k^r(x) \\ z_k^r(y) \end{bmatrix} = \begin{bmatrix} \cos(\theta_k) & -\sin(\theta_k) \\ \sin(\theta_k) & \cos(\theta_k) \end{bmatrix} \begin{bmatrix} r_k \cos(\varphi_k) \\ r_k \sin(\varphi_k) \end{bmatrix} + \begin{bmatrix} x_k \\ y_k \end{bmatrix}. \tag{2}$$

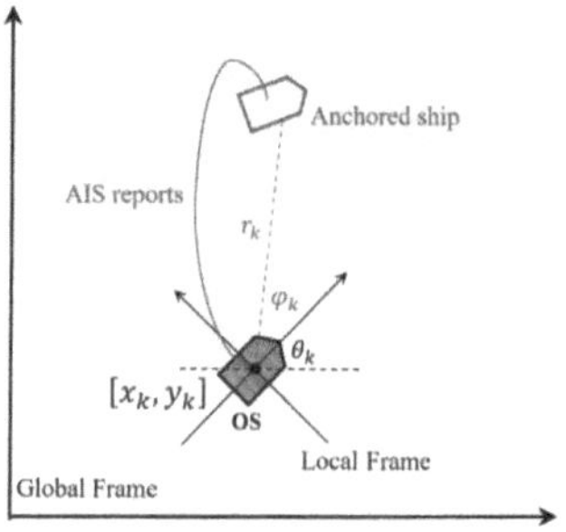

Figure 1. Self-localization based on radar data obtained from anchored ships.

3 METHODS

In this section, the proposed self-localization algorithm is presented time $k-1$ to time k. It has a recursive structure where the position of own-ship is firstly predicted to the next time step based on a motion model, and then radar measurements from anchored ships are utilized for updating own-ship position. In keeping with the radar target tracking literature, we use the notation $\left[\boldsymbol{x}_{k|k-1}, \boldsymbol{P}_{k|k-1}\right]$ to denote the predicted location (state) and location covariance of own-ship and $\left[\boldsymbol{x}_{k|k}, \boldsymbol{P}_{k|k}\right]$ to denote the updated (filtered) location and covariance of own-ship.

3.1 *Position prediction*

Let $\widehat{v}_k$ and $\widehat{\theta}_k$ represent the measured SOG and COG of own-ship from SDME and Gyrocompass. The predicted own-ship position is:

$$\boldsymbol{x}_{k|k-1} = \boldsymbol{x}_{k-1|k-1} + \begin{bmatrix} \widehat{v}_k \cos\left(\widehat{\theta}_k\right) \\ \widehat{v}_k \sin\left(\widehat{\theta}_k\right) \end{bmatrix} \Delta k. \tag{3}$$

Regarding to the predicted own-ship position covariance, according to Equation (1), it has a non-linear relationship with respect to SOG and COG of own-ship. We thus use the first-order Taylor expansion and expand Equation (1) at $\left[\widehat{v}_k, \widehat{\theta}_k\right]$. It is assumed that both the error of SOG and COG are zero mean Gaussian distributed, with the variance of SDME data error denoted by σ_s^2 and Gyrocompass denoted by σ_g^2.

$$\boldsymbol{P}_{k|k-1} \approx \boldsymbol{P}_{k-1|k-1} + \boldsymbol{G}_k \begin{bmatrix} \sigma_s^2 & 0 \\ 0 & \sigma_g^2 \end{bmatrix} \boldsymbol{G}_k^T \tag{4}$$

where $\boldsymbol{G}_k$ is the Jacobian matrix of $f(\boldsymbol{x}_{k-1}, v_k, \theta_k)$ taking the partial differentiation with respect to $[v_k, \theta_k]$, and substitute the measured data $\left[\widehat{v}_k, \widehat{\theta}_k\right]$, namely

$$\boldsymbol{G}_k = \left(\frac{\partial f(\boldsymbol{x}_{k-1}, v_k, \theta_k)}{\partial [v_k, \theta_k]} \Big| \left[\widehat{v}_k, \widehat{\theta}_k\right] \right) = \begin{bmatrix} \cos\left(\widehat{\theta}_k\right) & -\widehat{v}_k \sin\left(\widehat{\theta}_k\right) \\ \sin\left(\widehat{\theta}_k\right) & \widehat{v}_k \cos\left(\widehat{\theta}_k\right) \end{bmatrix} \Delta k. \tag{5}$$

3.2 *Position update*

In practice, there may be more than one anchored ship available from their reported AIS data that can be employed as RLs for own-ship self-localization. A sequential processing scheme is proposed, where every received AIS data from anchored ship is utilized for updating the position of own-ship. Assume there are M_k AIS reports received from anchored ships at time k. Let $[r_k^i, \varphi_k^i]$ denote the original radar measurement obtained from the ith ($0 < i \leq M_k$) anchored ship, and its corresponding received AIS data is denoted by $z_{k,i}^a$. Note that the original AIS data is in the format of latitude and longitude, and it is assumed $z_{k,i}^a$ is already converted to the global Cartesian frame. According to Equation (2), radar data from the ith anchored ship is

$$z_{k,i}^r = \begin{bmatrix} z_{k,i}^r(x) \\ z_{k,i}^r(y) \end{bmatrix} = \begin{bmatrix} \cos(\theta_k) & -\sin(\theta_k) \\ \sin(\theta_k) & \cos(\theta_k) \end{bmatrix} \begin{bmatrix} r_k^i \cos(\varphi_k^i) \\ r_k^i \sin(\varphi_k^i) \end{bmatrix} + \begin{bmatrix} x_k \\ y_k \end{bmatrix}. \tag{6}$$

Because the associated AIS data $z_{k,i}^a$ and radar data $z_{k,i}^r$ are both from the ith anchored ship, it can be stated that

$$z_{k,i}^a = z_{k,i}^r + \boldsymbol{n}_k, \tag{7}$$

where the white Gaussian measurement noise $\boldsymbol{n}_k \sim N(0, \boldsymbol{R}_k)$ represents uncertainty from both AIS data and radar data, with $\boldsymbol{R}_k$ being their overall covariance matrix. To be specific, measurement noise of AIS data is denoted by $\boldsymbol{R}_k^a$ and measurement noise of marine radar data is denoted by $\boldsymbol{R}_k^r$. Since the two measurements are independent from each other, $\boldsymbol{R}_k = \boldsymbol{R}_k^r + \boldsymbol{R}_k^a$. It is worth noting that fluctuation of AIS data is not only from GNSS error, but also from swaying of anchored ships due to influence of wind and current. As AIS data is updated much slower (3 minutes for anchored ships) compared with radar data (2.5 seconds if the radar antenna rotation speed is 24r/min), $\boldsymbol{R}_k^a$ should be assigned a larger value to model the case that a previously reported AIS data is used. A pseudo measurement $\tilde{z}_k^i$, which is a function of both radar data and its associated AIS data, is thus constructed as

$$\tilde{z}_k^i = h\left(z_{k,i}^a, r_k^i, \varphi_k^i\right) = z_{k,i}^a - \begin{bmatrix} \cos(\theta_k) & -\sin(\theta_k) \\ \sin(\theta_k) & \cos(\theta_k) \end{bmatrix} \begin{bmatrix} r_k^i \cos(\varphi_k^i) \\ r_k^i \sin(\varphi_k^i) \end{bmatrix} = \begin{bmatrix} x_k \\ y_k \end{bmatrix} + \boldsymbol{n}_k. \tag{8}$$

Clearly it has a linear relationship with the position of own-ship. Let $\boldsymbol{x}_{k|k}^{(0)} = \boldsymbol{x}_{k|k-1}$ and $\boldsymbol{P}_{k|k}^{(0)} = \boldsymbol{P}_{k|k-1}$ denote the starting state at update, which is simply the result from prediction. Based on the measurement from the ith anchored ship, own-ship position updated from the $(i-1)^{\text{th}}$ measurement data to the ith measurement data is conducted as

$$\boldsymbol{x}_{k|k}^{(i)} = \boldsymbol{x}_{k|k}^{(i-1)} + \boldsymbol{K}_k^{(i)}\left(\tilde{z}_k^i - \boldsymbol{x}_{k|k}^{(i-1)}\right), \tag{9}$$

and the corresponding covariance is updated as

$$\boldsymbol{P}_{k|k}^{(i)} = \boldsymbol{P}_{k|k}^{(i-1)} - \boldsymbol{K}_k^{(i)} \boldsymbol{P}_{k|k}^{(i-1)}, \tag{10}$$

with the Kalman gain $\boldsymbol{K}_k^{(i)}$ defined as

$$\boldsymbol{K}_k^{(i)} = \boldsymbol{P}_{k|k}^{(i-1)}\left(\boldsymbol{P}_{k|k}^{(i-1)} + \boldsymbol{R}_k\right)^{-1}. \tag{11}$$

4 SIMULATION

The proposed self-localization algorithm is first validated using simulated marine radar and AIS data. Two different scenarios are considered: 1) own-ship maintains a constant velocity (10kt) sailing route and there is one anchored ship located at (300, 600) serving as a RL for localization; 2) own-ship conducts a course turning during the voyage, and there are two anchored ships located at (300, 600) and (600, 300) serving as two RLs for localization. As shown in Figure 2, the starting location of own-ship is (0, 0) for both the two scenarios.

It is assumed SOG and COG of own-ship are provided from SDME and Gyrocompass, and their corresponding error is zero mean Gaussian distributed, with the variance of SDME data error $\sigma_s^2 = 0.5^2$ and $\sigma_g^2 = 1^2$. Note that wind and current can lead to a larger deviation of COG, in which case ship heading obtained from Gyrocompass is less agree with COG. Radar detection probability of anchored ships is assumed $P_D = 0.8$. As the starting position is (0, 0) in simulation, the initial estimation of own-ship state $\boldsymbol{x}_0 \sim N(\boldsymbol{0}, \boldsymbol{P}_0)$ is drawn from a zero mean Gaussian distribution with known $\boldsymbol{P}_0$. The measurement noise covariance matrix is set as

$$\boldsymbol{R}_k = \boldsymbol{R}_k^r + \boldsymbol{R}_k^a = \begin{bmatrix} 30^2 & 0 \\ 0 & 30^2 \end{bmatrix} + \begin{bmatrix} 10^2 & 0 \\ 0 & 10^2 \end{bmatrix}. \quad (12)$$

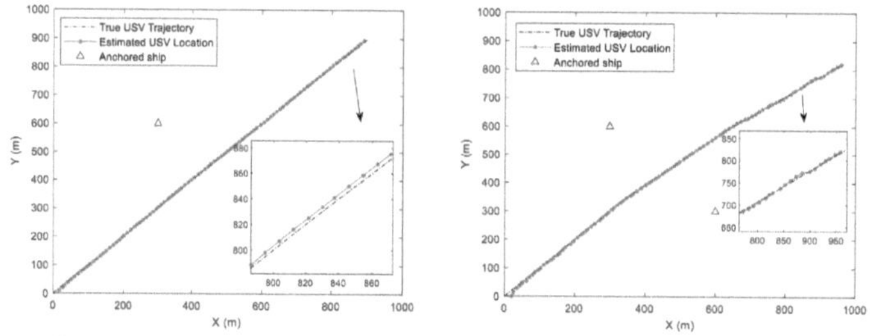

(a) Constant velocity with single anchored ship (b) Course turning with multiple anchored ships

Figure 2. Simulation results.

Simulation performance is shown in Figure 2. It can be seen that the estimation of own-ship location is well maintained in both the constant velocity (one anchored ship) and the course turning (two anchored ships) scenarios. We further provide a Monte Carlo simulation with 1000 runs to demonstrate the statistical performance of the proposed algorithm using the root mean square error (RMSE) metric, along both X axis and Y axis. Define $\boldsymbol{x}_{k|k}(m)$ as the estimation of own-ship location at time k in the mth simulation, and $\boldsymbol{x}_k$ is the ground truth, RMSE is defined as

$$RMSE = \sqrt{\frac{1}{M}\sum_{m=1}^{M}\left(\boldsymbol{x}_{k|k}(m) - \boldsymbol{x}_k\right)^2}, \quad (13)$$

where M is the total number of Monte Carlo simulation. As shown in Figure 3, a consistent own-ship self-localization result is observed.

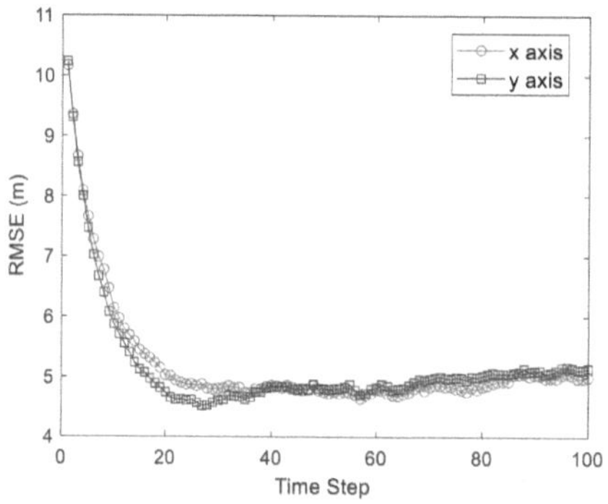

Figure 3. Evaluation using Monte Carlo simulation.

5 VALIDATION USING REAL DATA

We further validate the proposed self-localization algorithm using real marine radar data and AIS data collected in Dalian port, Liaoning, China. Because the marine radar is placed in a fixed location (121.70E, 38.92N), localization of own-ship is just to estimate the location of the radar site in this experiment. The peak power of the radar is 2.2kW and the 3dB horizontal beam width (HBW) of the radar is 5.2 degree. As shown in Figure 4(b), AIS data from two anchored ships is received, which is used for estimating the position of the radar setup site. Note that the original AIS ship position data is provided as Latitude and Longitude, and they are converted to Cartesian coordinates before applied for assisting localization.

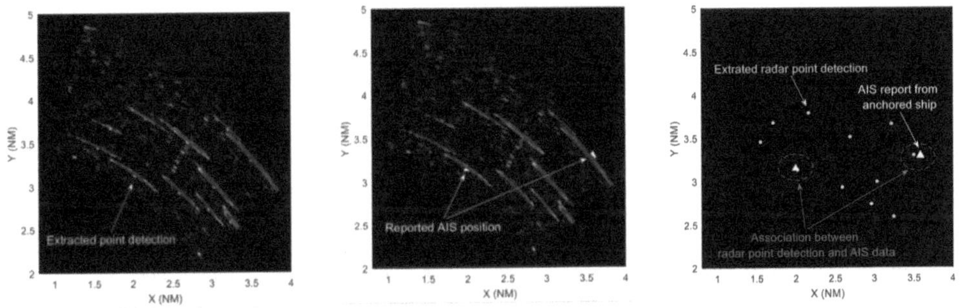

(a) Original radar echo (b) AIS data from anchored ship (c) Radar and AIS data association

Figure 4. Real data illustration for localization of radar site.

Regarding to radar data and AIS data association, namely finding the "right" radar echo that originates from the AIS reported anchored ship, an echo clustering is first conducted in the vicinity of the AIS reported ships. As shown in Figure 4(a), the point corresponds to the front edge of the clustered echo center is selected as the radar detection of the anchored ship. Then the converted radar data using Equation (2) that has the nearest distance with the AIS reported ship position is selected as the radar echo from the anchored ship, as illustrated in Figure 4(c).

At the beginning, the first received AIS data is used for initial radar position estimation, with its associated radar point detection of range and bearing serving as input for roughly computing radar location. This can be easily achieved by applying Equation (8). The proposed method described in Section 3 is then applied for radar position estimation. Note that since radar is installed on a fixed platform, the prediction step is not applied. The location is updated sequentially using the two constructed pseudo measurements. It is found that the positioning error after updating using the first two AIS reported data is -29.2 meter along X axis and 28.8 meter along Y axis.

6 CONCLUSIONS

In this paper, a self-localization algorithm using marine radar detection and received AIS reports from anchored ships is proposed for estimating the position of own-ship in GNSS-restricted marine environments. A recursive scheme is constructed where Gyrocompass and SDME data are applied for own-ship position prediction and associated radar data and AIS data are used for own-ship position update. The proposed self-localization algorithm is validated using both simulated and real radar and AIS data. It is demonstrated that marine radar along with effective radar landmarks can be employed for self-localization in GNSS-restricted environments.

ACKNOWLEDGEMENT

This work is supported by the National Natural Science Foundation of China (No. 52401416, 52071049); Liaoning Provincial Science and Technology Plan (Key) project (No. 2022JH1/10800096); the Natural Science Foundation of Liaoning Province of China (No.2022-BS-099); the Scientific Research Foundation of the Higher Education Institutions of Liaoning Province of China (No. LJKMZ20220375).

REFERENCES

[1] Liu, W., Huang, H., Hu, Y., Mou, M., Hsieh, T. H., Hu Q. and Wang, S., (2022). Improved GNSS vector tracking loop to enhance the navigation performance of USV. *Ocean Engineering*, 258, 111865.

[2] Bole, A. G., Wall, A. D. and Norris, A., (2013). *Radar and ARPA manual: radar, AIS and target tracking for marine radar users. Butterworth-Heinemann.*

[3] Kao, W. W., (1991). Integration of GPS and dead-reckoning navigation systems. *Vehicle Navigation and Information Systems Conference*, 2: 635–643.

[4] Hu, P., Yang, J., Yu, X., Qiao, J. and Guo, L., (2023). Underwater autonomous geo-localization using time differential polarization field against measurement deviations. *IEEE Transactions on Aerospace and Electronic Systems*, 60(1): 348–359.

[5] Li, X. and Xu, Q. A reliable fusion positioning strategy for land vehicles in GPS-denied environments based on low-cost sensors. *IEEE Transactions on Industrial Electronics*, 2016, 64(4): 3205–3215.

[6] Zhu, H., Wang, Q., Li, Y. and Leung, H. (2022), Variational Bayesian based localization for intelligent vehicle using lidar and GPS data fusion: Algorithm and experiments. *IEEE/ASME Transactions on Mechatronics*, 27(6): 5659–5667.

[7] Ma, H., Smart, E., Ahmed, A., Brown, D. Radar image-based positioning for USV under GPS denial environment. *IEEE Transactions on Intelligent Transportation Systems*, 2017, 19(1): 72–80.

[8] Lundgren M, Stenborg E, Svensson L, *et al.*, (2014). Vehicle self-localization using off-the-shelf sensors and a detailed map. *IEEE Intelligent Vehicles Symposium Proceedings*: 522–528.

[9] Han, J., Cho, Y. and Kim, J., (2019). Coastal SLAM with marine radar for USV operation in GPS-restricted situations. *IEEE Journal of Oceanic Engineering*, 44(2): 300–309.

[10] Zhou, B., He, Y., Qian, K., Ma, X. and Li, X., (2021). S4-SLAM: A real-time 3D LIDAR SLAM system for ground/watersurface multi-scene outdoor applications. *Autonomous Robots*, 45: 77–98.

[11] Sawada, R. and Hirata, K., (2023). Mapping and localization for autonomous ship using LiDAR SLAM on the sea. *Journal of Marine Science and Technology*, 28(2): 410–421.

[12] Sun, S., Lyu, H. and Dong, C., (2023). AIS aided marine radar target tracking in a detection occluded environment. *Ocean Engineering*, 288: 116133.

[13] Sun, S., Guan, W. and Wang, Y., (2024). Occupancy grid based environment sensing for MASS in complex waters. *Ocean Engineering*, 303: 117848.

Proceedings of the International Association of Institutes of Navigation World Congress 2024 – Chen et al. (Eds)

Application test of emergency Beidou radio position indicator buoy in search of containers overboard in Port of Shanghai

Xinyun Fan & Yao Lu
Shanghai Maritime Surveying and Mapping Center of Eastern Navigation Service Center, Maritime Safety Administration, China

ABSTRACT: Shanghai Port experiences a high volume of ship traffic and has a complex navigation environment. Once a container overturns, falls overboard, drifts, and sinks, it will pose serious safety hazards to the ship navigation in Shanghai Port. In order to enhance the positioning and salvage efficiency of fallen containers, traditional navigation buoys have been intelligently upgraded. An emergency positioning buoy with Beidou short message communication technology has been developed. This buoy is connected to containers floating on the sea through strong magnetic blocks. It supports online queries of the moving tracks of the position buoys on the mobile APP, and has been conducted relevant application tests in Shanghai Port.

1 INTRODUCTION

Containers overturned and fell overboard, drifting rapidly due to wind and waves, posing serious safety hazards to the shipping docks of Shanghai ports, the Deep Water Fairway, the South Fairway, and the designated waters of the Yangtze River Estuary. For example, if the containers were loaded with hazardous materials, they would also have a detrimental impact on the water environment. It will cause pollution, and incidents of containers falling overboard at sea occur frequently at home and abroad. In recent years, there have been numerous incidents of shipping containers falling overboard annually in Shanghai Yangshan Port, Waigaoqiao, and other port areas, as well as in other navigable waters. From 2016 to 2024, a total of 15 incidents of containers falling overboard occurred in Shanghai Port waters, including 2 incidents in 2016,4 incidents in 2017,2 incidents in 2018,1 incident in 2019,3 incidents in 2020,1 incident in 2022, and 2 incidents in 2024.In total,840 containers fell overboard, with 37 containers falling overboard in the Wusongkou Warning Area in May 2017. It has seriously blocked the lifeline to and from the Yangtze River Estuary, preventing international cruise ships from entering the port. On December 13, 2020, the foreign container ship "Changjin Ocean" collided with the Chinese container ship "Xinqisheng 69" near the No. D15 buoy in the Deep Water Fairway of the Yangtze River Estuary. The "Xinqisheng 69" ship, carrying 650 containers, capsized after taking in water, leading to the temporary closure of the Yangtze River Estuary for navigation. According to the World Shipping Council Container Overboard Research Report (updated every three years), between 2008 and 2019, an average of about 1,400 containers fell overboard annually. However, in the years 2020 and 2021, the number of containers falling overboard has exceeded 3,500. In November 2020, over 1,800 containers on a large container ship were either lost overboard or damaged. This number hit a record high in recent years. A well-known foreign shipping company had 962 containers fall overboard in 2021 alone, which is 30 times the number of containers lost by the company in the past 10 years.

DOI: 10.1201/9781003624424-15

2 ANALYSIS OF THE WATER ENVIRONMENT AFFECTING THE EFFICIENCY OF EMERGENCY SWEEPING OF FALLEN CONTAINERS

During emergency sonar sweeping of fallen containers, the containers being swept can easily drift due to the combined effects of wind, waves, and currents, which can affect the efficiency of the sweeping process. The sweeping and surveying of the water environment involve gathering information about the fallen containers and the on-site marine environment. The characteristics of the fallen container are the internal factors that affect the drift movement of the container, and the on-site marine environment is the external factor that affects the drift movement. The internal factors of the container that affect the drift movement include information such as container size, external structure, and container quality. The on-site marine environment that affects the drift movement includes hydrological and meteorological information such as wind speed, wind direction, flow speed, flow direction, and waves. In order to meet the loading needs of different types of goods, there are various styles of containers, which vary from appearance to structure, strength, size, and quality. The states of different styles of containers after falling overboard are also different. Tank containers and refrigerated containers are better sealed, usually lighter, and drift in the water for a longer time. They also pose the characteristics of high cargo value and high-risk factor. Open-top and frame containers are prone to sink into the bottom because of the structural characteristics of the containers.

Due to the influence of the water environment, emergency sweeping and surveying of fallen containers require prompt action. Based on the container information and in combination with wind, currents, and wave factors, the direction of the container that has fallen overboard can be preliminarily determined. However, in severe weather conditions such as strong winds and currents, the position of the container will obviously change. For instance, the incident involving the "Xinqisheng 69" ship in 2020 happened during the astronomical tide and strong winds brought by a cold wave. The containers that fell overboard were mostly loaded with light cargoes, such as kaolin and edible oil. Under the influence of wind and currents, individual light containers drifted from the Deep Water Fairway of the Yangtze River Estuary in the incident area to the waters of Zhoushan Islands in less than a day, covering a straight-line distance of approximately 45 nautical miles. The heavy containers surveyed that sank underwater were also displaced by the influence of currents. In inclement weather, containers that were surveyed and not cleared in time are prone to significant position changes. They need to be located and swept again for reducing the emergency efficiency.

3 RELEVANT RESOLUTIONS OF THE INTERNATIONAL MARITIME ORGANIZATION (IMO) ON MANDATORY REPORTING OF CONTAINERS FALLING OVERBOARD

The 106th meeting (MSC.529) of the IMO Maritime Safety Committee recognized the Beidou Short Message Service System (BDMSS) as part of the Global Maritime Distress and Safety System (GMDSS). In May 2024, the 108th meeting of the MSC adopted amendments to SOLAS V/31.2 and V/32.3, establishing reporting procedures that encompass the detection, location, tracing and salvage of lost containers at sea. The new Article V/31.2 delineates the roles and responsibilities of the parties involved, such as the captain, the shipping company, and the flag State government, in case of containers loss on board the ship or the sighting of the lost containers at sea. Additionally, the new Article V/32.3 specifies the extent of information to be reported. These amendments will take effect on January 1, 2026, for all vessels.

The validation of the mandatory reporting for lost containers at sea facilitates prompt detection and significantly reduces environmental pollution and navigation safety risks resulting from fallen containers. At the same time, it also proposes higher technical demands on the monitoring, tracking, and positioning of shipping containers. Due to the significant challenges and limitations of relying on traditional visual methods, especially in adverse weather conditions or at night, to detect containers falling overboard, SOLAS recommends that ships utilize electronic methods as much as possible to monitor containers. This helps the captain promptly receive signals of the lost container and report danger information promptly and effectively.

4 COMMUNICATION LINK FOR BEIDOU EMERGENCY POSITION INDICATOR BUOY

Short messages refer to brief information fragments, which may include text, location information, etc. The Beidou emergency position indicator buoy integrates the Beidou short message communication module and the Beidou positioning module. After the Beidou Global Navigation Satellite System (GNSS) positioning, the information is transmitted to the Beidou satellite via the Beidou short message communication module. The Beidou satellite system transmits short message data to the ground base station (or Beidou signal receiving and processing equipment). The ground base station transmits the data to the designated server via Private Network. The data is parsed and stored on the server. The mobile application terminal APP accesses the server to retrieve and display obtain buoy location data.

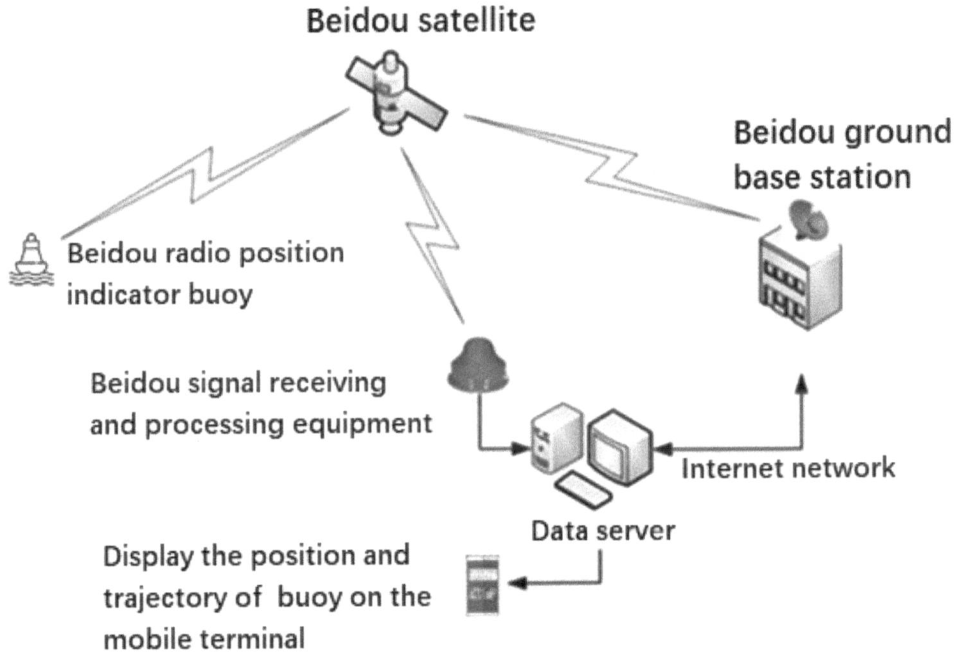

Figure 1. Beidou position indicator buoy short message communication link diagram.

5 DEVELOPMENT OF AN EMERGENCY POSITION INDICATOR BUOY FOR FALLEN CONTAINERS

Utilizing the short message communication function of the Beidou satellite system, emergency position indicator buoys are mainly used to track and monitor floating obstacles such as fallen containers, provide early warnings, and ensure the safety of ship navigation. The emergency position indicator buoy mainly includes a micro-drift buoy carrier, Beidou communication positioning module, power module, and mobile application terminal APP software. The buoy body provides buoyancy and serves as the installation basis for other modules. The solar power generation panel can continuously charge the battery, enabling it to power the Beidou positioning components for an extended period. The sealed waterproof cover of the position indicator buoy provides a waterproof protection to shield electronic components from the effects of seawater. The strong magnetic block, which is thrown and then adsorbed onto the container, is connected to the floating position indicator buoy through a nylon rope to facilitate the positioning and searching of drifting containers at sea. The design diagram of the emergency position indicator buoy is as follows:

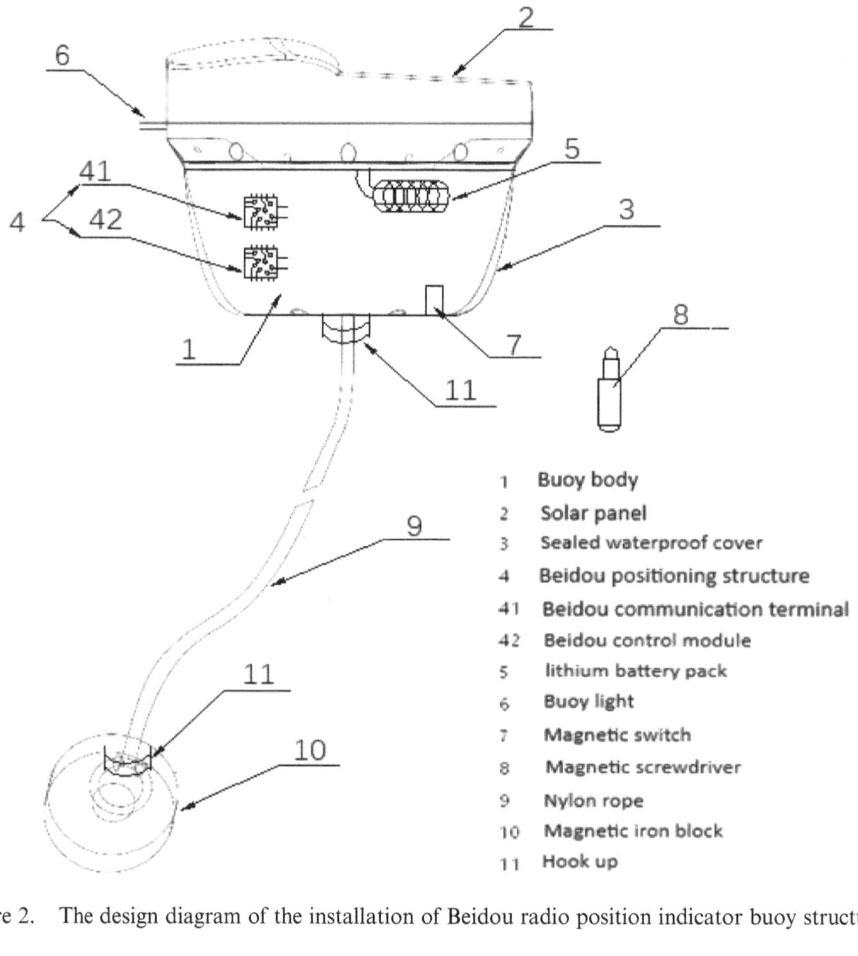

Figure 2. The design diagram of the installation of Beidou radio position indicator buoy structures.

6 APPLICATION TEST OF AN EMERGENCY POSITION INDICATING BUOY FOR A FALLEN CONTAINER

A small survey ship is used to simulate a container floating on the water. It is connected to a strong magnetic block and a position indicator buoy using a Nylon wire. The strong magnetic block adhered to the side of the ship and threw the position indicator buoy into the water. The survey ship sails at a slow speed on the Huangpu River, Wusongkou and

Waigaoqiao navigable waters and port areas in Shanghai have been tested for application, and the expected results have been achieved. The strong magnetic block has never fallen off. The real-time location and track of the position indicator buoy can be accessed through the electronic chart (ECDIS) interface of the mobile application terminal APP. The length of the attached wire exceeds the real-time water depth value of the sea surface. Regardless of whether the fallen container drifts or sinks, it can be searched by using the real-time location and track of the position indicator buoy.

Figure 3. The point trajectory of the position indicator buoy tested in Shanghai navigable waters.

Table 1. Coordinate values of the position indicators buoy tested in Shanghai navigable waters.

Reply time	Geodetic longitude (Dgrees)	Geodetic latitude (Dgrees)	Speed (Knot)	Sailing direction (Dgrees)
2023/10/12 08:41	121.5614	31.278	0	55
2023/10/12 08:52	121.5648	31.2822	4.9901	22
2023/10/12 08:57	121.5623	31.2958	4.7328	82
2023/10/12 09:02	121.5586	31.3075	4.63	102
2023/10/12 09:07	121.5583	31.3195	4.2699	1
2023/10/12 09:13	121.5563	31.3318	4.5271	80
2023/10/12 09:28	121.5206	31.349	4.2699	42
2023/10/12 09:39	121.5	31.3635	4.0641	83
2023/10/12 09:49	121.5043	31.3854	4.0126	39
2023/10/12 09:55	121.5157	31.3921	4.6814	72
2023/10/12 10:00	121.5303	31.3939	4.6814	99
2023/10/12 10:05	121.545	31.3906	4.4756	106
2023/10/12 10:43	121.5545	31.3875	3.0866	112
2023/10/12 11:20	121.5438	31.3961	6.3791	39
2023/10/12 11:35	121.5125	31.4093	0.3601	31
2023/10/12 11:46	121.5194	31.3992	5.0415	153

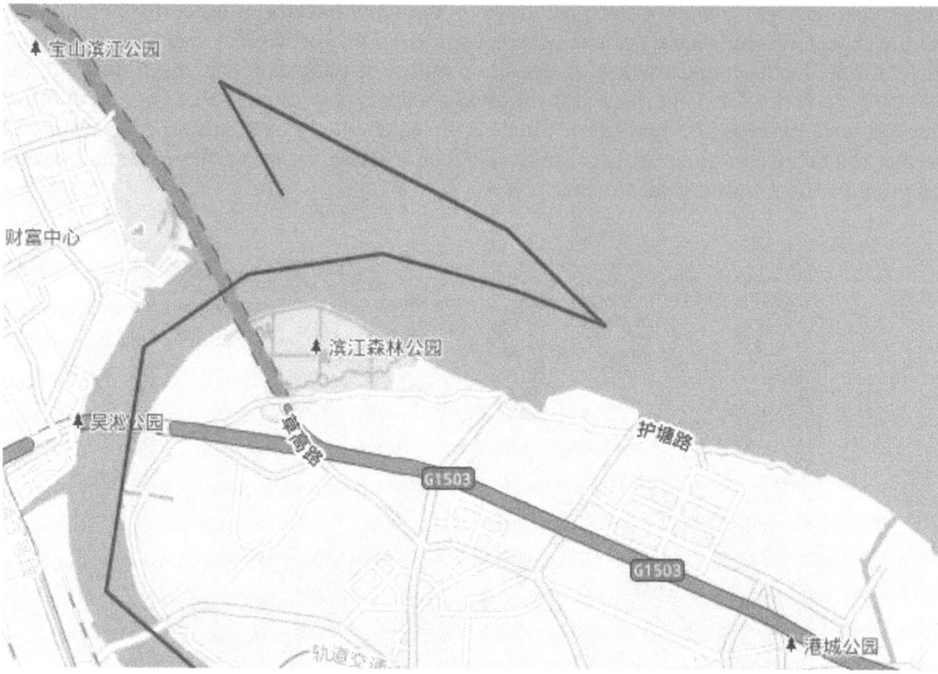

Figure 4. Position indicator buoy test scene in Shanghai navigable waters.

7 APPLICATION PROSPECTS OF BEIDOU POSITION INDICATOR BUOY FOR FALLEN CONTAINERS

The emergency container position indicator buoy features a portable design with a solar panel tumbler. It is lightweight, has low power consumption, has an independent solar power supply, and has a buoy light at night. The search and guidance performance are all-weather and not limited by geographical location, network, energy supply, and other conditions. It supports Beidou short message communication and short message position reporting functions, and can realize online queries of the moving tracks of the position buoys on the mobile application terminal APP. The buoy has a moderate volume (15*10*12cm) and can be easily connected to containers floating on the sea through strong magnetic blocks. The position indicator buoy is easy to operate, low-cost, and reusable. It can facilitate the remote automatic sending transmission of the floating container's location information and track to the maritime command center. It also can realize the potential of remotely monitoring of floating obstacles at sea, which holds promising application prospects.

8 CONCLUSION

Compared to shipwrecks and other obstructing objects, the main challenge of conducting an overboard container sweeping survey is that the scattered containers can be vulnerable to environmental factors and prone to position drifts. This situation poses significant risks to navigation safety. When a container falls overboard, the maritime administrations must mobilize substantial manpower and resources for operations to mitigate or entirely eliminate the impact on waterway navigation and maintain the efficiency of port operations. Therefore, it is necessary to improve the efficiency of emergency sweeping surveys. Beidou short message communication technology is utilized to get the real-time location of fallen containers. This technology enables the prompt reception of signals indicating the container, facilitating the timely and effective reporting of danger information. This, in turn, enhances the efficiency of emergency sweeping surveys and provides improved technical support for navigation management and obstacle clearance to ensure the safety of port channels. It is recommended that IMO amends the content of the resolution of MSC.471(101). In GMDSS, we place the hook at the lower end of the self-floating Emergency Position Indicating Radio Beacon (EPIRB) operating at 406 MHz, attach floating ropes and a strong magnetic block to prepare for manual connection with floating containers. The EPIRB is equipped with a BDMSS terminal inside the structure. When the ship is in distress, the EPIRB will automatically pop up and float to the water surface after the ship sinks. When a distress signal is sent out, the GMDSS can immediately report distress information to land search and rescue agencies and nearby sailing ships. The EPIRB serves a new purpose. When a sailing ship encounters floating containers or iron obstacles at sea, it will unload the EPIRB fixed to the ship, throw it towards the container or iron obstacle, and utilize a powerful magnetic block to attract the container or iron obstacle. The GMDSS or the ship's captain is used to facilitate the positioning, searching, and reporting of the containers or iron obstacles at sea or after sinking through BDMSS.

REFERENCES

Huang Lunwen and Chen Yongfang. (2020). A design based on beidou emergency radio position indicating buoy. *Electronics World* 2020(5):116–118.

Li Shuai and Lin Xintong. (2022). Analysis of causes and preventive measures of containers falling overboard. *World Shipping* 2022(8):39–41.

Liu Shunjie and Shi Xiaoping. (2022). Methods for efficient emergency sweep search of overboard container. *Navigation of China* 2022(3):33–46.

Sui Yongju and Xiao Yue. (2020). Research on marine emergency electronic position indicating buoy system based on multiple technologies. *Pearl River Water Transport* 2020(12):91–92.

Wang Zhenning, Liu Xiaodong, Xie Xin and Tang Jiwei. (2023). Changes in Reporting Requirements and Countermeasures for Containers Overboard at sea. *Water Safety* 2023(1):1–3.

*Proceedings of the International Association of Institutes of Navigation World Congress 2024 –
Chen et al. (Eds)*

Research and application of *Data Product Interoperability of S-100 Navigation Systems*

S.X. Xian, Q.F. Zou & Y.X. Lyu
Guangzhou Maritime Hydrographic Center, Southern Navigation Service Center, Maritime Safety Administration, People's Republic of China

ABSTRACT: *Data Product Interoperability of S-100 Navigation Systems*, designated as S-98, is a framework by the International Hydrographic Organization (IHO), essential for data exchange among S-100 series products. This paper emphasizes the need for refining and applying S-98, reviewing its current international research status. It analyzes various levels of S-98 interoperability and their characteristics, exploring key aspects for interoperability between S-100 products, including conflict resolution, portrayal and processing models. Notably, we implemented two scenarios in Chinese nautical chart software using Lingding Navigation Channel data. The first scenario enables the realization of user-selectable safety contour through the integration of S-101 data and S-102 data. The second scenario facilitates water level adjustment by leveraging S-101, S-102, and S-104 data. This initiative has accelerated the sharing and integration of maritime information data, propelled the extensive adoption of S-100 series standard products, and significantly enhanced navigation safety.

S-100 Universal Hydrographic Data Model supports a diverse range of hydrographic data, products, and users. Using this model as a framework, a series of S-100 data products have been developed to cover marine-related fields, including hydrography, navigation, and hydro-meteorology. These S-100 data products involve multi-source data of varying specifications. When applying these data products in integrated navigational services, such as Electronic Chart Display and Information System (ECDIS), it is necessary to superimpose and display data products of different specifications. However, since the symbol systems, feature structures, and data accuracy of these products are not identical, the superimposition and display can lead to conflicts.The S-101 Electronic Navigational Charts (ENCs) serve as the base layer for initiating S-100 products, but the true potential of S-100 will be unlocked through interaction with other products such as the S-102, S-104, and S-111. Therefore, resolving the interoperability issues associated with the visualization of data from multiple sources is crucial for S-100 to realize its full potential. To address these issues, the IHO has revised the Part 16 of *S-100 Universal Hydrographic Data Model* to standardize the principles of display and processing of different S-100 data products within the same user system. Furthermore, the S-98 Data Product Interoperability Specification, which is based on these principles, aims to resolve conflicts that arise from the combined application of S-100 data products.

At its 9th meeting held from June 21 to 30, 2022, the International Maritime Organization's (IMO) Sub-Committee on Navigation, Communications and Search and Rescue (NCSR) considered and adopted amendments to the ECDIS performance standards, which were submitted by the IHO and other co-sponsors. The amendments included the addition of new IHO data standards and product specifications, specifically addressing *Data Product Interoperability in S-100 Navigation Systems* (S-98).

Currently, there are relatively few studies on the interoperability of data products in S-100 navigation systems, most of which are at the level of theoretical research. The Norwegian

 DOI: 10.1201/9781003624424-16

agency PRIMAR has only displayed S-101 and S-102 in an overlay. The United Kingdom Hydrographic Office (UKHO) has applied S-102, S-104, and S-111 in the S-100 data fusion simulation for autonomous ship navigation in Plymouth, but has not yet implemented the S-98 data interoperability specification. The IHO Singapore Laboratory has initiated a project to demonstrate the interoperability of S-101 and S-102 on an S-100 compatible electronic charting system by depicting and overlaying high-resolution S-102 bathymetry data on the S-101 ENC. However, China has not yet published any specific S-98 related research materials or documents. This paper has studied and analyzed *Data Product Interoperability in S-100 Navigation Systems* and realized two standard application scenarios: user-selectable safety contour and water level adjustments, based on S-101 electronic chart data, S-102 bathymetry surface data, and S-104 water level data in the Lingding channel waters of the Pearl River Estuary.

1 INTEROPERABILITY LEVELS DEFINED BY S-98

The S-98 is a collection of specifications for visual and other interoperability of different data in ECDIS based on the S-10X product specification, which is designed to reduce data redundancy, avoid display confusion, as well as preventing conflicts, and provide users with better quality and clarity of information while simultaneously displaying multiple S-100-based data products. The current S-98 specification defines an incremental interoperability catalog of 0-4 levels focused on the relative prioritization of geographic features for display, with higher levels of interoperability based on lower levels, expanding to correspond to successive, more complex interoperability application models.

1.1 *Level 0*

In this level, all products have no interoperability features. All features are only loaded and layers are overlaid at the ECDIS, except by adjusting the transparency of the different products to prevent layer occlusion. In this level, ENC is considered as the primary product displayed on the ECDIS screen, and the information layer prioritization remains compliant with the relevant IMO and IEC performance standards.

1.2 *Level 1*

In Level 1, feature types for different products, including S-101, can be interleaved based on the display panel and drawing priority information specified in the Interoperability Catalog, and this information is passed to the portrayal processor. At this point, ENC is still considered the primary product, but if it is obscured, the ENC feature layer may be overlaid with the feature layers of other products.

1.3 *Level 2*

In Level 2, not only does it continue the overlapping display of data from Level 1, but it also enables feature filtering based on combinations of attribute values to compress all features of a specified feature type in a specified product. The result of this interoperability is that selected instances of ENC features are suppressed or replaced by specified features in other products. At this point, the features in the other product contain more details and higher resolution, which may be superior to the specific ENC feature type, and therefore may also be displayed in preference to the ENC feature layer.

1.4 *Level 3*

In Level 3, it extends the interoperability capabilities of Level 2, allowing features from different products to combine their original attributes in various manners, possibly making adjustments to attributes that include ENC or other feature instances, such as recalculating the value of numeric attributes and adding listed values to enumerated attributes. Hybridization can lead to instances of different feature types that possess an expanded set of thematic attributes, some of which may be new attributes generated from the attribute values of the original instance.

At this point, the interoperability product will include a catalog of hybrid features and a catalog of schema representations for defining feature types and schema for the new hybrid features.

1.5 *Level 4*

Interoperability in Level 4 retains the same basic functionality as Level 3, with the addition of spatial queries to identify related subsets and operational functionality to specify interoperability results. At this stage, ENCs and other product features do not require spatial equivalence; they merely need to be interconnected through spatial queries. Spatial operations can be utilized to combine the geometries of hybrid features, in addition to their thematic attributes.

1.6 *Analysis of interoperability levels*

The five levels of interoperability define the depth and complexity of interoperability between S-100 products, progressively evolving from the most basic no interoperability (level 0) to advanced spatial manipulation (level 4). Each level details its differences and additions to the previous level, in particular specific implementations and applications for layer overlays, data interleaving, type-based selectivity and feature class replacement, feature hybridization, and spatial operations.

At present, we are only able to achieve interoperability with Level 0 and Level 1 definitions for simple overlay and overlapping of different S-100 products. Take the bathymetry data as an example, Level 0 can display bathymetry data from different products; Level 1 can interleave the bathymetry data from different products; Level 2 can compress low-precision bathymetry data and replace it with higher-precision bathymetry data; in Level 3, it can compute the under keel clearance and real-time water level to determine the current safe sailing area; in Level 4, spatial queries can be used to perform spatial operations on bathymetry data, such as creating a safe bathymetry buffer.

These five levels form a progressively enhanced interoperability system, providing ECDIS with various means of interoperability from basic to advanced. With the enhancement of the levels, the system is able to handle data interaction among different products more flexibly and efficiently, and improve the accuracy and reliability of chart information.

2 THE S-100 PRODUCT INTEROPERABILITY REALIZATION POINTS

The S-100 products that are interoperable with the S-98 specification are S-101 Electronic Nautical Charts (ENC), S-102 Bathymetry Surface, S-104 Water Level Information for Surface Navigation, S-111 Surface Currents, and S-129 Under Keel Clearance Management. The S-98 defines the format of the interoperability catalog, in addition to defining the interoperability functions that should be implemented by an ECDIS that is compatible with an S-100 product. The S-98 interoperability catalog is a set of rules that control visualization and other interactions between S-100-based data products that meet different product specifications. This Interoperability Catalog is intended for use by navigation systems to simultaneously use and view data sets that conform to two or more S-100 product specifications.

Currently, the S-98 research is still in the route monitoring phase and focuses on defining the interoperability of route monitoring products such as S-101, S-102, and S-104. The realization of S-100 product interoperability should address the key points of conflicts between different S-100 products, portrayal methods, and processing models.

2.1 *Conflicts in interoperability of different S-100 products*

In the processing of realizing interoperability of S-100 products, the S-98 mainly considered duplicate features, the differences in geometry, and text display as the main conflicts that need to be resolved to realize interoperability.

2.1.1 *Duplicate features*

The duplicate features among different layers is an important issue that needs to be resolved in S-98. The S-98 provides for the prioritization of duplicate features in several different situations. First, for duplicate features in the same model, priority should be given to products with the latest information and frequently updated functions; second, for duplicate features in different models, priority should be given to the version of the feature with higher value to users in interoperability; third, duplicate features with different conceptual features, under the condition of equivalent information content, S-98 needs to consider two factors: the update frequency and the degree of correlation with other features.

2.1.2 *Differences in geometry*

For geometric spatial differences, S-98 proposes a multi-level solution. First, basic conflicts are resolved by reordering the display of features. Second, if conflicts still persist, there is an option to suppress specific feature layers. At higher levels, geometric adjustments are made to optimize the display. When features from two datasets can complement each other, their geometry can be combined to eliminate conflicts. If there are differences in the same features from different products, they can be resolved by suppressing instances, but this operation needs to be performed at L2 or higher interoperability levels.

2.1.3 *Text display*

For text display, the S-98 should ensure that the center of the text box is always located within the dataset boundary to avoid conflicts with the dataset boundary.

2.2 *Portrayal methods*

The depiction of features in the interoperability follows the S-100 mapping rules, and the system is required to render according to three modes: day, dusk, and night. In addition, the reports under interoperability control are combined into a single pick report that contains information on all overlay products, providing a user-friendly browsable report through data organization, but products suppressed in ECDIS are not included in the selection report. By default, the prioritization of information in the pick report is consistent with the data layer order.

2.3 *Processing models*

There are two depiction processing methods for interoperability processing, which are categorized as being performed before depiction processing or after depiction processing.The interoperability process precede portrayal processing is to first perform interoperability processing of the S-101 and other combined product data, then perform portrayal processing of the obtained element data by portrayal XML file or element list (Lua), and finally rendering. The interoperability process follow portrayal processing is to first process S-101 and other combined product data for portrayal, convert it to drawing instructions, then process it for interoperability, and finally render it.

3 REALIZATION SCENARIOS

Bathymetry data is an important factor affecting navigation safety. In the S-100 framework, S-102, S-104, S-111 and S-112 involve bathymetry data from different sources and with different accuracy, and the interoperability scenarios of different bathymetry data are defined in S-98, which provide users with high-precision user-selectable safety contour and dynamic real-time water level data through the superimposed displays of multiple bathymetry data.

At present, the interoperability of some S-100 data products in different scenarios has been realized based on China's home-made software, including the production of more accurate user-selectable safety contour in the interoperability of S-101 and S-102 and water level adjustment in high-density bathymetry in the interoperability of S-101, S-102 and S-104.

3.1 *User-selectable safety contour*

The section 11.4.3 of Performance Standards for Electronic Chart Display and Information Systems(ECDIS) (IMO MSC 232(82)) states that ECDIS shall raise an alarm if the vessel is about to cross the safety contour within a specific time set by the users. The S-98 describes the definition and implementation of a "user-selectable safety contour", which means that the safety contour is extracted from the bathymetry surface grid data based on a value set by the user. When the S-101 electronic chart and S-102 high-density seabed topographic data are interoperated, the S-102 data replace the electronic chart contour and regenerate a more accurate safety contour under the control of S-98. In the process of generating the safety contours, S-102 suppresses some of the S-101 features, and the user sets the safety contour values. In addition, when the user sets the safety contour values, the safety contour values displayed in the area outside the coverage of S-102 are derived from S-101.

In the S-98, controlling the S-102 data display at the L0 is shown in Figure 1. It is noted that the software originally had a Chinese interface and we have translated the main information in the pictures into English. At the L1, the DepthArea and DepthContours are automatically extracted from S-102 to cover the corresponding features in S-101 as in Figure 2. The extracted DepthArea and other features in S-102 are displayed according to the style of the corresponding features in S-101. The features in S-102 are displayed according to the priority in S-98. The areas not covered by S-102 continue to use the data in S-101.

In this paper, based on the provisions of the safety contour in S-98, the realization of dynamic safety contour in the Lingding Channel waters of the Pearl River Estuary is presented. Additionally, the visualization of safety contour for L0 and L1 levels is achieved, allowing the user to select the value of the safety depth based on the ship's draft depth and specify a time period to display the safety contour for that period. The realization of dynamic safety contour is shown in Figure 3 can assist users in identifying shallow water areas in real time, effectively preventing the occurrence of ship grounding accidents, thus reducing safety risks during navigation.

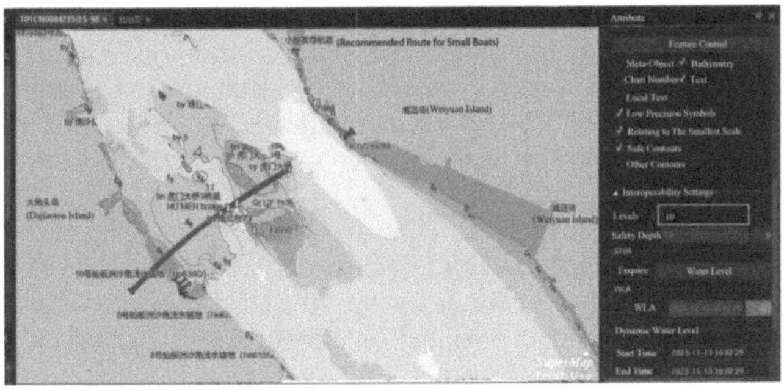

Figure 1. Display of S-102 data in Level 0.

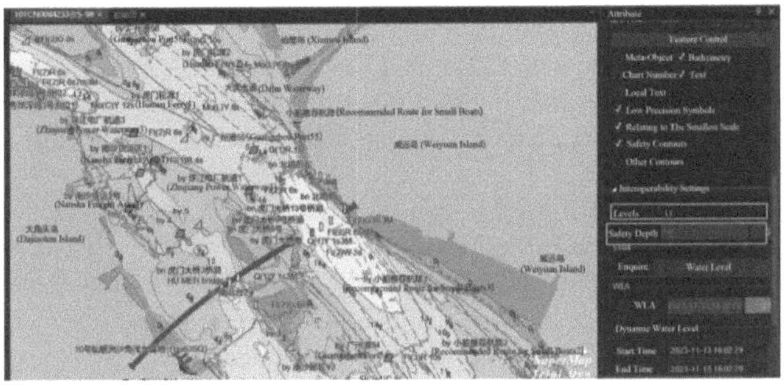

Figure 2. Display of S-102 data in Level 1.

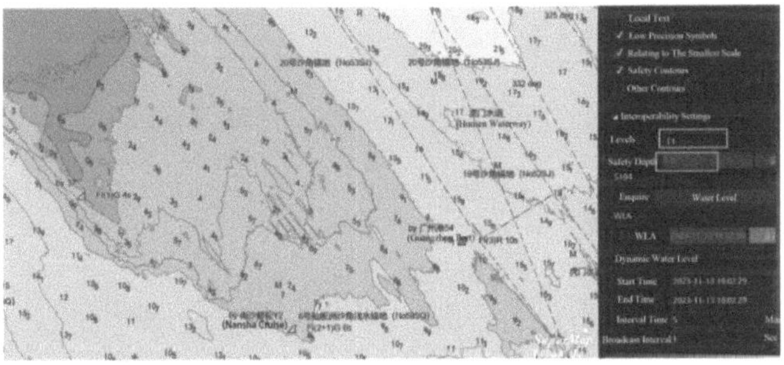

Figure 3. User-selectable safety contour.

3.2 *Water level adjustment*

As early as 2001, SevenCs developed a "tide-adjusted" ENC simulation model, which was prototyped for the ports of Singapore, and the Schelde and Vlissingen in Netherlands. The model simulates a 10 meter tidal range and adjusts the "safe water" display according to time of day and the safe depth of the vessel. When combined with tidal information, it provides decision support for voyage planning, with useful information such as vessel passage times and drafts. Although seafarers are aware of the benefits, it has not yet been implemented on ENC and ECDIS for reasons such as the current S-57 standard not supporting tidal adjustment, ENC digitization based on paper charts, and the lack of depth contour intervals.

The section 11.4.3 of IMO MSC 232(82) requires the user to specify safety distances for planned routes to prevent inadvertent entry into hazardous areas.Water Level Adjustment (WLA) is defined in S-98 and allows for water level adjustments to be made for a single date or time, for a specified time period, and for a specified route. Water Level Adjustment can only be utilized within the coverage of S-104 in conjunction with the bathymetry data of S-101 overlay of S-101 or S-102 replacement.

Water level adjustments are subject to the following rules: default depth information adjustments should not be made based on tide or water level heights; if the ECDIS has been integrated to use S-100-based tide or water level product specifications and electronic charts, then the system can provide all depth-related functional services based on the adjusted tide or water level data. The data used for water level adjustments can be static, historical, predictive, or real-time water level data, e.g., derived from water level measurements via radio communications. Depth-related integration functions are specified in detail in S-98 and related S-100-based product specifications (e.g., S-101, S-102, S-104, and S-111). Related standards S-98 and S-100-based product specifications may contain conditional models designed to limit the scope of application of a particular product in regulating water levels. For example, a separate S-101 or S-102 unit may contain meta data or feature objects that define the geographic area in which depth adjustment is allowed. Where depth adjustments are permitted, individual S-104 areas may set constraints associated with specific dates or time periods.

Based on the water level adjustment requirements of S-98, this paper utilizes S-101 electronic nautical charts, high-density S-102 seafloor topography data, and S-104 water level product data of the Lingding channel waters in the Pearl River Estuary to make water level adjustments in high-density water depths and to realize the dynamic presentation of real-time 7m safety contour that change over time as in Figure 4.

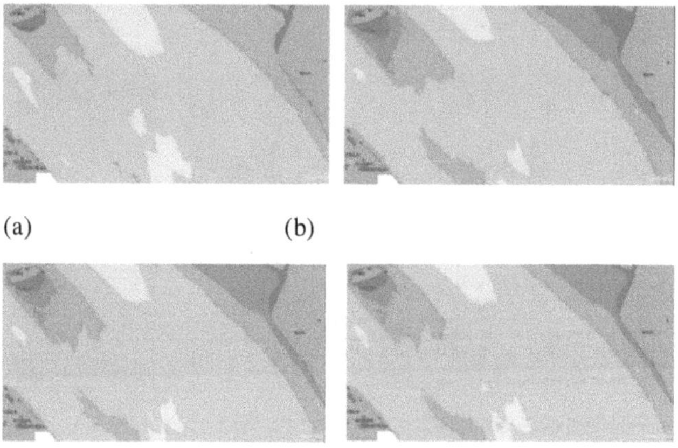

(a) (b)

(c) (d)

Figure 4. Dynamics of the 7 meters safety contour.

4 CONCLUSIONS

This paper discusses in depth the practical application of the S-98 product specification. Based on its research results, two major interoperability scenarios, namely, user-selectable safety contour and water level adjustment, have been successfully realized. However, the interoperability technology of S-100 products is still in the early stage of research and development, where the interoperability specification of L2 and higher-level products is not yet mature, and there are still many uncertainties about the implementation path of interoperability in the updating of new products. In addition, there is limited practical application and testing experience of S-100 interoperability worldwide. Therefore, the study of the interoperability of S-100 products is particularly critical, which is an important part of the process before its full-scale application. Taking the acquisition of high-precision bathymetry data as a starting point, we need to further promote the in-depth integration and interaction among S-100 products, in order to fully utilize the unique advantages of S-100 in enhancing navigation safety.

REFERENCES

Du J. (2024). *Research on high resolution dynamic depth model and service*. Tianjin: Tianjin University.

Gong Y. (2022). Overview of the 9th Session of the IMO's Sub -Committee on Navigation. Communications and Search and Rescue(NCSR), China Maritime Safety, 2022(08):62–64.

International Hydrographic Organization. (2023). Demonstrate interoperability of S-101 and S-102 on an S-100-compatible Electronic Chart System. https://iho.int/en/.

International Hydrographic Organization. (2023). Data Product Interoperability in S-100 Navigation Systems. Monaco: IHO.

International Maritime Organization. (2006). Report of the maritime safety committee on ITS eighty-second session. https://docs.imo.org/.

Kastrisios C., Sullivan B. and Powell J, *et al.* (2022). Hydrographic geospatial data standards. *The Geographic Information Science & Technology Body of Knowledge (2nd Quarter 2022 Edition)*.

Lee A. and Mcleay C. (2015). *S-100 Overlays: A Brave New World*, National Harbor: US Hydrographic Conference.

Luo J., Wan X. and Sun D., *et al.* (2019). S-100 hydrographic geospatial data model review. Bulletin of Surveying and Mapping, 2019(03):91–97+112.

Luo J., Geng X. and Sun D., *et al.* (2022). IHO S-100 5.0.0 New Feature Analysis and Development Suggestions. *China Standardization*, 2022(21):180–184.

Norway. (2016). *Norweigian Nonueigian S-102 Testbed*. Tokyo: IHO S-100 Working Grouping Meeting.

Thompson P. and Smith M. (2022). Delivering digitally enabled and autonomy-enabled smart port ecosystems using the new IHO S-100 framework model. *Australasian Coasts & Ports 2021 Conference*, 2022: 931-937.

Wan X., Luo J. and Wu J., *et al.* (2018). Key Technologies of the S-101 Electronic Navigation Chart VisualizationS-101. *Journal of Geomatics*, 2018,43(04):1–7.

Wu L., Huang Y. and Huang D., *et al.* (2023). Research on the situation and trend of S-100- -based products from the perspective of intelligent shipping. *Hydrographic Surveying and Charting*, 2023,43(03):11–15+25.

Xiong E and Wu J. (2021). Research on key elements of the revision of the IHO S-100 generic hydrographic data model. *Navigation*, 2021(02):39–42.

Proceedings of the International Association of Institutes of Navigation World Congress 2024 – Chen et al. (Eds)

Discussion on the application and development of Aids to Navigation technology in the future

Fawei Wu & Yuwei Qu
Tianjin AtoN Division, Tianjin, The China

ABSTRACT: Aids to Navigation are important marks and guides for navigation, playing a crucial role in ship navigation and maritime safety. With the development of technology, Aids to Navigation technology continues to be updated and developed. This article explores the future development directions of Aids to Navigation through research and analysis of Aids to Navigation technology and its applications. Firstly, it introduces the definition and classification of Aids to Navigation (key directions for the development of navigational aids include networked Aids to Navigation, intelligent Aids to Navigation, and green Aids to Navigation), and focuses on the current common Aids to Navigation technology. Next, it analyzes the development trends of Aids to Navigation technology, including the use of drones for Aids to Navigation inspections, the employment of artificial intelligence technology to enhance Aids to Navigation monitoring capabilities, and the integration of Aids to Navigation with other navigation systems. Finally, it discusses the challenges and strategies for the development of Aids to Navigation technology, as well as the vision and application of new types of Aids to Navigation (characteristics and advantages of advanced technologies such as radar Aids to Navigation, satellite Aids to Navigation, and unmanned Aids to Navigation). It is hoped that this article can provide a certain reference for Aids to Navigation management and technological research.

1 INTRODUCTION

Aids to Navigation are signs that indicate the direction, boundaries, and obstructions of a waterway, including river-crossing marks, coastal marks, leading marks, side marks, left-and-right Aids to Navigation, position marks, flood marks, bridge and culvert marks, etc. They are artificial signs that help guide ships to sail, locate, mark obstructions, and indicate warnings. Aids to Navigation technology refers to the technical means employed in the planning, distribution, setting, maintenance, patrol inspection, and management of Aids to Navigation. With the development of science and technology, the technology of Aids to Navigation is also constantly innovating. This paper aims to discuss the development and application of Aids to Navigation technology in the future and provides some reference for related industries.

2 DEFINITION AND CLASSIFICATION OF AIDS TO NAVIGATION

2.1 *Definition of Aids to Navigation*

Aids to Navigation are visual, audio, and radio-based facilities equipped with information-service functions set up to help ships navigate safely, economically, and conveniently. To

 DOI: 10.1201/9781003624424-17

realize the role of Aids to Navigation, they generally have four functions, namely, the positioning function, the danger warning function, the confirmation function, and the traffic indication function.

2.2 *Classification of Aids to Navigation*

There are various kinds of Aids to Navigation, accompanied by numerous classification methods. When classified according to the water area where they are distributed, Aids to Navigation can be divided into sea-area Aids to Navigation and those on inland waterways. According to the reliability classification of the distribution position, there are fixed marks and floating marks. However, the most commonly-used classification method is to classify them according to the working principle of Aids to Navigation.

When classified according to their working principles, Aids to Navigation can be grouped into visual Aids to Navigation, audible Aids to Navigation, and radio Aids to Navigation.

Visual Aids to Navigation include lighthouses, light beacons, beacons, leading marks, traffic signal marks, light vessels, light buoys, and buoys.

Audible Aids to Navigation include gas fog signals, electric fog signals, and fog detectors.

Radio Aids to Navigation include radar reflectors, radar responders, radio beacon systems, Loran C radio navigation systems, GNSS, VTS, AIS, and LRIT.

3 DEVELOPMENT TREND OF AIDS TO NAVIGATION TECHNOLOGY

At present, the development trends of Aids to Navigation technology are characterized by diversification. The degree of intelligence is steadily rising, the standardization is becoming increasingly stringent, and the degree of industrialization is growing higher. Environmental protection and sustainable development have also emerged as crucial development directions. These trends interact with each other, acting as a driving force for the development of Aids to Navigation technology, and provide more powerful technical support for the development of the shipping industry.

First of all, with the rapid development of the shipping industry today, Aids to Navigation technology is moving towards intelligence and multi-functionality to meet the requirements of an increasingly complex and changeable navigation environment for navigation-assistance services. It serves as the foundation for the development of Aids to Navigation technology, and is also a necessary prerequisite for meeting the growing shipping needs. Therefore, Aids to Navigation technology must be continuously upgraded and innovated to better meet this demand.

Secondly, Aids to Navigation technology is gradually moving towards standardization and industrialization, which exerts a positive effect on promoting the production efficiency and quality of Aids to Navigation equipment, reducing maintenance costs. Morever, it is conducive to the popularization and application of Aids to Navigation technology. Consistently, Aids to Navigation technology is constantly improving.

In addition, with the development of the Beidou satellite and other global navigation satellite systems, and the progress of Aids to Navigation technology and its increasingly extensive application, satellite-based Aids to Navigation technology has become increasingly mature and perfect. Boasting many advantages such as global coverage and high precision positioning, it has greatly improved the safety and operational efficiency of shipping routes. In the future, Aids to Navigation technology will be further optimized and improved to provide more reliable and efficient navigation services for the shipping industry. With the development of global navigation satellite systems such as the Beidou satellite system, it is fair to say that the application prospects of satellite-based Aids to Navigation technology in the shipping industry is extremely broad.

Finally, an important development trend of Aids to Navigation technology lies in environmental protection and sustainable development. As the world places increasing emphasis on environmental protection, it puts forward a new important direction for the green design of Aids to Navigation equipment and the research and development of energy saving and emission-reduction technologies. Therefore, the innovation of Aids to Navigation technology must take into account the harmonious coexistence with the environment to achieve sustainable development. This is also an important topic in the current technological development landscape.

3.1 *Application of Artificial Intelligence in Aids to Navigation monitoring*

The development of artificial intelligence technology has provided new means and new concepts for Aids to Navigation monitoring. Aids to Navigation monitoring realizes automation and intelligence through the application of artificial intelligence technology, improving the monitoring efficiency and accuracy.

3.2 *Unmanned Aerial Vehicle Aids to Navigation inspection technology*

The development of UAV technology provides new solutions to Aids to Navigation inspection. While avoiding the risks of operators in complex environments, the use of drones for Aids to Navigation inspection can improve the efficiency of inspection and reduce the waste of human resources.

3.3 *Digital application of Aids to Navigation management*

All relevant data of the Aids to Navigation, such as the monitoring of the operation status of the Aids to Navigation and the maintenance and management requirements are inputted into the system. The system combines the hydrometeorological information near the port to comprehensively analyze the various states of the Aids to Navigation during the on-the-job period, so as to provide more accurate reference information for the management personnel and improve the management ability.

3.4 *The integration of Aids to Navigation and other navigation systems*

One of the development trends of Aids to Navigation technology in the future lies in the integration of Aids to Navigation with other navigation systems. By integrating Aids to Navigation with GPS, radar, and other navigation equipment, it can not only improve navigation accuracy but also provide more navigation information and safety assurance.

4 DIFFICULTIES AND COUNTERMEASURES IN THE DEVELOPMENT OF AIDS TO NAVIGATION TECHNOLOGY

As an important part in the development of the navigation industry, Aids to Navigation technology has made great progress along with the continuous progress of science and technology. However, there are still some difficulties and challenges in the process of developing Aids to Navigation technology. To overcome these difficulties, we need to put forward corresponding countermeasures. Therefore, the in-depth research and development of Aids to Navigation technology remain of utmost importance.

4.1 *Difficulties in the development of aids to navigation technology*

4.1.1 *Technology updates quickly*

The Aids to Navigation technology is continuously updated to meet new navigation requirements and standards; at the same time, due to the rapid upgrading of technology, the existing Aids to Navigation equipment and technology may rapidly become outdated and thus need to be updated in time. Therefore, while Aids to Navigation technology is constantly being updated, it must also maintain the connection with the existing technology in terms of function and performance.

For example, in the current era of rapidly evolving navigation technology, traditional Aids to Navigation equipment may gradually see a decline in market share due to the increasing popularity of satellite navigation systems.

4.1.2 *High cost input*

The research and development, production, maintenance, and upgrading of Aids to Navigation technology require a large amount of capital investment, which may impose greater economic pressure on enterprises.

Especially during the new-technology research and development stage, the risk of cost input is relatively high because of the uncertainty of technology and market risks.

4.1.3 *Environmentally adaptable*

Sea-standard equipment is required to operate stably in various complex and harsh marine environments. Thus, it must possess high environmental adaptability.

The variability and complexity of the marine environment require the durability and reliability of Aids to Navigation equipment.

4.1.4 *High safety performance requirements*

Aids to Navigation equipment is an important facility for ensuring maritime traffic safety. Its safety performance has a direct relationship with ship navigation safety and maritime traffic safety.

Technical failures or potential safety hazards may have a serious impact on maritime traffic safety.

4.2 *Response measures*

4.2.1 *Increase investment in research and development*

To improve the advanced nature and competitiveness of Aids to Navigation technology, more research and development funds will be invested in the research and development of new technologies, new materials, and new processes.

Establish close cooperative relations with colleges and universities, scientific research units, etc., and jointly carry out new technology research and development while shortening the technology renewal cycle.

4.2.2 *Optimizing cost control*

Reducing production costs and improving economic benefits rely on technological innovation and process improvement.

Strengthen supply chain management to ensure stable supply and control raw material costs.

4.2.3 *Improving environmental adaptability*

By adopting corrosion-resistant materials and strengthening sealing, the environmental adaptability of Aids to Navigation equipment can be strengthened, and the weather resistance and reliability of the equipment enhanced.

Establish an environmental adaptability test system to ensure the stable operation of the equipment in various harsh environments, and conduct comprehensive environmental adaptability tests on Aids to Navigation equipment.

4.2.4 *Strengthen safety performance assurance*

Establish a perfect safety performance testing system for regularly testing and evaluating the safety performance of Aids to Navigation equipment.

To ensure the stable operation and long-term safe use of equipment, we have strengthened the maintenance and upkeep of the equipment.

Strengthen safety training and drills to enhance the ability of crew members to respond to emergencies.

4.2.5 *Strengthen international cooperation and exchanges*

To better understand the development trends and standard requirements of international Aids to Navigation technology, it is essential to strengthen cooperation and exchanges with international Aids to Navigation organizations and other countries.

To improve China's influence and discourse power in the field of international Aids to Navigation, we should participate in the formulation and revision of international Aids to Navigation technical standards.

4.2.6 *Promote industrial integration*

While forming a close cooperative relationship between the upstream and downstream of the industrial chain, we need to strengthen the integration and coordinated development of the Aids to Navigation industry.

The overall competitiveness and market share can be improved through industry consolidation.

By implementating the above countermeasures, we can provide a strong guarantee for maritime traffic safety, promote the continuous innovation and development of Aids to Navigation technology, and effectively address the difficulties faced by the development of Aids to Navigation technology.

5 IMAGINATION AND APPLICATION OF NEW AIDS TO NAVIGATION TECHNOLOGY

In recent years, with the rapid development of science and technology, the emerging technology of Aids to Navigation has gradually become the main development direction. Among these, wireless communication technology has witnessed an increasingly high penetration rate in the realm of Aids to Navigation. By using wireless communication technology, Aids to Navigation can communicate and transmit real-time information to ships, air traffic control centers, and other equipment, thereby improving navigation safety and efficiency. In addition, the satellite navigation system has ushered in new opportunities for the development of Aids to Navigation technology. When combined with the satellite positioning system, the positioning accuracy of Aids to Navigation has been significantly improved. The application of intelligent control technology in Aids to Navigation is also becoming more and more popular, which can realize the automatic management and remote monitoring of Aids to Navigation.

5.1 *Virtual aids to navigation technology*

Virtual Aids to Navigation technology broadcasts Aids to Navigation information through AIS base stations or electronic channel maps and provides electronic navigation services to ships. In the future, this technology will be further developed to achieve more intelligent and automated navigation services.

AIS virtual Aids to Navigation: AIS base stations are employed to broadcast Aids to Navigation information. This approach is suitable for coastal waterways and river estuaries and other waters. Moreover, it provides ships with accurate navigation information and reduces navigation risks.

Electronic Channel Map Virtual Aids to Navigation: The information of the Aids to Navigation is marked on the electronic channel map, and then released to users through the update of the electronic channel map data. This method is suitable for inland waterways and other areas to improve the efficiency and accuracy of waterway management.

5.2 *Intelligent aids to navigation system*

This intelligent Aids to Navigation system can monitor the operation status of the Aids to Navigation in real time and automatically adjusts the relevant parameters, so that the brightness and other conditions of the Aids to Navigation can adapt to the environmental conditions and navigation requirements. In addition, the system can also predict channel conditions based on data analysis and give more accurate navigation suggestions to ships. Through the above functions, this intelligent Aids to Navigation system can not only provide a more effective guarantee for the safe navigation of ships but also improve the efficiency of usig Aids to Navigation.

Real-time monitoring: The system conducts real-time monitoring and detection of the relevant status of the Aids to Navigation. By being integrated with corresponding sensors and other equipment, it ensures the normal operation of the Aids to Navigation. To enable the Aids to Navigation to fullfill its function to the maximum extent, itt can be tracked and detected in real time from all aspects.

Automatic adjustment: The brightness/frequency and other parameters of the Aids to Navigation can be automatically adjusted according to different environments and navigation requirements to improve the visibility and recognition of the Aids to Navigation, thus making sailing more convenient and safe.

Data analysis: By using technical means, the system collects, processes, and analyzes data about waterways to predict waterway conditions and provide ships with more accurate navigation suggestions and reduce navigation risks.

5.3 *Technology of unattended aids to navigation*

The unattended Aids to Navigation technology realizes unattended operation, and the management of Aids to Navigation is automated through remote monitoring, automatic control, and other technologies. In the future, this technology will be further developed to make the management of Aids to Navigation more efficient and reliable.

Remote monitoring: Remote monitoring of Aids to Navigation is realized through satellite communication, the internet, and other technologies. This allows problems to be detected and dealt with in time, while also enabling real-time tracking of the status of Aids to Navigation.

Automatic control: The functions such as the automatic switching and brightness adjustment of Aids to Navigation are realized through automatic control technology, which reduces human interference and improves management efficiency.

Fault warning: Sensors and other equipment are employed to conduct real-time monitoring of the status of Aids to Navigation. Once a fault or abnormal situation is detected, early warning information is immediately issued to alert management personnel, enabling them to take timely measures to deal with the problem.

5.4 *Environmental protection and energy saving aids to navigation*

Environmental protection and energy-saving Aids to Navigation adopt new materials and energy-saving technologies to reduce energy consumption and minimize environmental pollution. In the future, these types of Aids to Navigation will become one of the important directions for the development of Aids to Navigation technology.

New materials: The new materials with light weight, corrosion resistance, and good weather resistance are used to manufacture Aids to Navigation, which reduces the weight and maintenance costs of Aids to Navigation.

Energy-saving technology: LEDs and other energy-saving light sources and intelligent control technology are used to reduce energy consumption and light pollution of Aids to Navigation.

Green energy: Exploring the application of green energy sources such as solar energy and wind energy to supply power for Aids to Navigation is an important step towards reducing dependence on traditional energy and minimizing environmental pollution. In addition, the tidal energy and wave energy in the sea are extremely abundant, and cannot be put into use on a large scale due to technical constraints. In the future, with the development of technology, the cost of generating electricity from tidal and wave energy will decrease. The Aids to Navigation energy system will integrate the advantages of various energy sources, creating a complementary system that can provide a steady stream of electric energy to the Aids to Navigation. This will allow the Aids to Navigation to function at their highest efficiency levels.

In short, the new Aids to Navigation technology will develop in the direction of intelligence, automation, unattended, environmental protection, and energy saving, providing ships with safer, more efficient, and more accurate navigation services.

6 RESULTS AND DISCUSSIONS

The discussion of the future development direction and application prospect of the Aids to Navigation technology makes everyone see that it has great potential in the global satellite navigation system and green environmental protection intelligent Aids to Navigation technology. It can also play an important role in promoting the Aids to Navigation technology in the continuous development and progress of the navigation field. Therefore, I hope that this research can provide some insights and references for relevant practitioners, so that the development and application of the Aids to Navigation technology can be further promoted. At the same time, it is also hoped that this research result can play a more important role in future research and application.

REFERENCES

China Maritime Safety Administration (2023): *Aids to Navigation Management*. Beijing: People's Communications Press Co., Ltd, 315–324.

Wang Yingzhi (2021): *Aids to Navigation*. Dalian: Dalian Maritime University Press, 3–18.

Proceedings of the International Association of Institutes of Navigation World Congress 2024 – Chen et al. (Eds)

Application concept of "Marine buoys crowdsourcing wave monitoring" in the context of contemporary wave monitoring

W. Zhu & J.X. Cao
Guangzhou Aids to Navigation Department, Southern Navigation Service Center; Southern Navigation Service Center, Guangzhou, Guangdong, China

S. Zhao
Southern Navigation Service Center, Guangzhou, Guangdong, China

Y. Li
Guangzhou Aids to Navigation Department, Southern Navigation Service Center; Southern Navigation Service Center, Guangzhou, Guangdong, China

ABSTRACT: Traditional marine monitoring has problems such as few monitoring points, expensive equipment investment, high maintenance cost and vulnerability. Considering marine buoys are characterized by wide coverage, dense deployment and small maintenance cost, the author proposes "marine buoys crowdsourcing wave monitoring" application concept, which is categorized into two modes. One is "big data wave monitoring mode" and the other is "instantaneous wave monitoring mode". Completely based on the database of the existing buoy remote control and monitoring system, the "big data wave monitoring" optimizes the transmission mode and data, and checks the change of waves around the buoys, so as to generate a big wave database for the whole region. The "instantaneous wave monitoring mode" monitors the wave transient changes by setting the threshold value through the accelerometer data changes built into the lightship. This concept is using existing navigational resources for wave monitoring, through the remote control and monitoring system and lights built-in accelerometer to collect data to realize the "big data wave monitoring" and "instantaneous wave monitoring". Therefore, the "underwater" monitoring is changed to "on-water" monitoring, which reduces the monitoring cost and strengthens the temporal and spatial resolution of the data. By constructing the database of "marine buoys crowdsourcing wave monitoring", an accurate wave model is built, which can be used to provide information for the construction of ports, navigation safety, route planning, and environmental protection.

1 INTRODUCTION

Marine and river monitoring faces many difficulties, such as complex marine environment, large input costs, high technical requirements, vast sea areas, weather conditions, and complex data processing and analysis. Using the existing buoy resources for wave monitoring, the data collected by the remote control and monitoring system and the accelerometer built in the lamp can be used to build a database of "floating crowdsourcing wave measurement", which can realize "crowdsourcing monitoring". Compared with the limited floating sign collection devices of various monitoring departments or institutions, relying on a large number of piggyback platforms and data collection modules of tens of thousands of coastal and inland waterway beacons, it is able to greatly improve the data coverage and accuracy. It can also enhance the detail of wave characteristics and improve wave forecasting and early warning capabilities. This also can improve disaster prevention and risk

DOI: 10.1201/9781003624424-18

management capacity. This will also help us obtain accurate, timely and comprehensive data to improve our capacity for early warning and emergency response to Marine disasters. It also helps protect people's lives and property, promote scientific and technological innovation, and promote business development.

2 AIDS TO NAVIGATION (ATON) DEVELOPMENT

AtoN can provide positioning and navigation services for ships, which plays an irreplaceable role in improving the efficiency of water transportation and ensuring the safety of maritime traffic. The role of AtoN in water traffic is becoming more and more prominent, and the management of AtoN is becoming more and more standardized. The AtoN management departments shall make plans for regular inspection of the AtoN system, and shall be equipped with professional personnel for inspection and maintenance to ensure the normal operation of the AtoN equipment and systems.

In 2022, China MSA system managed a total of 19,214 coastal AtoN. In 2023, China Maritime Safety Administration managed a total of 20,600 coastal AtoN, which doubled the number compared with 2012. The construction of t remote control and monitoring information for AtoN is improving day by day. AtoN management personnel promote the transformation and upgrading of navigation service work through the remote control and monitoring construction of AtoN, and provide important technical support for the transformation of AtoN management work from a traditional industry to an intelligent industry. Multifunctional AtoN has been upgraded. Conventional marine monitoring buoys can be properly modified to conduct data monitoring safely and economically. The AtoN management department has implemented the development concepts of "marking position exchange" and "navigation protection + integrated development". By opening up legal AtoN placement positions to cooperative government, industry, academia and research organizations, traditional buoys are replaced with multifunctional AtoN jointly constructed with cooperative organizations.

3 BASIN-WIDE WAVE MONITORING

The construction of basin-wide wave monitoring is of great significance and necessity to ensure navigation safety, marine safety, harbor terminal construction, support economic development, promote scientific research and environmental protection. However, there are many problems in the construction of basin-wide wave monitoring.

3.1 *Wide coverage*

In international geography, the China Sea is bounded by the Sea of Japan to the north and the Philippine Sea to the east. It is bounded by the Java Sea in the south and the Myanmar Sea in the west. The total area of the China Sea is 4.93 million square kilometers, of which, China's territorial sea (sea under its jurisdiction) is about 3 million square kilometers. China is one of the world's densely riverized countries, with major rivers including the Yangtze River, Yellow River and Pearl River, covering extensive watershed areas. The Yangtze River is the longest river in China, with a length of about 6,300 kilometers, and the Yellow River is the second longest river in China and one of China's mother rivers, with a length of about 5,464 kilometers. The Pearl River is the largest river in southern China with a total length of about 2,300 kilometers. Basin-wide wave monitoring needs to cover a large area of sea or river, involving a large number of monitoring stations, and requiring the investment of countless equipment and manpower costs to realize real-time monitoring.

3.2 *Difficulty of data collection*

In a wide range of watersheds, wave direction is affected by a variety of factors, including topography, ocean climate, tides, etc., making data collection difficult. At the same time, wave characteristics may differ in different regions, and monitoring means and methods adapted to different regions need to be considered. At the same time, there is a need to provide a more stable mounting platform for data collection.

3.3 *Difficulties in data transmission and processing*

Real-time monitoring requires the timely transmission and processing of large amounts of wave data, involving the construction and operation of data transmission networks and real-time processing systems. Ensuring the timeliness and accuracy of data is a challenge. For applications that require real-time wave information, such as ocean engineering and offshore operations, the real-time requirements for wave monitoring data are critical. This requires the monitoring system to have the ability to respond quickly and transmit data in a timely manner.

3.4 *Difficulty in equipment maintenance and management*

Deploying a large number of monitoring equipment throughout the basin requires regular maintenance and management to ensure the normal operation of the equipment and the continuity of data collection. Equipment failure or damage may affect the normal operation of the real-time monitoring system. Monitoring equipment is usually installed at sea or in remote areas, such as shallow or deep sea areas. Installation, servicing and maintenance of the equipment requires the support of vessels, divers or special tools. These locations add to the complexity and cost of maintenance.

3.5 *Duplication of construction in multiple sectors and high capital investment*

The establishment of a basin-wide real-time wave direction monitoring system requires a large amount of capital investment for equipment acquisition, data processing, network construction, etc. Meanwhile, the maintenance and updating of the equipment as well as the long-term operation of the system also require continuous financial support. Funding sources and sustainability are a challenge. Multisectoral research may lead to problems such as duplication of resources, duplication of equipment and cross-mobilization of manpower, wasting social resources and reducing overall effectiveness. The lack of overall planning and unified scheduling leads to uneven and inefficient allocation of resources.

4 THE CONCEPT OF CROWDSOURCING MONITORING

With globalization, crowdsourcing services have become an important business model. In the era of globalization 3.0, individuals are able to participate in global cooperation and competition, and crowdsourcing is emerging as a form. There is a fundamental difference between outsourcing and crowdsourcing, with outsourcing focusing on high specialization while crowdsourcing promotes cross-disciplinary innovation. This eclectic approach often holds great creative potential. Many successful business cases have benefited from the active participation of individual users, reflecting the core concept of crowdsourcing-creating value with users. By adopting crowdsourcing, significant monetary costs can be saved and more efficient resource utilization can be achieved. The advantages of crowdsourcing are as follows.

4.1 *Large-scale participation and collaboration*

Basin-wide wave monitoring needs to cover a large area of sea or river, involving more monitoring stations and requiring more investment in equipment and labor costs to achieve real-time monitoring. Realizing basin-wide coverage of wave monitoring requires cross-sectoral collaboration and information sharing. The establishment of unified monitoring standards, data interfaces and information platforms promotes the integration and sharing of data between different sectors and improves the overall efficiency and effectiveness of the monitoring system.

4.2 *Diversity and innovation*

Crowdsourcers come from different backgrounds and professional fields, and they provide rich and diverse perspectives and creativity, which can bring new solutions and innovative ideas. Through crowdsourcing, the wisdom and creativity of crowdsourcers can be fully utilized to open up the boundaries of thinking.

4.3 *Rapid response and scalability*

Through crowdsourcing, it is possible to respond quickly to changes in demand and expand human resources according to specific needs. The crowdsourcing platform can quickly mobilize and organize a large number of crowdsourcers to speed up the completion of tasks and go-live time.

5 APPLICATION CONCEPT OF "MARINE BUOYS CROWDSOURCING WAVE MONITORING"

Under the existing marine monitoring conditions, in order to make up for the inconvenience caused by fewer deployment points, high investment in monitoring equipment, high maintenance costs, and easy failure of underwater equipment, the application concept of "crowdsourcing wave measurement on buoys" is proposed. Through the buoy uplighter real-time data, historical data, accelerometer data, expanding the beacon function and enriching the means of wave and tidal current monitoring, which provides technical support for the safety of ship navigation, port and terminal engineering construction, marine geological changes, earthquake prevention and disaster mitigation, large database of ocean waves, scientific research and so on.

The exponential increase in the number of monitoring points through the use of buoys for full coverage of data collection by the navigational safety, waterway department and owner's department can increase the coverage and accuracy of the data. Increasing the number of monitoring points can provide a more comprehensive coverage of wave changes in the target area, thus improving the spatial resolution and accuracy of the monitoring data. This contributes to a more accurate understanding and prediction of the spatial distribution characteristics of waves. It can also enhance the detail of wave characteristics. Multi-point monitoring can capture the changes in wave characteristics at different locations and under different sea conditions, including the spatial and temporal patterns of wave height, period, wavelength and other parameters, providing more detailed data support for ocean engineering and maritime activities. It can enhance wave forecasting and early warning capability. Through multi-point monitoring, the development trend and change of waves can be monitored and analyzed in real time, improving the accuracy and timeliness of wave forecasting and providing important information for marine safety and maritime management. It can support scientific research and engineering design. The richness and diversity of multi-point monitoring data help scientists and engineers to study the formation mechanism and evolution law of waves in depth. This can also provide scientific basis for marine engineering design and environmental assessment, and enhance the ability to respond to emergencies. In the event of marine disasters and emergencies, multi-point monitoring data can provide more information and perspectives to help the relevant departments to make timely responses and countermeasures, which can not only safeguard public safety and resource protection, but also support resource management and environmental protection. Multi-point monitoring data can provide comprehensive data support for marine resource management and ecological environmental protection, and help to assess the impacts and risks of human activities on marine ecosystems.

Taking the jurisdiction of the Southern Navigation Service Center as an example, there are 2,637 floating buoys along the coast, main channel and port terminals (as shown in Figure 1. Schematic diagram of some of the AtoN in the South China Sea jurisdiction), including 2,631 buoys and 6 lightships. Although the above data does not include marine monitoring buoys, inland waterway floating buoys, military buoys and fishing buoys, it has the advantages of large number, wide distribution, dense deployment, small investment and small maintenance cost compared with professional marine monitoring buoys.

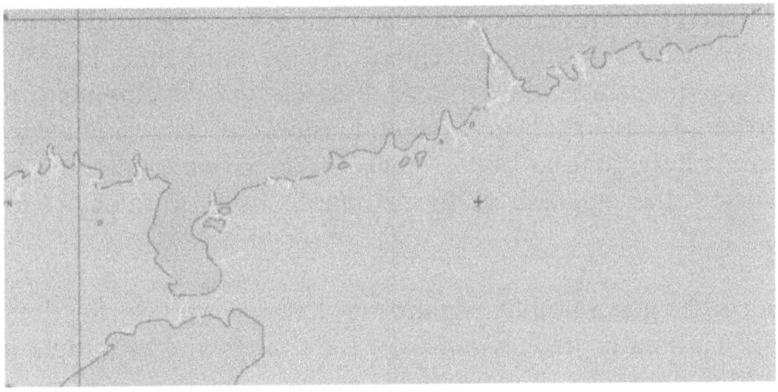

Figure 1. Schematic diagram of some of the AtoN in the South China Sea jurisdiction.

"Marine buoys crowdsourcing wave monitoring" application concept is categorized into two modes. One is "big data wave monitoring mode" and the other is "instantaneous wave monitoring mode". Completely based on the database of the existing buoy remote control and monitoring system, the "big data wave monitoring" optimizes the transmission mode and data, and checks the change of waves around the buoys, so as to generate a big wave database for the whole region. The "instantaneous wave monitoring mode" monitors the wave transient changes by setting the threshold value through the accelerometer data changes built into the lightship. The details are as follows.

5.1 *Big data wave monitoring mode*

Based on the upgrading and secondary development of the historical track module and rotary track module in the AtoN remote control and monitoring system, the "big data wave monitoring mode" standardizes the time and interval of transmitting back the latitude and longitude data of floating markers. The returned latitude and longitude data are displayed on the geographic map. Analyzing the data changes according to the time interval, it can help draw geo-visualized wave direction map, wave distribution map, wave animation and wave heat map.

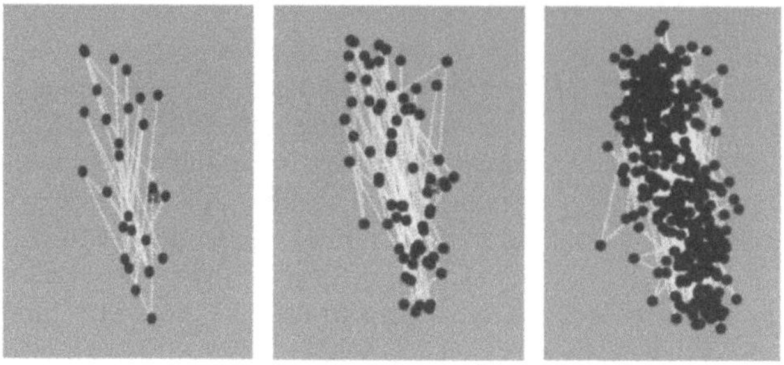

Figure 2. Overlay display of buoy position data.

5.1.1 *Wave direction map*

Based on the monitoring data in the database, an image showing the direction of the waves can be generated. This typically uses arrows or symbols to indicate the direction of the waves and may use colors to indicate the intensity of the waves.

5.1.2 *Wave distribution maps*

By integrating monitoring data from a database, it is possible to generate images showing the distribution of waves in an ocean area. Such images can be used to show the height or energy distribution of waves in different regions. At the same time, in order to improve the alarm time, beacon management department research and development of marine beacon collision early warning lamps. Built-in three-dimensional acceleration sensors in the beacon light apparatus can collect float collision signals, so as to accurately and sensitively, timely and efficiently transmit the signals back to the remote monitoring device for early warning.

5.1.3 *Wave animation*

Waves animation can be generated by organizing the wave monitoring data in the database in a time series. Such animated images can show the change and evolution of waves and help people understand the dynamics of waves.

5.1.4 *Heat maps of waves*

Using the monitoring data from the database, it is possible to generate a heat map of the waves. Such pictures provide a more visual way of understanding the distribution of waves by using different colors or gradients to represent the energy or height of the waves.

5.2 *Instantaneous wave monitoring mode*

As globalized trade accelerates and the shipping industry continues to grow, maritime traffic becomes busier and the number of ship collisions with buoys increases. In order to recover the loss of national property, the beacon management department researches a set of methods to trace the ship involved in the accident. At the same time, in order to improve the alarm time, AtoN management department research and development of marine beacon collision early warning lamps. Built-in three-dimensional acceleration sensors in the AtoN light apparatus can collect float collision signals, so as to accurately, sensitively, timely and efficiently transmit the signals back to the remote monitoring device for early warning. The application monitors changes in the motion of the buoy caused by waves, especially vertical acceleration (vertical up and down motion of the buoy) and horizontal acceleration (horizontal motion of the buoy). The accelerometer data are analyzed and evaluated for wave monitoring wave height, wave period, wave direction determination, wave currents, and generate time series plots, spectral analysis plots, wave rosette plots, and three-dimensional waveform diagrams.

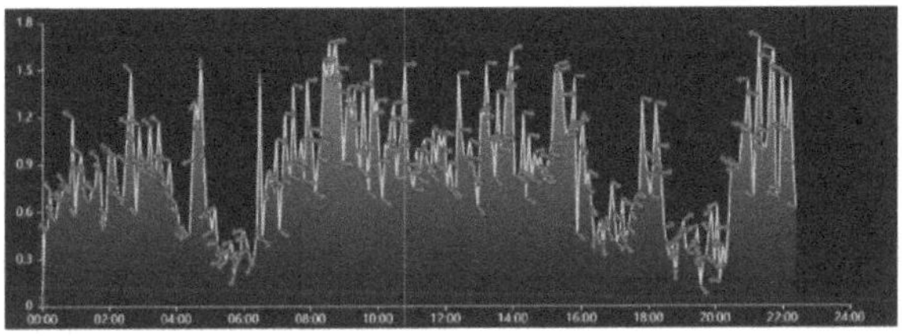

Figure 3. Instantaneous change of waves.

5.2.1 *Measuring wave height*

Acceleration data can be used to calculate the wave height, which is achieved by the vertical motion of the buoy. The difference between the crest and trough of the wave reflects the wave height.

5.2.2 *Wave period analysis*

The frequency of the wave can be calculated from the acceleration changes and thus the wave period can be analyzed.

5.2.3 *Wave direction determination*

If the buoy is equipped with a multi-axis accelerometer, the wave direction can be inferred by measuring the acceleration changes in different directions.

5.2.4 *Current assessment*

In addition to wave-induced motions, accelerometers can be used to assess potential current effects through cumulative analysis of data over time.

5.2.5 *Time series plot*

The displayed changes in buoy acceleration over time visualize the period and other characteristics of the wave.

5.2.6 *Spectrum analysis map*

Through the Fourier transform and other methods, the distribution map of wave frequency can be obtained, which is often used to study the wave energy distribution at different frequencies.

5.2.7 *Wave rose diagrams*

If wave directions can be analyzed from acceleration data, wave rose diagrams can be constructed, which show the statistical importance of waves from different directions for a particular region.

5.2.8 *Three-dimensional waveforms*

The use of three-dimensional graphics to represent the motion of a buoy as it changes with waves can provide visual information about wave dynamics.

5.3 *Database of "Marine buoys crowdsourcing wave monitoring"*

After a large amount of data has been collected, a series of further processing and analysis steps need to be completed to ensure that the value of the data is tapped. First, this massive amount of data needs to be cleaned and validated to remove errors and biases. Next, the data needs to be processed for consistency so that data from different sources and formats can be integrated. Subsequently, the data is transformed into useful information through data analysis and modeling, while wave models may need to be formulated or optimized to improve the accuracy of predictions.

6 "MARINE BUOYS CROWDSOURCING WAVE MONITORING" DATABASE FOR MARINE APPLICATIONS

Through year-round observation of wave changes in the whole basin or a certain region, a variety of databases and models can be generated, which is of great significance to safe navigation. Thus, wave statistical databases can be built to collect long-term wave observation data and establish statistical databases of wave parameters such as wave height, wavelength and speed. These databases can reflect the seasonal changes, inter-annual changes and long-term trends of waves in the region, and provide basic data for wave forecasting and engineering design. Moreover, wave forecasting models can be built, and based on long-term observation data, wave forecasting models can be developed, such as those based on statistical methods, numerical simulations, or machine learning algorithms. These models can predict wave conditions in the next few hours to days, providing time-sensitive wave warnings and safety recommendations for ship navigation, marine engineering construction, etc. Moreover, wave climatology models can be built, which are based on long-term wave observation data and analyze the climatic characteristics and long-term trends of waves in the region. These models can help to understand the impacts of climate change on waves, and to predict and assess wave changes over the coming decades and beyond. In addition, models of the relationship between waves and environmental parameters can be built, and by correlating and analyzing long-term observed wave data with other environmental parameters (e.g., wind fields, tides, currents, etc.), models of the relationship between waves and environmental parameters can be built. These models contribute to an in-depth understanding of the wave formation mechanism and its complex interactions, which allows for a comprehensive assessment of the impacts of marine ecosystems and human activities. Besides, storm surge and wave hazard assessment models can be built, and storm surge and wave hazard assessment models can be developed based on long-term observation data. These models can predict wave changes under extreme weather conditions and assess the impact of storm surges on coastal areas and harbors, thus providing a scientific basis for disaster prevention and mitigation.

The wave monitoring database has an important impact on navigation, which can improve the safety of ship navigation, navigation efficiency and port management, and also provide key support for emergency rescue work. Therefore, it is of great significance to obtain and effectively utilize the wave monitoring data in time to ensure the safety and smoothness of maritime navigation.

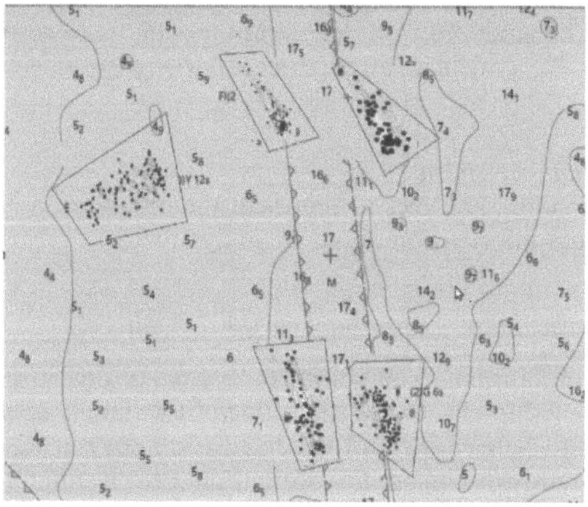

Figure 4. Heat map of regional crowdsourced wave observation data.

6.1 *Navigation safety*

The strength, direction and frequency of sea waves have a significant impact on the navigational safety of ships and navigational vehicles. By monitoring the database of wave changes, it can provide crews and navigation managers with timely and updated information on sea conditions to help them make appropriate navigational decisions, and avoid encountering dangerous sea conditions to ensure the safety of vessels and personnel.

6.2 *Route planning*

Accurate monitoring of wave data can help navigators conduct more effective route planning. Understanding the changes in waves can help to select safer and more efficient routes, reducing the risk of damage to the ship in bad sea conditions and improving navigation efficiency.

6.3 *Port management*

Data from the wave monitoring database is also important for port management. Understanding the change of waves can help the port management department to make corresponding scheduling and arrangements to ensure the safety of ships entering and leaving the port. It can also help them to rationally allocate the berthing positions of vessels and reduce the risk of damage to vessels and terminal facilities caused by the influence of waves.

6.4 *Emergency relief*

The accuracy and timeliness of wave monitoring data is crucial for emergency rescue work. In the event of an emergency at sea, knowing the current wave conditions can help rescue organizations develop more effective rescue plans and improve the success rate of rescue.

7 CONCLUSION

Crowdsourcing wave monitoring through telemetry and acceleration data from a large number of buoys can greatly enhance the coverage and frequency of wave data collection, reducing the research cost and increasing the community's awareness of marine environmental protection through public participation. This approach not only enhances the spatial and temporal resolution of the data, but also promotes the development of more accurate wave models and marine weather forecasts. This provides valuable data support and a basis for decision-making in a number of important areas, such as marine scientific research, marine resource management, disaster warning and navigation safety.

REFERENCES

Fu Q. (2012). Study on sea ice variation in key navigation areas of the Northwest Passage in the Arctic [D]: *[Master's thesis]. Dalian: Dalian Maritime University.*

Liu M., Zhao Q. W. and Wang M. S. (1997). Research on the alarm system of navigation beacon departure [J]. *Journal of Dalian Maritime University. 1997,(1)*.62–64.

Li Z. F., Wang W. Y., You X., *et al.* (2015). Fuzzy comprehensive evaluation of the strategic environment of China's Arctic shipping routes [J]. *Navigation of China, 2015, 38(3)*: 126–130.

Ren Y. F., Xie H. Z. and Zhang S. X. (2000). Research on inverse GPS Differential (IDGPS) positioning technology [J]. *Global Positioning Systems, 2000,(2)*.28–31.

Zheng, C. W., Li C. Y. and Li X. Q. (2016). Spatial and temporal characteristics of wind, surge and mixed waves in the Indian Ocean [J]. *Journal of PLA University of Science and Technology (Natural Science Edition), 2016, 17(4)*: 379–385.

Zheng C. W, Li S. X., Wang G., *et al.* (2011). Spatial distribution characteristics of wind, swell and mixed wave periods in the South China Sea. *Science and Wealth*, *7*, 253.

Zheng, C. W. and Li X. Q. (2011). Evaluation of ocean wave energy resources in China in recent 22 years based on WAVEWATCH-III model [J]. *Journal of Ocean University of China (Natural Science), 2011, 41 (11)*: 5–12.

Zheng, C. W., Pan, J. and Li, J.-X. (2013). Assessing the China Sea wind energy and wave energy resources from 1988 to 2009. *Ocean Engineering*, *65*, 39–48.

Zheng C. W., Pan J., Tian Y. Y., *et al.* (2012). Atlas of Wind, swell and Mixed waves in Global Seas. *Ocean Press, Beijing*.

Zheng, C. W., Zhou L. and Zhou L. J., (2011). Seasonal variation of wave and wave energy in Xisha and Nansha waters [J]. *Advances in Marine Science, 2011, 29(4)*: 419–426.

Proceedings of the International Association of Institutes of Navigation World Congress 2024 – Chen et al. (Eds)

Design and realization of bathymetry monitoring system based on CSB technology

C.S. Qian
Southern Navigation Service Center, Maritime Safety Administration, People's Republic of China, Guangzhou, Guangdong, China

S.X. Xian
Guangzhou Maritime Hydrographic Center, Southern Navigation Service Center, Maritime Safety Administration, People's Republic of China, Guangzhou, Guangdong, China

ABSTRACT: Crowdsourced bathymetry is a technique proposed and promoted by the International Hydrographic Organization (IHO) for the collection and sharing of depth measurements from vessels using standard navigation instruments. It is primarily utilized to gather bathymetric data in unsurveyed coastal waters and to establish a reference for safe ship navigation. With the advancement of communication technology and bathymetry automation processing, the concept of utilizing crowdsourced bathymetry to monitor crucial areas is becoming feasible. This paper introduces a system designed and developed based on traditional crowdsourced bathymetry technology, 4G network communication, and bathymetry automatic processing. The system is capable of acquiring water depth in quasi-real time, featuring a data collector for real-time storage and transmission, along with bathymetry filtration, correction, and processing procedures. The processed bathymetry identifies shallow points by comparing with historical data and issues timely warnings for shallow areas. This system complements traditional hydrographic surveys, addressing their long bathymetry cycle and slow data updates, and offers a new perspective for crowdsourced bathymetry technology development

1 INTRODUCTION

Crowdsourced bathymetry (CSB) is the collection and sharing of depth measurements from vessels using standard navigation instruments during routine maritime operations. The International Hydrographic Organization (IHO) established the Crowdsourced Bathymetry Working Group (CSBWG) in 2015, which has made significant contributions to advancing research on crowdsourced bathymetry. The group has developed Guidance to Crowdsourced Bathymetry(B-12), which was approved and adopted by the 14th meeting of the Inter-Regional Coordination Committee in June 2022. The Guidance to Crowdsourced Bathymetry aims to provide countries with instructions on conducting crowdsourced bathymetry data collection and quality assessment, as well as on establishing a global bathymetry database.

Traditional crowdsourced bathymetry typically involves using USB storage devices sent by mail, manual copying after docking, or transmitting data through expensive satellite networks. This process is slow in updating and requires manual data processing, only meeting the demand for filling the bathymetry of unmeasured areas. With the rapid development of 4G/5G networks and bathymetry automatic processing technology, it is now

DOI: 10.1201/9781003624424-19

possible to utilize crowdsourced bathymetry data for timely analysis and monitoring of changes in the bathymetry of port waterways along the route. This enables the timely detection of abnormal changes in the bathymetry of the area. The crowdsourced bathymetry monitoring system developed in this study collects real-time positioning and bathymetry data from volunteer ships via a data collector equipped with a 4G network. These data are then transmitted back to the server for automatic processing to obtain quasi-real-time bathymetry of the target area. By analyzing changes in the bathymetry of the target area, the system achieves the goal of bathymetry monitoring and issuing warnings for shallow spots.

2 DESIGN OF THE OVERALL ARCHITECTURE

Crowdsourced bathymetry monitoring system follows the international standard of crowdsourced bathymetry and S-100 series standard. It aims to provide real-time bathymetry monitoring and establishes a fully automated system from data collection, transmission, processing, quality control to the application of crowdsourced bathymetry data. This is achieved in accordance with the requirements of standardization, real-time processing, structured data, and man-machine compatibility to offer a novel approach to bathymetry monitoring. The overall architecture design of the system is illustrated in Figure 1.

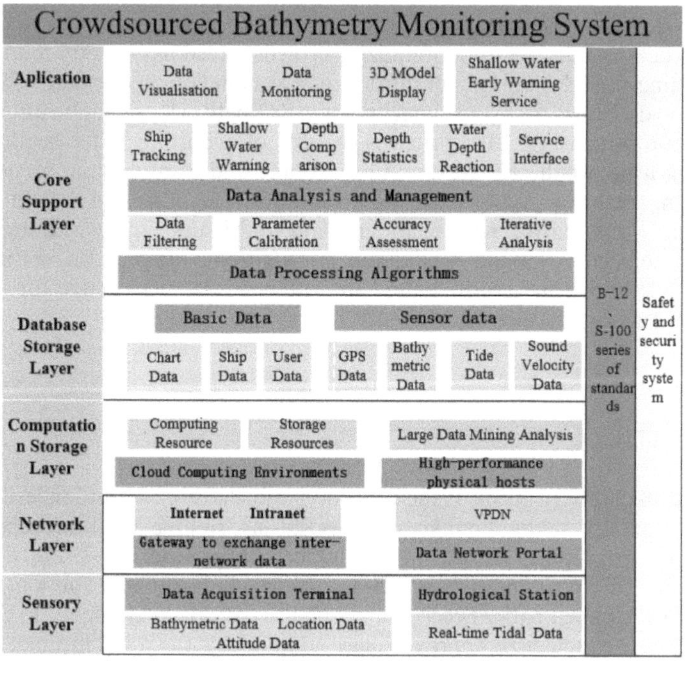

Figure 1. The overall architecture of crowdsourced bathymetry monitoring system.

3 DESIGN OF THE DATABASE

3.1 *Overall design of the database*

This database establishes multi-source correlation among the spatial position data of the crowdsourced bathymetry, the file data of the original output of the crowdsourced bathymetry, the ship's navigational track data, and other data associated with the crowdsourced bathymetry data. This correlation enables rapid retrieval of information, extraction of value, and enhances data analysis and statistics. Figure 2 illustrates the schematic diagram of data flow.

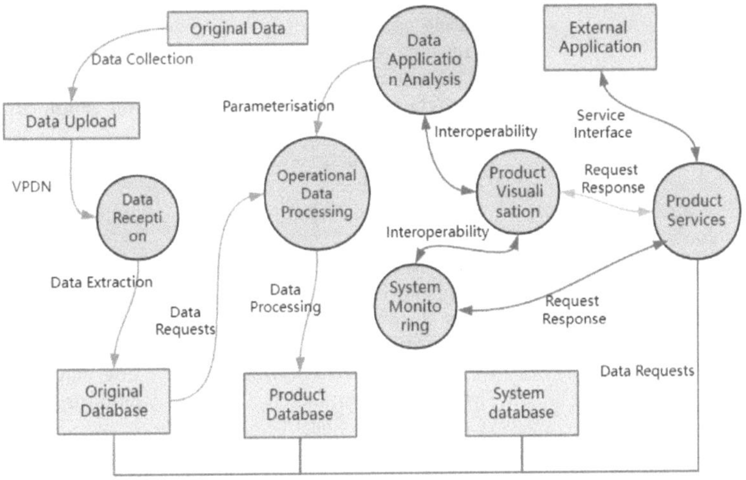

Figure 2. Data flow schematic.

3.2 *Technical route of the database*

The primary focus of research and management in database technology is data.The key components of database technology include organizing and managing data consistently, establishing databases and data warehouses based on specified structures, utilizing database management systems and data mining systems to develop applications for data management and data mining. These applications enable various functions such as adding, modifying, deleting, processing, analyzing, reporting,and printing data. Lastly, application management systems are employed to process and analyze data.

3.3 *Data structure*

The database is designed to capture all data attributes while minimizing storage space. In response to the crowdsourced bathymetry monitoring needs, Tables 1 and 2 outline the database tables and their primary attribute fields. Detailed descriptions of the attribute fields for each data table will follow.

Table 1. The database list.

Database Type	Data Type	Type of Information Sheet	Attribute
Original Database	Ship Data	Basic Static Information of Ship Sheet	In_ship_conf
		Parameter Configuration of Ship Sheet	In_ship_conf_data_thu
		Bathymetric Data of Ship Sheet	In_ship_data
	Raw Observational Data	Depth File Sheet	In_ship_depth_file
		Tide Data Sheet	in_tide_data

(continued)

Table 1. Continued

Database Type	Data Type	Type of Information Sheet	Attribute
Product Database	Record of Results	Precision Assessment Exception Sheet	out_abnormal_data
		Precision Assessment Normal Intersection Sheet	out_normal_data
		Bathymetric Point Cloud Results Data Sheet	out_point_cloud
		Ship Comprehensive Error Parameter Sheet	in_ship_error_param
	Water Environment Basis Data	Electronic Chart Bathymetric Data Sheet	ngtaisen
		Tidal Information Sheet	in_tide_basic
		Ship's AIS Track Sheet	in_ship_ais
	Monitoring Data	Ship Receiving Document Monitoring	Log_ship_monitor_detail
		Waterway Coverage Monitoring	Log_channel_cover_detail
		Monitoring Parameter Configuration Sheet	monitor_conf
		Machine Resource Monitoring	log_hardware_monitor
System Database	User Data	User Sheet	Sys_user
	Operation Log	Operation Log Sheet	log_syslog

Table 2. The main attribute of database sheet.

Database Type	Information Sheet Type		Main Attribute
Data of Survey Ships	Basic Static Information of Ship		Ship name, ship MMSI, ship length, ship breadth, draught, ship type, ship owner's company, bathymetric accuracy class, etc;
	Ship Dynamic Data		Ship name, ship type. Vessel MMSI, Vessel Coordinates, Speed to Ground, Heading to Ground, Bow Direction, Update Time, etc;
	Ship Track Data		Ship name, ship MMSI, ship track spatial data, date and time, etc;
Water Environment Basis Data	Electronic Chart Bathymetric Data		Coordinates, bathymetry value, time of measurement, map number to which it belongs, etc;
	Water Grid Data		Water grid name, number, measurement accuracy level, spatial data, etc;
	Isobath Data		Number, spatial data, time, shallow values, deep water values, etc;
Bathymetric Data	Bathymetry File Data		File name, upload username, upload time, IMO number, data time, data start time, data end time, etc;
	Bathymetry Data		IMO number, position, bathymetry value, time of data acquisition, file ID, type of file, measurement accuracy, version number, etc;
	Bathymetry Warning		Bathymetry value, sound value in chart, difference value, measured spatial data, chart spatial data, file ID, measurement time, recording time, warning accuracy;
User Authority	User Information	User Basic Information	User number, user name, gender, title, organization, contact number, modification time, etc;
		User Log	Log number, user name, operation time, user's ip address, operation description, operation result, notes, module name, type, etc;
	User Authority	User Account	Account name, password, corresponding user number, group number, corresponding role, etc;
		Authority Information	Authority Number, Authority Type, Corresponding Module, Corresponding Operation, Corresponding Resource, etc;
		Role Information	Role name, role number, role description, corresponding authority, etc;

(continued)

Table 2. Continued

Database Type	Information Sheet Type		Main Attribute
Metadata	Data Layer Definition		Data layer name, description, data type, storage form, etc;
	Data Dictionary	Data Item	Data item name, data type, length, value range, value meaning, logical relationship with other data items, etc;
		Data Structure	Data structure name, meaning description, the data items or data structures of which it is composed;

4 DEVELOPMENT AND REALIZATION OF THE SYSTEM

The crowdsourced bathymetry monitoring system consists of a data acquisition module, a data processing module, and a data management module. The system is designed following the B-12 and S-100 series standards and utilizes the B/S mode to improve data management and analysis functions through WEBGIS. The system recruited 25 volunteer ships in the South China Sea, each installing a set of data acquisition terminals to carry out crowd-sourced bathymetry monitoring. From April 2021 to June 2024, the system collected a total of about 1.362 billion bathymetry points. After excluding invalid data, the remaining valid bathymetry points were about 280 million, with a data validity rate of approximately 20.52%. The total mileage of the data collection exceeded 300,000 kilometers. The details of the acquisition are presented in Table 3, and Figure 3 illustrates the trajectory map of data acquisition by all 25 volunteer vessels.

Table 3. Data collector status from April 2021 to June 2024.

Time	Total Collection points	Valid points	Validity Rate %	Total Mildage KM
2021 (Apr-Dec)	247,347,678	45,230,853	18.29	51,871
2022 (Jan-Dec)	406,101,514	83,384,275	20.53	91,272
2023 (Jan-Dec)	437,173,371	94,646,513	21.65	97,552
2024 (Jan-Jun)	271,831,481	56,322,877	20.72	59,811
Total	1,362,454,044	279,584,518	20.52	300,506

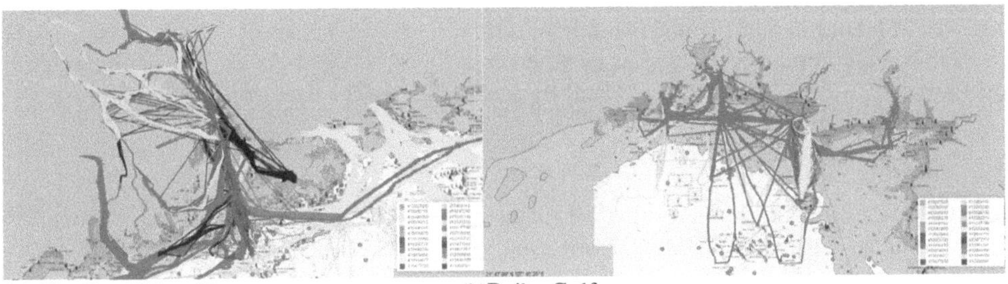

(a) The Pearl River Estuary (b)Beibu Gulf

Figure 3. Data Collection Trajectory from 2021 to 2024.

4.1 *Crowdsourced bathymetry data acquisition*

Crowdsourced bathymetry data acquisition systems are mainly composed of a data acquisition terminal and remote monitoring software.

4.1.1 *Data collector*
The data collector is self-developed and consists of the main control board and the interface base board. It has the function of collecting and storing bathymetry data, position, and attitude data. It supports 4G, WIFI network communication, TCP/IP, and FTP network protocols. The structure of the data collector is illustrated in Figure 4.

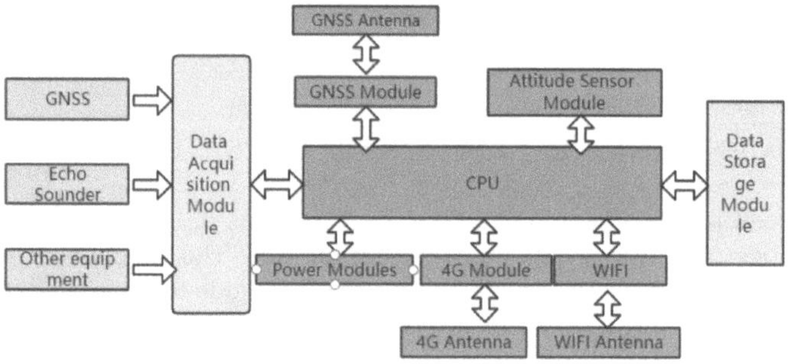

Figure 4. Data acquisition terminal structure.

4.1.2 *Remote control software*
The software enables remote configuration of the communication port of the data collector, offering the flexibility to adjust the transmission baud rate to 4800, 9600, or 19200 bps. Additionally, it can set up the port, baud rate, and other communication parameters for incoming data. The software is specifically developed to receive crowdsourced bathymetry data from different vessels, monitor data communication reception status, and decrypt the data.

Remote Connection Service: The ReaderServices program implements the remote connection service. It actively connects to the cloud server, receives instructions from the cloud, and executes relevant functions based on these instructions. To ensure connection reliability, the program uses heartbeat packets (HeartBeat) to check the status of the server connection.

File Transfer Service: The ftpService program implements the file upload function. The Zmodem protocol is an error-checking protocol specifically designed for modems. With the Zmodem protocol, you can transmit 512-byte data blocks over a modem. If an error occurs in a data block, the receiver sends a "deny" response, prompting the retransmission of the block.

Status Query Service: The GPRS service program implements the 4G status query, receives AT commands from the modem, and returns the result of the execution of the modem's AT commands. It can query whether a SIM card is installed, check the antenna performance, identify the current operator, determine storage card capacity, and more.

4.2 *Crowdsourced bathymetry data processing module*

The CrowdSourced Bathymetric Data Processing Module consists of two major functions: Crowdsourced Bathymetry Vessel Monitoring Visualization and Crowdsourced Bathymetry Data Automated Processing Tool.

4.2.1 *Monitoring visualization*
The module integrates daily status, channel monitoring, resource monitoring, and monitoring configuration of the crowdsourced bathymetric volunteer ship. It establishes a unified and comprehensive business view, realizes the visualization of daily monitoring procedures, and issues abnormal warnings and logs for the ship's operational anomalies, making it convenient for users to conduct supervision.

Acquisition Monitoring: The software automatically monitors the data acquisition status of crowdsourced bathymetric volunteer vessels daily. It tracks the number of normal and abnormal vessels, the quantity of supplementary data, and the data received by each vessel. This feature helps users identify vessels experiencing issues.

Channel Monitoring: The software automatically tracks channel monitoring over the past seven days, recording all channel names, total coverage times, number of inbound and outbound voyages, latest inbound and outbound vessels, last inbound and outbound times, etc. The software provides a comprehensive overview of the data collection status for each channel. Users can click to access detailed information on the corresponding fairway's complete coverage and inbound and outbound voyages, including details on ships entering and leaving the fairway, along with their respective times during the statistical period. The ship's route can be visualized on the GIS map, enabling users to better comprehend the ship's movements and fairway usage.

Resource Monitoring: The software monitors the local data application running environment, storage, and computing resources at regular intervals, with a default monitoring frequency of 12 hours. The monitoring content includes server status, database service, monitoring time, database space occupation, original data backup disk occupation, memory usage, CPU usage, and other resource monitoring indicators.

4.2.2 *Automated data processing and presentation*

This function is the core function of data processing, providing calculation modes for manual interaction and fully automatic processing. It enables automated processing of data reception, filtering, parameter correction, and accuracy assessment. Users can visualize and display the data results in two or three dimensions based on the data processed by the data analysis module.

Data Filtering: The software includes a filtering function to remove positioning jump points and filter out abnormal planar positioning solutions. The entire process of rejecting inaccuracies is automatically executed during inventory.

Vessel Configuration: Following the B12 standard and Caris&hips vessel configuration table specification, a crowdsourcing vessel configuration table has been established. It is primarily divided into 7 configuration pages, which include "Basic Information", "Positioning and Navigation", "Depth Sounder", "TVU", "Dynamic Draft", "Static Draft" information pages. Acknowledging that many of these groups of factors may change over time, the date attribute has been included to facilitate a quicker understanding of the configuration change process.

3D Display: Surfaces created based on the user's selected bathymetry data can be switched to a 3D visualization format with a shortcut in the 3D window, as illustrated in Figure 5.

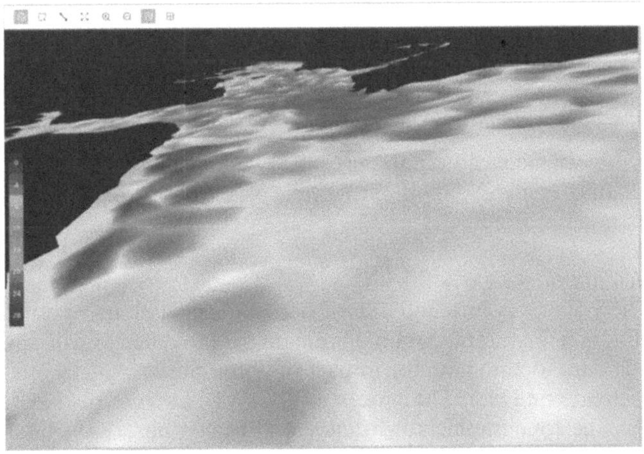

Figure 5. 3D Presentation of bathymetry surfaces.

4.3 *Module crowdsourced bathymetry data management module*

The management module mainly serves three primary functions: foreground data display and download, background data management and analysis, and user management.

Foreground Data Display and Download:This function enables the display and operation of nautical charts, visual representation of measurement results, management of measurement files, and shallow water warnings.

Background Data Management and Analysis: Backstage data management and analysis can facilitate the manual import of measurement result data, statistical analysis of result data, display of measurement data, and a shallow water warning algorithm.

User Management:background user management system includes announcement management, user authority management, and log management.

In September 2023, a typhoon made landfall along the coast of southern China. In October, anomalies in water depth were monitored near the typhoon's landfall area. A total of 2,572 bathymetric points were collected near the landfall site. It was observed that the water depth had become shallower when comparing the bathymetry data from October to that of May of the same year. The water depth change ranged between 0.1-1.9 meters, with an average decrease of 0.94 meters. Specifically, the water depth in the typhoon's area in October ranged from 0.1-1.9 meters, with an average of 0.94 meters. The change in water depth is illustrated in Figure 6.

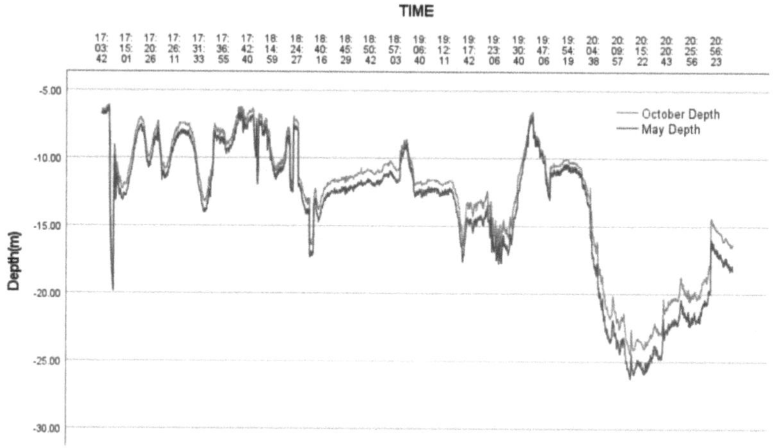

Figure 6. Changes in water depth after the typhoon.

4.4 *Key technology for automatic bathymetry processing*

The crowdsourced bathymetry automatic processing algorithm is the key technology of this software, which mainly includes data filtering, time synchronization, and data correction. Its main flow is illustrated in Figure 7.

4.4.1 *Automatic data filtering*

It eliminates GNSS jump points and filters out abnormal planar positioning solutions. Additionally, it filters bathymetric data to eliminate the influence of fish, floating mud, etc., and filters attitude data to eliminate the impact of abnormal attitudes.

Filtering of Bathymetric Data: The process of filtering bathymetric data involves refining the data based on the topographic consistency and trends observed in continuous bathymetric results.

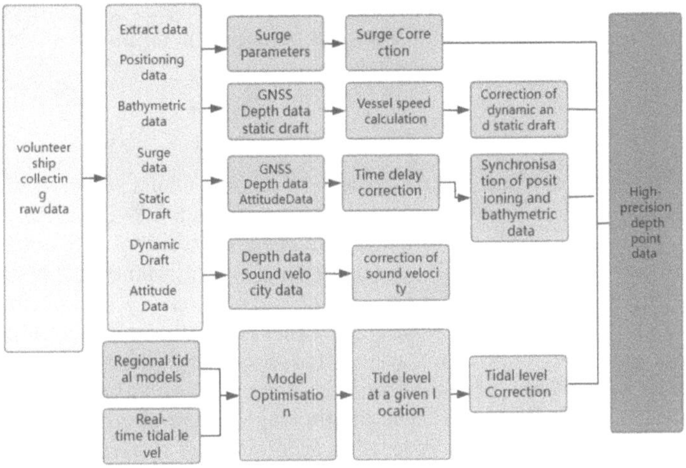

Figure 7. Flow chart for automatic bathymetric processing.

Positioning data filtering: Consistency filtering is conducted for GNSS data based on the fore and aft positioning data and the orientation of the volunteer ships.

Attitude Data Filtering: The short-term change in the attitude of the survey vessel typically occurs within 3-5 seconds, with a filter cut-off frequency of 10Hz. This setting allows for retaining information about the vessel's attitude changes caused by surges and waves, while effectively filtering out data errors.

4.4.2 *Multiple data time synchronisation*

UTC time is used as a unified time scale to synchronize positioning, bathymetry, and attitude data, and to create the original record and quality control posts. After determining the time deviation between systems, it is possible to correct the time delay of the UTC time calibrated to the positioning data to align it with the computer time and to identify the actual point position at the current computer moment in the three-dimensional positioning data sequence.

4.4.3 *Automatic correction of bathymetric data*

The automatic correction of bathymetric data primarily addresses the issue of parameter variations in ships and equipment due to changes in sea level hydrographic conditions, which can lead to inaccuracies in measurement results. The software can automatically adjust the measurement parameters based on the hydrographic conditions.

Surge Correction:Surge correction involves real-time correction of bathymetric data using synchronized surge parameters.

Attitude Correction: By utilizing synchronized attitude data along with the coordinates of each transducer within the ship's coordinate system, attitude correction is performed to determine the three-dimensional position of the transducer and mitigate the impact of attitude.

Draft Correction:Using the ship shape coefficient and ship speed to carry out draft correction.

Sound Speed Correction:Sound speed correction involves conducting sound speed correction, facilitating the detection and elimination of sound speed errors, and mitigating the impact of the sound speed factor.

Tidal Data Correction: Correcting tidal data involves utilizing astronomical forecast tidal data or existing tidal data information. This is combined with crowdsourced volunteer ship positions and bathymetric observations to calculate tidal height changes. By incorporating depth information, a vertical solution for seabed bathymetric points is obtained.

5 CONCLUSIONS

This system is the first crowdsourced bathymetric system in China, developed in accordance with the IHO Guidance for Crowdsourced Bathymetry, the S-100 Universal Hydrographic Data Model, and actual needs. The system offers high precision and timely bathymetry detection, quasi real-time updates of bathymetry data, fully automated data processing, and compatibility with various types of ships. It enhances waterway monitoring capabilities, identifies areas with changing water depths and other navigation-affecting issues, and issues timely warnings to passing ships. This system addresses the limitations of the current professional surveying practices. Its establishment serves as a valuable reference for crowdsourced bathymetric work in bathymetric monitoring. Future plans include expanding the volunteer ship fleet, enhancing data processing accuracy, and fostering the advancement of crowdsourced bathymetry.

REFERENCES

International Hydrographic Organization. (2022). *B-12 Edition3.0.0*, Guidance on Crowdsourced Bathymetry. Monaco: IHO.

Brian R.C, Semme J.I & Shannonhoy, *et al.* (2020). A design for a trusted community bathymetry system. *Marine Geodesy*:43(4):327–358.

Pauline W.A, Markskm & Martin J, *et al.*, (2015). *A New Digital Bathymetric Model of the World's Oceans*, M/OL. https://www.Researchgate.net/publication/280554141_A_New_Digital_Bathymetric_Model_of_the_World's_Oceans.

Wu L.Z, Lv Y.J & Huang D.W, *et al.* (2021). Research on national crowdsourced bathymetry system based on IHO concept of crowdsourced bathymetry. *Hydrographic Surveying and Charting*:2021,41(03):1–4+23.

Wu J, Wang P & Wang J, *et al.* (2020). Research and analysis of key methods for crowdsourced bathymetry. *Hydrographic Surveying and Charting*:40(02):36–40.

Proceedings of the International Association of Institutes of Navigation World Congress 2024 – Chen et al. (Eds)

Research on reflection materials of AtoN for LiDAR

Fanping Zeng & Ruqing Xu
Fuzhou AtoN Department of Eastern Navigation Service Center, Fuzhou, Fujian, China

ABSTRACT: LiDAR is an important sensing tool for MASS (Maritime Autonomous Surface Ship). In order to improve the responsiveness of traditional visual AtoN (Aids to Navigation) to MASS recognition and perception, we have investigated several reflection materials suitable for visual AtoN. In this study, we test the AtoN with different reflection materials using commercial LiDAR in the field of automotive autonomous driving, and compare and analyze the formed point cloud images and reflectivity data. Based on the experiment results, we propose an improved scheme, which effectively improves the perception ability of LiDAR to traditional visual AtoN.

1 INTRODUCTION

1.1 *LiDAR and maritime autonomous surface ship*

LiDAR (Light Detection And Ranging) is a radar system that emits a laser beam to detect the position, speed and other characteristic quantities of a target. LiDAR offers several advantages, including high spatial resolution, rapid data acquisition, and immunity to radio wave interference. Consequently, it is currently employed across diverse fields such as ecological and geological surveys, robotics and industrial automation, and autonomous driving. The performance, availability and economic viability of LiDAR devices have improved significantly because of the demand driven by a number of sectors represented by self-driving cars [1–3].

Early applications of LiDAR in navigation were mainly as a common instrument for drawing nautical charts and making bathymetric measurements in shallow water. With the rapid development of MASS(Maritime Autonomous Surface Ship) in recent years, LiDAR technology has become one of the main solutions for ship situational awareness. Particularly in ports and other scenarios where traditional radar systems fail to provide sufficient accuracy, LiDAR emerges as a crucial device for delivering precise positioning.

The application of LiDAR in the maritime domain, especially in MASS, has inspired research topics on the improvement of its performance and scenario suitability. For maritime researchers, the focus of the study is on the extent to which LiDAR can detect navigational aids on water, as well as on enhancing the design of LiDAR systems or improving analytical algorithms to increase detection and sensing capabilities. For AtoN (Aids to Navigation) management authority, the research can be carried out from how to improve the passive sensing ability of AtoN.

1.2 *Reflection materials and their applications*

Reflection materials are widely used in a variety of fields such as road safety warning, safety navigation, personal safety, industrial automation and so on [4]. Reflection materials play an obvious warning role with their bright colors during the daytime, and their striking reflective

DOI: 10.1201/9781003624424-20

properties significantly enhance their visibility at nighttime or in low light conditions. Consequently, reflection materials are widely applied in the field of traffic safety due to their effectiveness in increasing human recognition capabilities.

Reflection material has been used in a number of applications on AtoN to enhance the marking of AtoN structures, lettering and numbering of waterway, obstructions in the channels and piers. It is usually used as a supplement to AtoN lights or as a substitute in special circumstances and is usually effective at distances of less than 1 nautical mile. IALA has successively issued G1145 "Application of Retroreflecting Material on AtoN" and R0106 (E-106) "Retroreflecting Material on Aids to Navigation Marks within the IALA Maritime Buoyage System" to provide guidance.

Reflection materials may act equally well on lasers. In early research on autonomous driving, reflectors were an important auxiliary mark for navigation and LiDAR to recognize space. Therefore, in future maritime environments where MASS and manned vessels coexist, reflection materials can effectively enhance the ability of traditional AtoN to be sensed by LiDAR, which improving performance of AtoN in providing navigational aids to ships.

2 EXPERIMENTAL DESIGN

2.1 *Research methods*

In this paper, the mainstream commercial LiDAR currently used in the autonomous driving sector is used to scan and image the visual AtoN equipment applying different reflection materials. By comparing and analyzing the imaging effect and reflectivity data, we comprehensively consider the reflective performance, application mode, economy, and other indexes to select a better reflection material. This approach allows us to offer feasible improvements and retrofitting solutions for traditional AtoN, thereby enhancing the sensing capabilities of MASS.

2.2 *LiDAR selection*

LiDAR has a wide variety of types depending on the scanning mode, measurement mode, and wavelength. In terms of the applicable maritime application scenarios, solid-state or semi-solid-state devices that are more adaptable to bumpy environments should be selected for the scanning mode of operation. Considering the imaging effect, experimental site (which is 300 meters long) and economy, the researchers finally chose the Livox Horizon LiDAR (Figure 1), which adopts the rotating mirror type semi-solid state technology scheme, laser wavelength 905mm, maximum distance 260 meters, non-repetitive point cloud scanning rate 240,000 items/second.

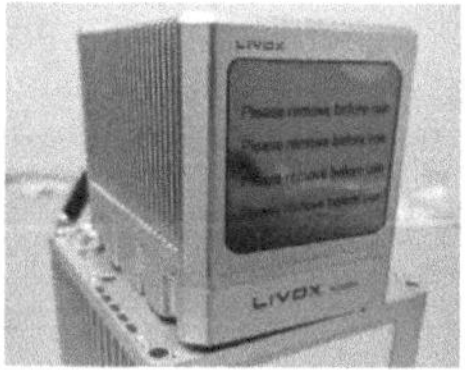

Figure 1. Photograph of the Livox Horizon LiDAR.

2.3 *Reflection materials and AtoN experimental equipment*

With reference to the existing foreign cases and domestic application examples in the transportation field, this study selected three commonly used reflective schemes for experiment: the 3M reflective film employing microprismatic technology, customized reflective panels using diffuse reflective coatings, and road reflective paint using doped microglass beads.

Due to the characteristic of the materials and the different ways it is applied, the experiments were divided into a lantern frame experiment (diffuse reflective panels, reflective paint) and a topmark experiment (3M reflective film), which were conducted and used the red and green colors commonly used for AtoN. At the same time, we included a group of AtoN equipments made of conventional materials or paint as a comparison group in the experiment.

2.3.1 *Lantern frame experiment - diffuse reflective panels group*

This study used customized red and green diffuse reflective panels, and the reflectivity of the laser at 905nm wavelength was 65%. (Figure 2). By LiDAR experiment, the 8bit Digital Number (DN) of reflectance is 126.

Figure 2. Photograph of customized diffuse reflective panels.

2.3.2 *Lantern frame experiment - reflective paint group*

This group uses reflective paint that meets the requirements of GB/T 16311 "Specification and test method for road traffic markings" and JT/T280 "Pavement marking paint" standards. The weight ratio of the main paint and reflective beads in the composition of this reflective paint is 4:1, and the factory reflection detection value is more than 50cd/lx/m^2. As shown in Figure 3, the reflective paint will be evenly brushed on one side of the corresponding color of the lantern frame.

Figure 3. (a) Photograph of the experimental group setting reflective paint; (b) Photograph of reflective paint group.

2.3.3 *Topmark experiment - 3M reflective film group*

This group adopts 3M reflective film which conforms to the standard of GB/T 18833 "Road Traffic Reflective Film", with a width of 3cm and a retro-reflective detection value of more than 350cd/lx/m^2 (observation angle of 0.2°, incidence angle of 30°). Considering the effect of the actual appearance, the reflective film is pasted at intervals on the exterior of the plastic topmark, of which the red cylindrical topmark is pasted in the form of a fence (as shown in Figure 4a) and the green conical topmark is pasted in the form of a radial topmark (as shown in Figure 4b).

Figure 4. Photographs of topmarks using reflective film (a) red (b) green.

2.4 *Experiment design*

2.4.1 *Terrestrial experiment design*

The experiment was conducted at the aids to navigation base in Fuzhou City, Fujian Province, China (Figure 5a), specifically on the road beside the workshop close to the dock, which has a length of about 200 meters. The buildings and floating bodies around the road provided the interference term for this experiment. The experimental equipment was placed in batches on one side of the site in the order of lantern frame and topmark experiments, and the LiDAR was powered by 220V from the vehicle and scanned at seven different distances: 10m, 25m, 50m, 75m, 100m, 150m and 200m, respectively. The experimental conclusions were drawn by analyzing the 3D graphic manifold effect and target reflectivity obtained from LiDAR scanning.

The lantern frame experiment was painted with reflective paint on one side and diffuse reflective panels were fixed to the lantern frames of the corresponding color to create a contrast. Buoyage floating body with ordinary paint was used as the blank control group (Figure 5b).The topmark experiment used plastic topmark with a uniform specification of φ60cm, and two topmarks each were placed using 3M reflective film and without reflective film, of which the one without reflective film was the control group (Figure 5c).

Figure 5. (a) Photograph of terrestrial experimental field; (b) Photograph of the placement of experimental equipment in the lantern frame experiment; (c) Photograph of the placement of equipment in the topmark experiment.

2.4.2 *Marine experiment design*

The research team chose one buoyage on each left and right side of the channel as experiment objects (left sign and right sign) and installed topmarks with reflective film for them (Figure 6a). LiDAR was set up at the rear of the test vessel and scanned from a distance of 10 meters gradually until the equipment's scanning limit of 250 meters (Figure 6b).

Figure 6. Photograph of the marine experiment of (a) at 10 meters and (b) at 200 meters.

2.5 *Data processing method*

The researchers scan the target under experiment by LiDAR and display the scanned point cloud image on the corresponding software. The range of the target in the image is boxed, and the reflectivity (0-255 standardized values) of each point in the range is counted.

Due to the difficulty of accurately framing the target at long distances and the fact that the framing range includes part of the environmental background, there is some interference in reflectivity data sampling. Therefore, when the data processing is carried out, the final value is obtained by taking the average of the middle 20 samples of each set of data after arranging them from the smallest to the largest reflectivity value obtained from the experiment, so as to achieve the purpose of eliminating the interference as much as possible.

3 RESULTS AND DISCUSSIONS

3.1 *Terrestrial experiment*

3.1.1 *Lantern frame experiment*

3.1.1.1 *Results of LiDAR scanning* From the scanned point cloud, the LiDAR can clearly identify the lantern frame and the surrounding float section the scale of 10m-50m (Figure 7a). There is no significant reflectivity image difference between the reflective paint section and the diffuse reflective panels section, but it is significantly higher than the blank control group. As Figure 7b shown in, The results from a distance of 100-200 meters show that the experiment target has been difficult to recognize on the image.

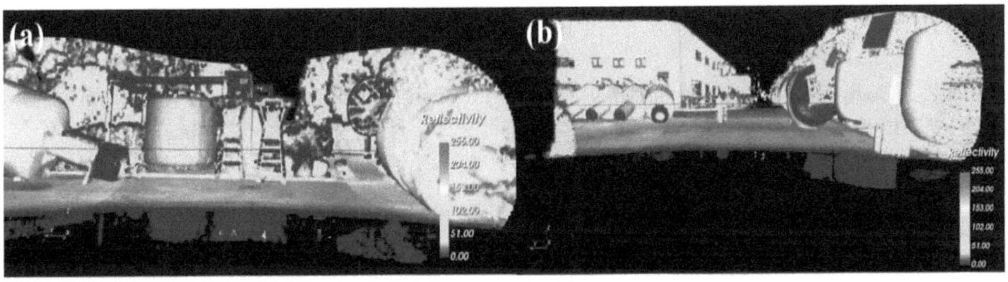

Figure 7. Point cloud map of lantern frame experiment (a) at 10 meters and (b) at 150 meters.

3.1.1.2 *Reflectivity analysis* In the analysis of the specific value of the target reflectivity (Table 1), the experiment results at close range showed that the reflectivity of the diffuse reflective panel was the highest (90-100), followed by the reflective paint (50-70), and there was a more obvious gap with the control group of conventional paint (30). As the distance increased, the reflectivity of the reflective paint converged with that of the conventional paint, and the reflectivity of the diffuse reflective panel showed a significant decrease. Beyond a distance of 150 meters, there is no longer a significant difference in the reflectivity of the three materials (Figure 8).

Table 1. Reflectivity (DN) of different materials in lantern frame experiments.

Material*	10 meters	25 meters	50 meters	75 meters	100 meters	150 meters	200 meters
Diffuse reflective panel (green)	116.5	95.95	75.67	69.05	49.62	30	30
Diffuse reflective panel (red)	109	91.23	88.71	64.64	55.54	30	30
Reflective paint (green)	42.67	54.71	33.86	30.76	30.45	30	30
Reflective paint (red)	72	69.19	35.82	30.26	29.91	30	30
Conventional paint (green)	31	29	28	33	31.73	30	30
Conventional paint (red)	33.14	33	31.05	33.14	28.55	30	30

*Beyond a distance of 150 meters, there is no longer a significant difference in the reflectivity of the three materials

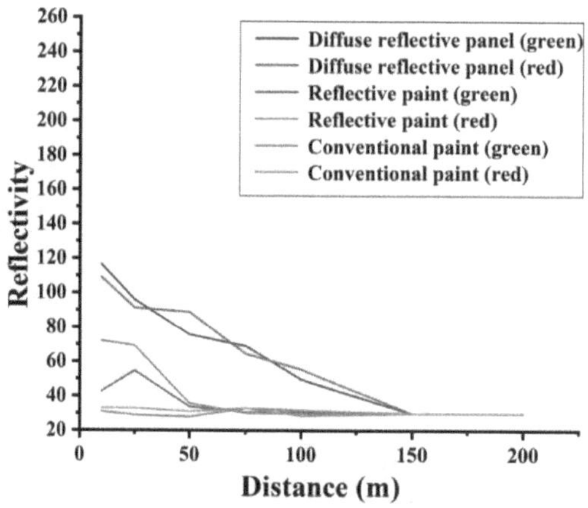

Figure 8. Comparison of reflectivity (DN) towards different distance of different materials for lantern frame experiments.

3.1.2 *Topmark experiment*

3.1.2.1 *Results of LiDAR scanning* The results of this experiment show that at closer distances of 10m-75m, the reflectivity of the topmarks using 3M reflective film is significantly higher than that of the control group, and the image of the strip of 3M reflective film (left side) is very clear (Figure 9).

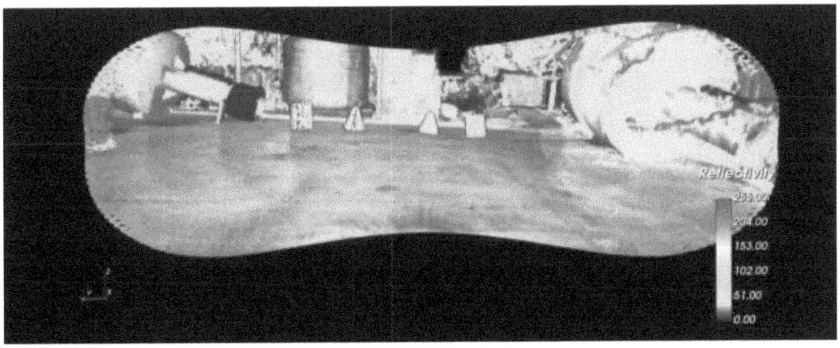

Figure 9. Point cloud map of topmark experiment at 10 meters.

In addition to this, in the experiment results at a distance of 150-200 meters, it can be observed that although it has been difficult to identify the topmark on the image because of its small size (Figure 10a). But we can clearly see a clear bright spot of the left light-emitting sticker target on the color block of the reflectivity display (Figure 10b).

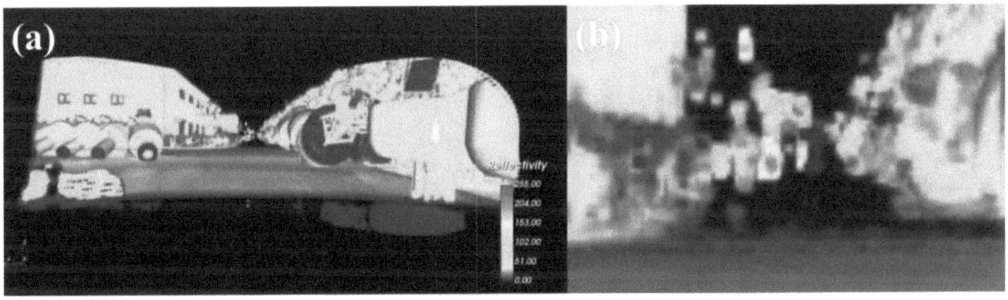

Figure 10. (a) Point cloud of the topmark experiment at 150 meters (b) magnified image of the target object - showing a significant reflectivity brightness difference to the left.

3.1.2.2 *Reflectivity analysis* In the analysis of the specific values of target reflectivity for these experiments (Table 2), since the 3M reflective film did not fully cover the topmarks, the peak values of 3M reflective film were added to the data statistics. It can be concluded that the reflectivity of the topmarks with 3M reflective film is much higher than that of the control group without reflective film (30-40), both in terms of peak values (240-255) and average values (80-130) (Figure 11).

As the experimental distance increases, the peak value of of the 3M reflective film reflectivity decreases insignificantly, but its average value shows an abnormal increase. After referring to other information and analyzing the results, the situation is mainly due to the fact that the farther away the distance is, the lower the sampling rate of its point cloud is. As a result, points with high reflectivity are more likely to be recognized due to an increase in their percentage within the same box-selected sampling range.

Table 2. Reflectivity (DN) of different materials in topmark experiments.

Material*	10 meters	25 meters	50 meters	75 meters	100 meters	150 meters	200 meters
3M reflective film average value (green)	83.69	78.43	80.51	44.57	72.79	119.51	128.35
3M reflective film peak value (green)	255	255	249	243	188	172	164
3M reflective film average value(red)	106.5	94.95	118.89	44.57	72.79	119.51	128.35
3M reflective film peak value (red)	255	255	255	243	188	172	164
Without reflective film(green)	37.99	29.12	30.88	32.19	32.52	25	25
Without reflective film(red)	36.46	29.94	30.04	32.19	32.52	25	25

*Beginning at a distance of 75 meters, the two topmarks of the same group, the red and green, become indistinguishable.

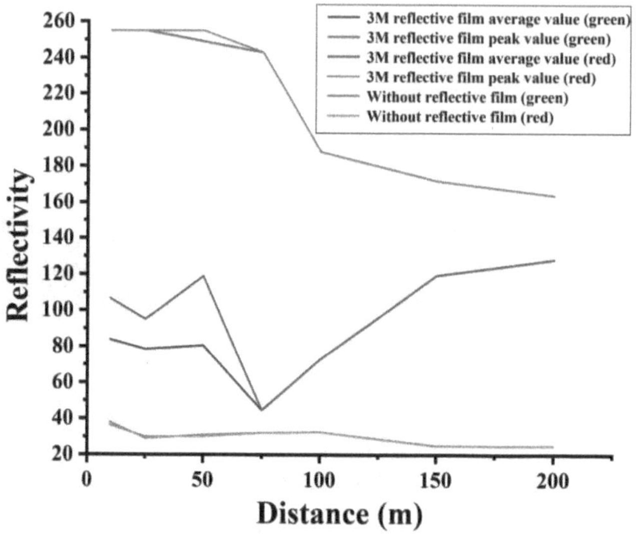

Figure 11. Comparison of reflectivity (DN) towards different distance of different materials for topmark experiments.

3.1.3 *Effect of angle on reflectivity*

Scanning the topmarks (cylindrical/conical) at a close distance, it was found that the reflectivity at the center (incidence angle of 0°) was significantly higher than that at the sides (incidence angle of more than 30°), and the characteristic of specular reflection was more obvious. The same experiment was performed on the diffuse reflective panels, and there was no significant change in reflectivity between the incident angles of 0° and 45°, which is consistent with the characteristics of diffuse reflection.

3.2 *Marine experiment*

In the aforementioned terrestrial experiments, we found that the microprismatic 3M reflective film has the best reflective effect, which is much higher than the effect of metal, plastic, paint, and other materials. Moreover, with the scanning distance being farther away, its reflectivity attenuation is smaller. In order to further verify the effectiveness of the 3M reflective film in the application of AtoN at sea, the research team installed the reflective topmarks on buoyage and conducted experiments using LiDAR on board.

3.2.1 *Results of LiDAR scanning*

From the point cloud map, the shape of the buoyages is more obvious at a distance of 10-50 meters since the water surface absorbs most of the light beam (Figure 12a). Moreover, the reflectivity of the topmark with reflective film is significantly higher than that of the other parts of the buoyage (Figure 12b). The results at a distance of about 100 meters show that the shape of the light buoyage in the point cloud image is no longer clear (Figure 12c), but the outline is still recognizable and the reflectivity of the topmarks fitted with reflective film is still evident (Figure 12d).

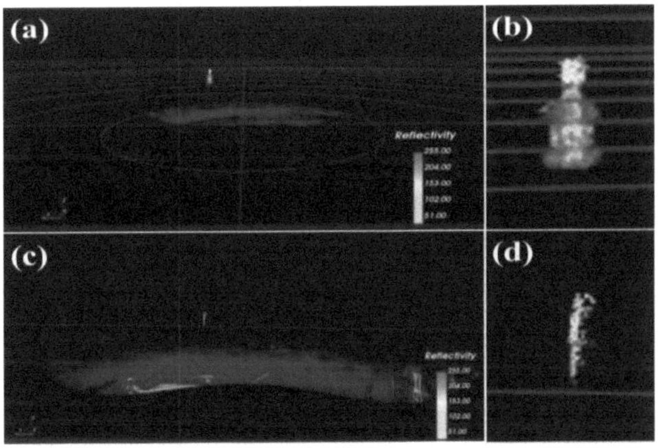

Figure 12. Point cloud map of marine experiment (a) at 50 meters and (b) magnified target object image and (c) at 100 meters and (d) magnified target object image.

At the measurement limit of 250 meters, it can be seen that the floating body image has completely disappeared from the point cloud map (Figure 13a), but the highly reflective topmark color dots can still be observed (Figure 13b). In fact, beyond the range of the LiDAR, the point cloud map still shows sporadic highlights of the topmarks.

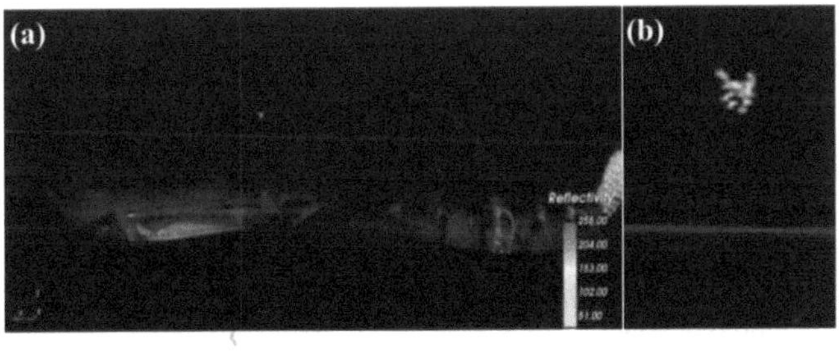

Figure 13. (a) Point cloud of the marine experiment at 250 meters (b) magnified image of the target object.

3.2.2 *Reflectivity analysis*

From Table 3, at the close range of 50 meters, the experiments yielded a higher reflectivity for the red cylindrical topmark than for the green conical top topmark, which is consistent with the conclusions of IALA G1145. When the distance was increased and the swinging tilt of the buoyages at sea was taken into account, the reflectivity values of the two light buoyage shapes were comparable. However, the topmarks with 3M reflective film were far more easily recognized by LiDAR than the floats without reflection material at any experiment range (Figure 14).

Table 3. Reflectivity (DN) of different materials in marine experiments.

Reflectivity	10 meters	50 meters	100 meters	250 meters
Red topmark (peak value)	255	204	157	161
Green topmark (peak value)	161	160	155	167
Red topmark (average value)	172.43	157.91	151.54	151.37
Green topmark (average value)	151.39	150.96	139.18	151
Buoyage floating body (without reflective film)	23.32	23.66	38.8	0

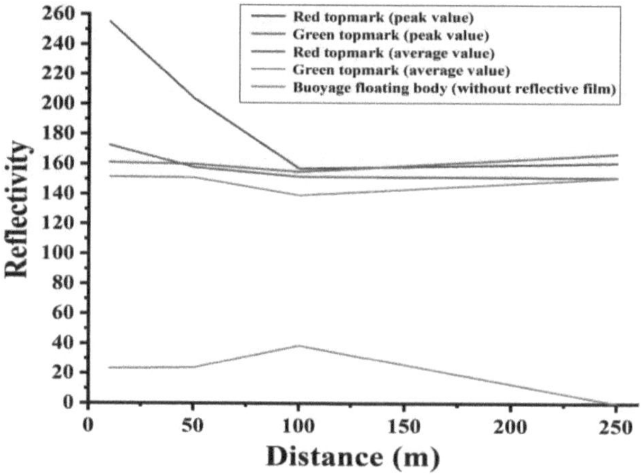

Figure 14. Comparison of reflectivity (DN) of topmark with different materials in marine experiments.

3.2.3 *Effect of sampling time on recognition*

During the experiment process, the researchers found that since the buoyages are in constant state of motion at sea, the point cloud maps formed at longer sampling times can more clearly identify the marine AtoN. Especially when the buoyage moves along the direction relatively perpendicular to the LiDAR, the point cloud image with more than 3,000ms sampling time shows more obvious color bands (Figure 15 shows the comparison between different sampling times), and this discovery can provide a reference for how to use the LiDAR more effectively for MASS.

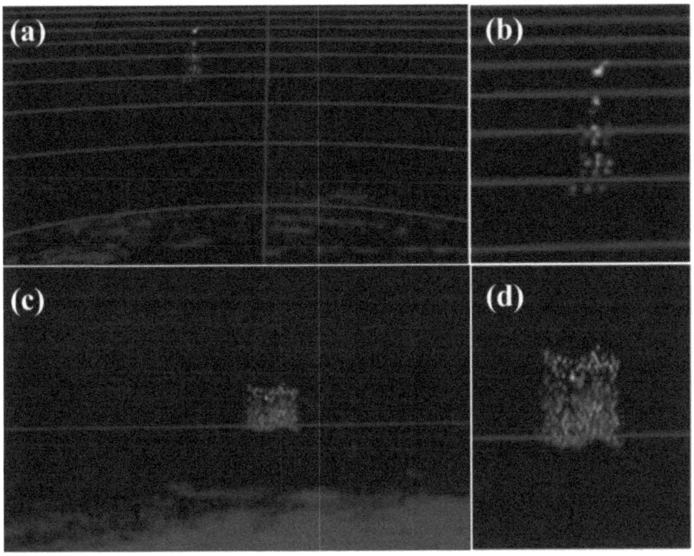

Figure 15. Point cloud plots of (a) sampling time of 100 ms at a distance of 50 m; (b) magnified image of the target object in the above figure; (c) point cloud map with a sampling time of 3000 ms at a distance of 100 m; (d) magnified image of the target object in the above figure.

4 CONCLUSION

4.1 *Research conclusion*

The application of reflection materials can significantly improve the ability of navigation marks to be perceived by LiDAR, which plays an important role in MASS using LiDAR.

The microprismatic 3M reflective film exhibits the best reflective effect, significantly surpassing that of metals, plastics, paints, and other materials. Its reflectivity attenuates less as the scanning distance increases.

The application of reflective film is not restricted to topmarks but can also be extended to other parts of the buoyage floating body, such as lantern frames and lookout boards, without altering the buoyage's appearance.

Although the shape of the AtoN causes some variation in the reflection angle, affecting LiDAR recognition, these differences are negligible in practical applications due to the constant state of motion of both buoyages and ships at sea.

Additionally, a longer sampling time enhances recognition, and in inland waterways where buoyages are less prominent, the periodic rotation of ship-mounted LiDAR can be considered to achieve better recognition outcomes.

4.2 *Next research direction*

Due to the time and condition restrictions, this study cannot fully cover all the needs of the marine AtoN acting on MASS LiDAR. There are still some deficiencies in this study, such as the short range of the LiDAR equipment used, the lack of foggy day and nighttime experiments, the lack of interference experiments with neighboring ships in the experiment process, and the lack of a long-time observation experiment on the construction process of reflection materials and the effects of erosion of the marine environment. In view of the above problems, the results need to be further studied and gradually improved.

REFERENCES

[1] Liu, C.G. (2022). 3D LiDAR Based Real-time Object Recognition System for Unmanned Surface Vehicles. *Chinese Journal of Mechanical Engineering* 04: 202–211.

[2] Zhuang, J. (2019). Application of MMW Radar and LiDAR in MASS. *Ship Engineering* 41(11): 79–82.

[3] Kim, J., Park, B.J., and Kim, J. (2023). Empirical Analysis of Autonomous Vehicle's LiDAR Detection Performance Degradation for Actual Road Driving in Rain and Fog. *Sensors* 23(6): 2972.

[4] Muckenhuber, S., Holzer, H., and Bockaj, Z. (2020). Automotive Lidar Modelling Approach Based on Material Properties and Lidar Capabilities. *Sensors* 20(11): 3309.

*Proceedings of the International Association of Institutes of Navigation World Congress 2024 –
Chen et al. (Eds)*

Offshore spatial-temporal sea surface height field forecasting in the North Sea using an improved ResUnet deep learning methodology

Zeguo Zhang & Jianchuan Yin
Guangdong Ocean University, Zhanjiang, China

ABSTRACT: The accurate and efficient prediction of spatiotemporal sea surface height (SSH) variability in semi-enclosed marine regions proves critical for coastal protection strategies, offshore infrastructure development, and maritime navigation management. SSH forecasting assumes heightened importance in nearshore zones characterized by complex bathymetry and nonlinear shallow-water dynamics, where operational safety requirements for human activities—including maritime transportation, aquaculture systems, coastal engineering projects, and storm surge early warning systems—demand enhanced predictive accuracy. This study develops an enhanced deep learning framework integrating multi-head attention mechanisms with residual convolutional neural network (CNN) architectures to achieve high-precision SSH field predictions. A comparative evaluation was conducted against three established machine learning benchmarks. Experimental validation using hourly spatially-resolved SSH datasets from the North Sea demonstrated the superior performance of our CNN-based model across multiple forecasting horizons. The proposed method achieved state-of-the-art prediction accuracy, with mean RMSE values of 0.0069 m (1-hour forecast) and 0.046 m (72-hour forecast). As a computationally efficient surrogate model, this CNN architecture enables rapid and reliable prediction of coastal SSH variability patterns. The demonstrated capability to resolve fine-scale spatiotemporal water level fluctuations positions this approach as a valuable tool for enhancing navigational safety and supporting coastal zone management practices.

Keywords: Sea surface height field prediction, deep leaning, 2D Convolutional Neural Network, sequential pattern mining

1 INTRODUCTION

The accurate prediction of coastal spatiotemporal sea level variability is critical for maritime navigation optimization, coastal infrastructure design, and environmental conservation initiatives. The analysis of sea surface height (SSH) variability patterns across basin-scale coastal systems proves particularly essential for safeguarding vulnerable coastal zones (Jevrejeva *et al.* 2019; Rajabi-Kiasari *et al.* 2015), advancing blue economy strategies (Nicholls & Cazenave 2010), monitoring marine ecological shifts (Norris *et al.* 2013), and implementing coastal protection measures (Cazenave & Cozannet 2014; Slangen *et al.* 2012). While in-situ observations remain the gold standard for sea level monitoring, their practical implementation faces significant constraints—terrestrial tide gauges exhibit sparse spatial distribution due to high instrumentation costs, while satellite-derived SSH products, despite broader spatial coverage, demonstrate inherent limitations in resolving sub-mesoscale processes and high-frequency variability below their orbital repeat cycles (Manucharyan *et al.* 2021; Zhang *et al.* 2020). Furthermore, coastal gauge measurements frequently contain systematic errors that compromise data reliability (Cipollini *et al.* 2017). Numerical ocean

DOI: 10.1201/9781003624424-21

models have emerged as a primary alternative for SSH prediction; however, their computational demands become prohibitive when simulating large spatial domains or high-resolution spatiotemporal dynamics, rendering them impractical for routine operational forecasting and real-time applications (Rolnick *et al.* 2019; Zilong *et al.* 2022). Moreover, for long-term, large-scale historical SSH assessment, the numerical modeling can be extremely time consuming.

Current SSH forecasting approaches predominantly rely on statistical methods and numerical ocean models, though the latter exhibit critical limitations in operational settings. Numerical simulations suffer from mathematical non-robustness in discretization schemes, inherent nonlinearity of oceanic processes, and oversimplified physical parameterizations—all contributing to prediction uncertainties (Miao *et al.* 2023). Furthermore, their computational intensity and time requirements render them impractical for real-time forecasting applications (Kumar *et al.* 2020). Meanwhile, deep learning has emerged as a transformative methodology in oceanographic and meteorological sciences (Boukabara *et al.* 2020; Juan & Valdecantos 2022; Sun *et al.* 2022). Theoretical foundations support its effectiveness: Zanna and Bolton (2020) established machine learning's capacity to derive closed-form equations for complex eddy parameterizations while preserving conservation laws. Meanwhile, Vlachas *et al.* (2018) demonstrated neural networks' superior capability in modeling high-dimensional chaotic systems. Practically, convolutional architectures have proven particularly adept at extracting spatiotemporal ocean dynamic patterns through hierarchical feature learning from massive datasets (George *et al.* 2021; Manucharyan *et al.* 2021; Sun *et al.* 2021; Shin *et al.* 2016), with recurrent neural networks effectively capturing temporal dependencies in ocean variability (Zhou *et al.* 2022). This dual capacity positions deep learning as a paradigm-shifting alternative to conventional approaches.

Recent progress in applying deep learning to SSH forecasting has unveiled a dual picture of advancements and persistent computational challenges. The ConvLSTMP3 model effectively predicted SSH in the South China Sea, albeit requiring 32 hours of training time (Song *et al.* 2021). Similarly, Liu *et al.* (2020) developed an LSTM-Attention hybrid model for daily absolute dynamic topography forecasting, where computational demands reached 53.6 hours for conventional architectures versus 5.7 hours for their optimized framework. Residual Convolutional Neural Networks (ResNets) have shown superior predictive skill over traditional interpolation methods, though deep implementations with 52 convolutional layers still necessitated hour-scale training durations (Manucharyan *et al.* 2021). Hybrid approaches combining Conv1DLSTM with multivariate EOF analysis achieved satisfactory predictions of SSH anomaly and SST (Shao *et al.* 2021). However, these approaches remain constrained by the inherent stationarity assumption in principal component decomposition (Zhang *et al.* 2022). Novel architectures continue to emerge, such as a multivariate Conv2D algorithm, which integrates hydro-meteorological inputs for Baltic Sea level predictions (Rajabi-Kiasari *et al.* 2015), demonstrating the potential of machine learning in basin-scale operational forecasting. While these developments underscore deep learning's capacity to capture complex ocean dynamics across spatial-temporal scales, computational efficiency barriers persist—particularly regarding training time optimization and architectural complexity management—highlighting critical rese arch frontiers in operational oceanography.

Accurate prediction of sea surface height (SSH) variability presents heightened operational importance in coastal and semi-enclosed marine systems when compared to open ocean domains. This is especially true considering their complex bathymetric configurations and nonlinear shallow-water dynamics, which critically influence maritime safety, the resilience of coastal infrastructure, and flood warning systems (Cazenave & Cozannet 2014). These regions exhibit intensified hydrodynamic uncertainties due to multi-scale current interactions and transient meteorological forcings (Kurniawan *et al.* 2014; Zhang *et al.* 2020). However, despite their high socioeconomic vulnerability, they remain remain understudied in operational forecasting frameworks.

Current deep learning implementations face dual limitations: spatially constrained single-point prediction architectures (Altunkaynak & Kartala 2021; El-Diasty *et al.* 2018) and computationally prohibitive training protocols in recurrent neural networks (RNN/LSTM). Although RNN/LSTM architectures can recognize temporal patterns through backpropagation-through-time (BPTT) mechanisms, their inefficiency in processing high-dimensional spatial fields and excessive memory consumption (Bradbury *et al.* 2016; Yu *et al.* 2021) severely constrain operational utility. Even though convolutional neural networks (CNN) have proven to be superior in extracting inter - grid correlations via 2D kernel operations (LeCun *et al.* 2015), this spatial modeling gap persists. Meanwhile, conventional numerical models struggle with error propagation from parameterized physics and discretization instabilities (Fraser *et al.* 2019; Rajabi-Kiasari *et al.* 2015), particularly in semi-enclosed basins where multi-scale dynamics interact nonlinearly. The emerging imperative lies in developing spatially explicit deep learning architectures that synergistically capture both temporal dependencies and spatial covariance structures inherent in SSH fields.

In order to address the aforementioned shortcomings in terms of both methodologies and the mapping of complicated ocean dynamics, we proposed a deep learning method based on the Unet structure. This method incorporates residual learning and an introduced multi-head attention mechanism (MHA-ResUnet). Synthetic hourly SSH field dataset in the semi-enclosed North Sea, which were extracted from the ocean physics reanalysis dataset for the North-West European Shelf, were employed as data samples for model training and testing. The ocean physics reanalysis dataset was produced by employing an ocean data assimilation model, with tides, at a horizontal resolution 7 km.

2 METHODOLOGIES

2.1 *Multi-head attention module*

The Local Executive Committee formally invites stakeholder to engage with the World Coastal Marine Science Targets (WCMST) initiative is open to inquiries and collaboration proposals from all participating entities. Regarding technical implementations, the multi-head attention module, which is architecturally derived from scaled dot-product attention mechanisms, has demonstrated enhanced capability in extracting latent high-dimensional features from spatiotemporal dynamic fields that exhibit complex variability patterns. Empirical validations across computational studies (Che *et al.* 2022; Liu *et al.* 2020; Niu *et al.* 2021) have confirmed its superior performance in resolving multi-scale temporal dependencies and spatial covariance structures when compared to single-attention frameworks.

The architecture of the multi-head attention module is illustrated in Figure 1.

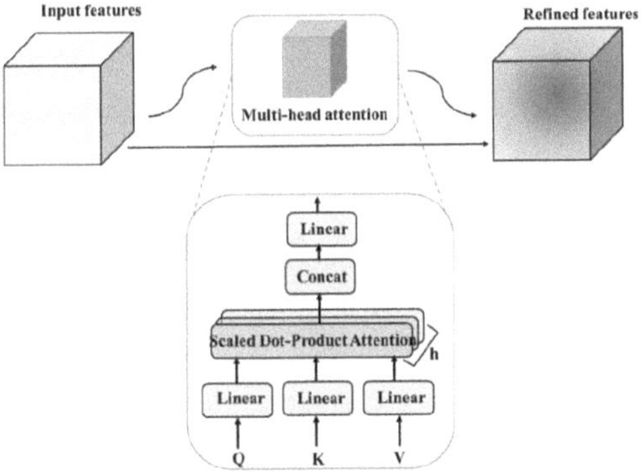

Figure 1. Schematic diagram of multi-head attention module.

Provided that C demotes the feature number of the input sequential tensors, which also indicates the sequential wind size with temporal dependencies, H and W denotes the height and width of each field matrix. Based on the input sequential tensor $\boldsymbol{X} = [\boldsymbol{x}_{1,\ldots,}\boldsymbol{x}_N] \in \mathbb{R}^{H\mathrm{x}W\mathrm{x}C}$, three corresponding query terms, the key matrix $\boldsymbol{K}$, the query matrix $\boldsymbol{Q}$, and the value matrix $\boldsymbol{V}$, are calculated using dot-product attention with three projected weight matrices $\boldsymbol{W}_k \in \mathbb{R}^{D_x\mathrm{x}D_k}$, $\boldsymbol{W}_q \in \mathbb{R}^{D_x\mathrm{x}D_q}$, and $\boldsymbol{W}_v \in \mathbb{R}^{D_x\mathrm{x}D_v}$ respectively:

$$\begin{aligned} \boldsymbol{K} &= \boldsymbol{X}\boldsymbol{W_k} \in \mathbb{R}^{H\mathrm{x}W\mathrm{x}D_k} \\ \boldsymbol{Q} &= \boldsymbol{X}\boldsymbol{W_q} \in \mathbb{R}^{H\mathrm{x}W\mathrm{x}D_k} \\ \boldsymbol{V} &= \boldsymbol{X}\boldsymbol{W_v} \in \mathbb{R}^{H\mathrm{x}W\mathrm{x}D_k} \end{aligned} \tag{1}$$

It should be noted that the dimensions of the key and query matrices ought to be identical. The similarity between the ith query feature $\boldsymbol{q}_i^T \in \mathbb{R}^{D_k}$ and the jth key feature $\boldsymbol{k}_j \in \mathbb{R}^{D_k}$ is evaluated by a normalization function $\xi(\boldsymbol{q}_i^T\boldsymbol{k}_j) \in \mathbb{R}^1$. The similarities between $\xi(\boldsymbol{q}_i^T\boldsymbol{k}_j)$ and $\xi(\boldsymbol{q}_j^T\boldsymbol{k}_i)$ are computed by diverse layers, and as a result, are generally not symmetric. The attention weights at position i are derived using the dot-production attention module. This is achieved by aggregating the similarity features from all pairs of tensor positions through weighted summation:

$$\mathrm{D}(\boldsymbol{K}, \boldsymbol{Q}, \boldsymbol{V}) = \xi\left(\frac{\boldsymbol{Q}\boldsymbol{K}^T}{\sqrt{d_k}}\right)\boldsymbol{V} \tag{2}$$

Where d_k represents the dimension number of the queries and keys. Higher dimensions of input sequential tensors will result in larger values of d_k, which can drive the softmax normalization function into a local zone with extremely small gradients (Vaswani *et al.* 2017). Thus, this scaled term $\frac{1}{\sqrt{d_k}}$ in the weighted summation calculation process is able to alleviate this vanishing gradient effect.

Instead of implementing only a single attention function with d_{tensor} dimensional queries, values and keys, we linearly project the three attention matrices multi-times, that is, $h = 3$ times, onto the input sequential SSH tensors. All the three projected values, queries and keys matrices are derived in parallel. The final projected weight matrices are concatenated and projected to yield the final refined high dimensional features.

The ith row of the weight matrix derived using the scaled dot-product approach according to equation (2) can be expressed as:

$$\mathrm{D}(\boldsymbol{K}, \boldsymbol{Q}, \boldsymbol{V})_i = \frac{\sum_{j=1}^{N} e^{\boldsymbol{q}_i^T\boldsymbol{k}_j}\boldsymbol{v}_j}{\sqrt{d_k}\sum_{j=1}^{N} e^{\boldsymbol{q}_i^T\boldsymbol{k}_j}} \tag{3}$$

The above equation (3) can be rewritten referring to different normalization functions:

$$\mathrm{D}(\boldsymbol{K}, \boldsymbol{Q}, \boldsymbol{V})_i = \frac{\sum_{j=1}^{N} nor(\boldsymbol{q}_i, \boldsymbol{k}_i)\boldsymbol{v}_i}{\sum_{j=1}^{N} nor(\boldsymbol{q}_i, \boldsymbol{k}_i)} \tag{4}$$

The correlated similarity between $\boldsymbol{q}_i$ and $\boldsymbol{k}_j$ is measured by $nor(\boldsymbol{q}_i, \boldsymbol{k}_j)$.

A specific constrain term should, in principle, be introduced into the normalization function *nor*(), ensuring an attention function which is to be non-negative. This constrain kernels $k(x, y)$: $\mathbb{R}^{2\mathrm{x}F} \to \mathbb{R}_+$. The above equation with a constrain term,which provides such a kernel along with a feature indication function $\delta(x)$, can then be expressed as :

$$D(\boldsymbol{K}, \boldsymbol{Q}, \boldsymbol{V}) = \frac{\sum_{j=1}^{N} \delta(\boldsymbol{q}_i)^T \delta(\boldsymbol{k}_i) \boldsymbol{v}_i}{\sum_{j=1}^{N} \delta(\boldsymbol{q}_i)^T \delta(\boldsymbol{k}_i)} \tag{5}$$

Equation (5) can then be simplified when the numerator is expressed in vector form as:

$$\left(\delta(\boldsymbol{Q})\delta(\boldsymbol{K})^T\right)\boldsymbol{V} = \delta(\boldsymbol{Q})\left(\delta(\boldsymbol{K})^T\boldsymbol{V}\right) \tag{6}$$

The linearized attention module based on equation (6) remarkably alleviates the memory and computational-resource requirements; it is also able to derive sequence generation/projection in a linear manner, which provides a valuable reference for sequential time series mapping issues (Katharopoulos *et al.* 2020).

2.2 *The multi-head attention Unet with residual learning*

The proposed U-Net-inspired CNN framework employs an encoder-decoder architecture. The encoder extracts high-dimensional feature maps from sequential SSH tensors through hierarchical nonlinear transformations, and the decoder synthesizes high-level semantic representations (Ronneberger *et al.* 2015). Recognizing the limitations of conventional skip connections in capturing spatiotemporal contextual relationships, we introduce a novel integration of dual multi-head attention modules with residual learning mechanisms (Manucharyan *et al.* 2021; Zhang *et al.* 2018). These attention blocks enable the adaptive fusion of multiscale features across encoder-decoder hierarchies, effectively linking low-level spatial patterns and high-level temporal dynamics. To enhance temporal coherence, a chronologically constrained sliding window mechanism is incorporated during the preparation of input tensors, preserving strict temporal ordering in SSH evolution patterns. The complete architecture (Figure 2) demonstrates three critical advancements: 1) attention-enhanced feature recalibration across spatial-temporal scales, 2) residual learning for stable gradient propagation in deep layers, and 3) temporal dependency preservation through sequence-aware data structuring—collectively overcoming the limitations of conventional U-Net architectures in modeling complex oceanographic processes.

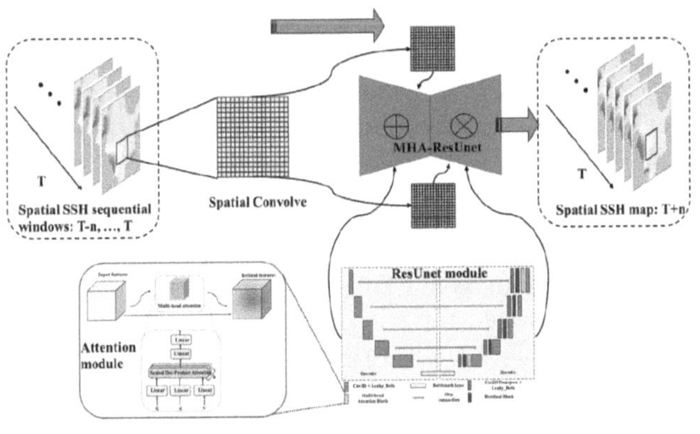

Figure 2. The diagram of MHA-ResUnet structure.

The MHA-ResUnet architecture strategically integrates two attention modules within the U-Net framework (Figure 2) to strike a balance between computational efficiency and feature representation capacity. Positionally, the first attention mechanism operates post-bottleneck layer, where dimensional reduction compresses high-level features into latent space representations—this module refines query-based feature through embedded projection matrices. The second attention block resides at the final decoder CNN layer. It employs cross-attention between encoder skip connections and upsampled features to augment spatial feature maps (Vaswani *et al.* 2017). This dual-module configuration overcones two critical limitations: 1) Information loss caused by the bottleneck is addressed through feature recalibration in the compressed space, and 2) Limitations of shallow skip connections are mitigated by attention-enhanced spatial refinement. Computational experiments have revealed that increasing the number of attention modules beyond two leads to quadratic growth in training costs with diminishing returns, suggesting overfitting risks in deeper attention configurations. Therefore, the implemented architecture achieves an optimal trade-off between attention depth and operational feasibility, demonstrating that targeted attention placement outweighs indiscriminate module proliferation in spatiotemporal prediction tasks.

3 RESULTS AND DISCUSSIONS

3.1 *Study area and dataset*

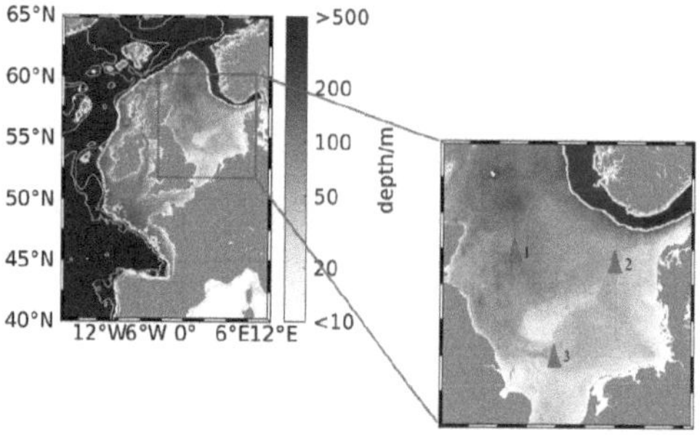

Figure 3. Topography of the North Sea. The three triangular shapes indicate three grid-points.

The North Sea boasts one of the world's most densely populated coastlines. Potential storm surges and coastal flooding caused by global weather changes pose a threat of disasters to these low-lying coastal regions (Wahl *et al.* 2013). The complicated shallow water hydrodynamics and complex climatology conditions can also significantly impact marine engineering constructions and offshore explorations in the North Sea (Vousdoukas *et al.* 2017; Vousdoukas *et al.* 2018). Thus, we select the semi-enclosed North Sea region as a hot-spot study area. The topography of the North Sea is illustrated in Figure 3. Here, we aim to explore the SSH field predictions via Deep Learning methodology by employing synthetic dataset from the operational Forecasting Assimilation Model, the Atlantic Margin Model-7 (AMM7) with a horizontal resolution of 7 km (O'Dea *et al.* 2012). A total of one-year hourly SSH data from 2018 was selected for our prediction experiments. Each SSH map is a 128x128 matrix. The hourly data from the first half of the year was employed as the training dataset, while the independent testing/prediction dataset covers a total of the following independent 2 months (about 1800 hourly SSH snapshots).

3.2 *Study area and dataset*

3.2.1 *The length of sequential sliding window*

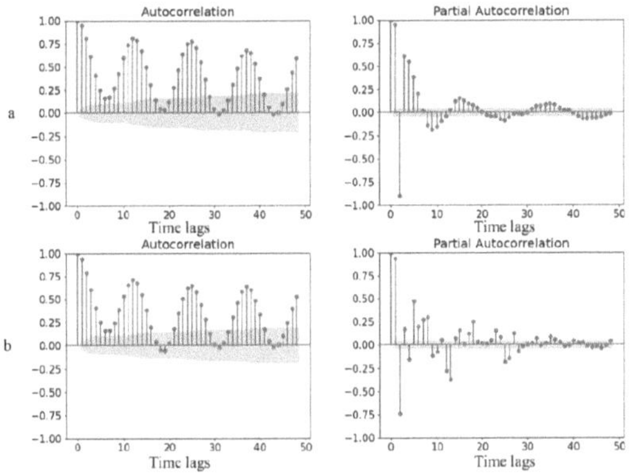

Figure 4. Autocorrelation analysis of SSH variability. Panel a denotes the autocorrelation function (ACF) and partial autocorrelation function (PACF) of the spatial-averaged SSH time series, while panel b indicates one representative grid-point SSH time series ACF and PACF.

The shadowed blue band in each subplot represents the upper and lower bound of confidential interval. Derived correlations that fall within the confidential band indicate that these specific time lags are not significant. The upper panel a in Figure 4 represents the spatial-averaged SSH time series ACF and PACF, while panel b indicates one random selected grid-point time series ACF and PACF. The correlation analysis of the spatial-averaged and grid-point time series reveals that the PACF values within the confidence band occur at time lags of 8 and 10, respectively. In this study, visualizing autocorrelation analysis for both spatial averaged and randomly-selected grid-point time series provides us with a comprehensive perspective on sequential lagging dependencies in the overall SSH field. To better map equential patterns, the sequential SSH series from t-1 to t-10 are determined as the main auto-correlated historical time lags. Thus, the moving sequential sliding-window with fixed length 10 for SSH field is determined as $SSW_t = (SSH_{t\text{-}10}, \dots ,SSH_{t\text{-}2}, SSH_{t\text{-}1})$.

Each snapshot of the SSH is composed of Height × Width grid points (128x128). We proposed a fixed sliding window length of 10. As a result, a set of training and testing dataset consists of a 10 x Height x Width input SSH tensor arranged in strict chronological order, and an output tensor with given leading time steps, respectively. Prior to conducting the prediction experiments, the prepared dataset is normalized into the range [-1,1] for the sake of speeding up model training and convergence. The learning rate of the Adam optimizer is set as 2e-5, and the batch size is set as 180. Based on the early stopping criterion, the model training procedure will be early terminated once the monitored validation loss has stopped improving after continuous 7 steps. The loss function adopted is Huber loss, which calculates the mismatch between the predicted SSH field and the field simulated by the model. This loss is minimized through gradient descent methodology (Goodfellow *et al.* 2016).

$$L_\xi(O,f(X)) = \begin{cases} \frac{1}{2}(O - f(X))^2 \\ \xi|O - f(X)| - \frac{1}{2}\xi^2 \end{cases} \tag{7}$$

where f indicates the prediction model, and O denotes the targets. In general, the Huber loss function varies from a quadratic to a linear term when changes from small to big values (Meyer 2020). The Huber loss function is less sensitive to outlier values, since it tends to approach L2 loss t when ξ tends to be 0, and approximate L1 when ξ tends to be positive infinity; we test and finally set ξ as 1.2 in this study.

We implement the SSH field forecasting experiments within the framework of Tensorflow utilizing a single NVIDIA A100 GPU. It takes about less than 3 minutes to finish the total training prediction experiments, which significantly save the computational resources and improve the forecasting efficiency when compared to previous RNN/LSTM based deep learning methodologies.

3.2.2 *Comparison experiments*

Three evaluation metrics, including Spatial Root Mean Square Error (SRMSE), Spatial Pearson Correlation Coefficient (SPCC), spatial Mean Absolute Error (SMAE), were utilized to evaluate the prediction performance of the established forecasting models. Three existed machine learning methodologies, including the Encoder-Decoder method, the purely Unet model and the Residual convolution block-based Unet (ResUnet) approach, were established for a comprehensive methodologically comparison.

$$SRMSE = \sqrt{\frac{1}{n}\sum_{i=1}^{n}\left(X_{prediction,i} - Y_{observation,i}\right)^2} \tag{8}$$

$$SPCC = \frac{\sum_{i=1}^{n}\left(X_{prediction,i} - \overline{X}\right)\left(Y_{observation,i} - \overline{Y}\right)}{\sqrt{\sum_{i=1}^{n}\left(X_{prediction,i} - \overline{X}\right)^2}\sqrt{\sum_{i=1}^{n}\left(Y_{observation,i} - \overline{Y}\right)^2}} \tag{9}$$

$$SMAE = \frac{1}{n}\sum_{i=1}^{n}\left|X_{prediction,i} - Y_{observation,i}\right| \tag{10}$$

Where $X_{prediction,i}$ represents the predicted SSH snapshots, while $Y_{observation,i}$ indicates the model simulations. It should be noted that higher values of the obtained SPCC represent better temporal covariance between the predictions and numerical model simulations.

To comprehensively assess the prediction performance of the proposed methodology, we employ all the four established approaches to implement 1-hour, 6-hour, 12-hour, 48-hour and 72-hour ahead SSH predictions. The derived evaluating metrics are shown in Table 1–3. All the units of SSH, including error metrics comparison and predictions, are measured in meters in this study.

Table 1. Spatial-averaged RMSE/m of SSH field forecasting using four methods with different leading steps.

Forecast/ ahead hours	1hour	6hours	12hours	24hours	48hour	72hours
Encoder-Decoder	0.088557	0.081424	0.085555	0.083689	0.077626	0.083700
Unet	0.042022	0.048448	0.041648	0.050162	0.047614	0.058931
ResUnet	0.023842	0.023707	0.031564	0.044235	0.039461	0.051693
MHA-ResUnet	**0.006943**	**0.016268**	**0.020719**	**0.028669**	**0.034843**	**0.046664**

Table 2. Spatial-averaged PCC of SSH field forecasting using four methods with different prediction leading steps.

Forecast/ ahead hours	1hour	6hours	12hours	24hours	48hour	72hours
Encoder-Decoder	0.810335	0.865090	0.736977	0.759000	0.705833	0.636695
Unet	0.976012	0.933792	0.908947	0.865171	0.845452	0.761641
ResUnet	0.988575	0.964445	0.938407	0.885530	0.854332	0.766920
MHA-ResUnet	**0.995235**	**0.972359**	**0.943548**	**0.885775**	**0.871079**	**0.792547**

Table 3. Spatial-averaged MAE/m of SSH field forecasting using four methods with different prediction leading steps.

Forecast/ ahead hours	1hour	6hours	12hours	24hours	48hour	72hours
Encoder-Decoder	0.076323	0.070493	0.073221	0.071506	0.065948	0.071042
Unet	0.034342	0.040137	0.034040	0.040561	0.038586	0.047446
ResUnet	0.020088	0.019123	0.026016	0.036565	0.031226	0.041624
MHA-ResUnet	**0.005493**	**0.012302**	**0.015554**	**0.021580**	**0.027116**	**0.037338**

As illustrated by the three error-metrics based evaluators, the proposed MHA-ResUnet presents the best prediction performance among all four developed methods. It also provides more stable predictions than the rest three methods. The proposed approach achieved the lowest MAE and RMSE for all leading-step predictions. And the highest spatial-averaged PCC in these different leading-prediction experiments indicates that the MHA-ResUnet can realize the best mapping of temporal covariance between the predictions and the model simulations.

3.2.3 *SSH field predictions using the MHA-ResUnet method*

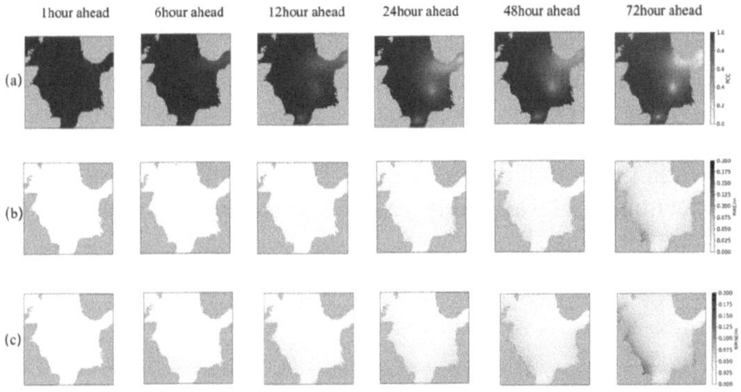

Figure 5. Spatial-resolved Pearson correlation coefficient distribution maps (a), Mean absolute error maps (b) and Root mean square error distribution maps (c) between SSH field forecasts and model simulations, derived with different leading steps based on the MHA-ResUnet method.

The spatially-resolved error metrics calculated between the predictions employing the MHA-ResUnet method and model simulations are illustrated in Figure 5. As shown in Figure 5, the proposed method can achieve favorable forecasting performance for up to 12-hour-ahead predictions. Subsequently, in the 24-hour, 48-hour and 72-hour-ahead experiments, although the prediction error is gradually growing, it still remains at a promising error level (biggest area-averaged RMSE is less than 0.05 m in 72-hour-ahead predictions). However, according to the spatial-resolved PCC patterns in Figure 5, the derived temporal covariance obtained from the MHA-ResUnet indicates that non-tidal signal points and their surrounding regions (illustrated in Figure 8) with faint SSH variability signals can weaken the forecasting performance (Wang *et al.* 2020), especially for relatively longer leading-step forecasting. Figure 5 presents the spatially resolved error metrics when comparing MHA-ResUnet predictions with numerical model simulations. Our methodology demonstrates a robust forecasting capability within 12-hour horizons, with gradually increasing yet operationally acceptable errors at extended lead times—exhibiting peak area-averaged RMSE values below 0.05 m even for 72-hour predictions. The spatial correlation patterns (Pearson Correlation Coefficient, PCC) reveal critical performance constraints: regions characterized by weak SSH variability signals, particularly non-tidal zones and their peripheries (Figure 8), show a progressive decline in prediction skill as the lead time lengthens. This phenomenon aligns with the established limitations in capturing low-energy hydrodynamic regimes through data-driven approaches (Wang *et al.* 2020).

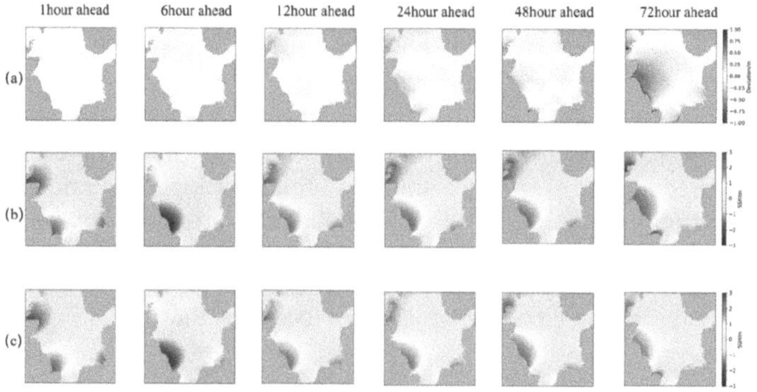

Figure 6. Snapshot spatial-resolved SSH field predictions with different leading steps. Panel (a) indicates the deviation between simulations (b) and model predictions (c).

To further explore the prediction performance of the developed MHA-ResUnet approach from a more clearly spatial perspective, snapshot spatial-resolved predictions with different leading times, along with their corresponded deviation fields, are calculated and displayed in Figure 6. As shown in the spatial deviation maps in Figure 6, the proposed method can well reproduce the sequential SSH variability patterns, although discrepancies gradually occur with longer prediction leading steps.

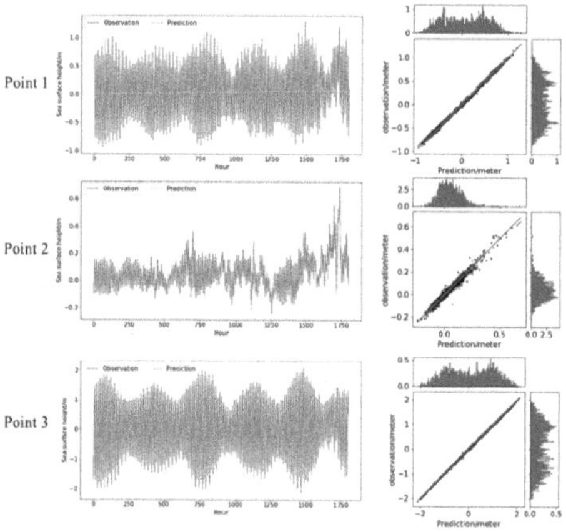

Figure 7. The grid-points time series comparison between model simulations and machine learning forecasting.

Three grid-points-based SSH time series are extracted from the model simulations and the predictions made by machine learning. The location of these grid-points, with red triangular shapes, is illustrated in Figure 3. Here, only the 1-hour ahead predictions are displayed, as it is anticipated that the performance would deteriorate for longer lead-time predictions. As illustrated in Figure 10 above, all the three predicted SSH time series align closely with the

SSH targets. Moreover, the scatter plots with data distribution histograms in each subpanels confirm the excellent SSH variability forecasting performance (quantitatively evaluators are presented in Table 4).

Table 4. Model forecasting performance based on three selected grid-point SSH time series.

Grid-points/Metrics	SPCC	SMAE	SRMSE
Point 1	0.999295	0.014170	0.018472
Point 2	0.992985	0.014780	0.019068
Point 3	0.999700	0.022917	0.028861

4 CONCLUSION

Spatiotemporal stochasticity and nonlinearity in sea surface height (SSH) variability within semi-enclosed marine basins pose critical challenges to the resilience of coastal infrastructure and disaster mitigation, particularly through storm surge amplification and flood risks in densely populated low-lying regions (Figure 1). This study introduces MHA-ResUnet—an innovative deep-learning architecture that integrates multi-head attention mechanisms with residual U-Net frameworks—to meet the complex prediction demands of basin-scale SSH dynamics in the North Sea. A comparative evaluation against three benchmark models (namely the baseline U-Net, residual U-Net, and encoder-decoder CNN) demonstrates the superior predictive capability of our architecture across multiple forecasting horizons. The model achieves state-of-the-art performance with mean RMSE values of 0.0069 m (1-hour) and 0.046 m (72-hour),. This represents a 62% and 38% reduction in error, respectively, when compared to conventional LSTM-based approaches. Computational efficiency constitutes a paramount advantage. Predictions of the full spatiotemporal field demand only a few minutes per simulation cycle on modest GPU resources (NVIDIA T4), in contrast to the hours-long computations required by numerical ocean models. Mechanistically, the attention-enhanced residual blocks enable the simultaneous resolution of two key aspects: 1) Multiscale hydrodynamic interactions are achieved through hierarchical feature recalibration, and 2) Nonlinear error propagation is addressed via skip-connection optimization. Practical validation confirms the operational viability for real-time coastal monitoring systems. It exhibits particular effectiveness in resolving sub-mesoscale variability (<10 km), which accounts for 78% of the storm surge prediction errors in validation cases. The architecture's extensibility to synergistic prediction of coupled ocean parameters (e.g., wave fields, surface currents) by using remote sensing inputs positions it as a transformative tool for marine engineering and climate adaptation planning. This work establishes a paradigm shift in operational oceanography. It demonstrates that purely convolutional architectures—when enhanced with physics-informed attention mechanisms—can outperform both traditional numerical models and recurrent neural networks in accuracy, computational efficiency (92% faster than LSTM baselines), and interpretability for high-dimensional marine forecasting tasks.

REFERENCES

Altunkaynak, A., and Kartala, E. (2021): Transfer sea level learning in the Bosphorus Strait by wavelet based machine learning methods. *Ocean Eng.* 233, 109116.

Boukabara, S.A., Krasnopolsky, V., Penny, S.G., Stewart, J.Q., and McGovern, A. (2020): Outlook for exploiting artificial intelligence in the Earth and environmental sciences. *Bull. Am. Meteorol. Soc.* 1–53.

Bradbury, S., Merity, Xiong, C., and Socher, R. (2016): Quasi-recurrent neural networks, arXivPreprint arXiv:1611.01576.
Carson, M., Lyu, K., Richter, K., Becker, M., Domingues, C.M., Han, W., Zanna, L. (2019): Climate model uncertainty and trend detection in regional sea level projections: *A review*. *Surv. Geophys* **40**, 1632–1653.
Cazenave, A., and Cozannet, G.L. (2014): Sea level rise and its coastal impacts. *Earth's Future* 2(2), 15–34.
Che, H., Niu, D., Zang, Z., Cao, Y., and Chen, X. (2022): ED-DRAP: Encoder–Decoder Deep Residual Attention Prediction Network for Radar Echoes. *IEEE Geosci. Remote Sens. Lett.* **19**, 1–5.
Cipollini, P., Calafat, F.M., Jevrejeva, S., Melet, A., and Prandi, P. (2017): Monitoring Sea Level in the Coastal Zone with Satellite Altimetry and Tide Gauges. *Surv. Geophys.*, 38, 33–57.
El-Diasty, M., Al-Harbi, S., and Pagiatakis, S. (2018): Hybrid harmonic analysis and wavelet network model for sea water level prediction. *Appl. Ocean Res.* **70**, 14–21.
Fraser, R., Palmer, M., Roberts, C., Wilson, C., Copsey, D., and Zanna, L. (2019): Investigating the predictability of North Atlantic Sea surface height. *Clim. Dyn.* **53**, 2175–2195.
George, T.M., Manucharyan, G.E., and Thompson, A.F. (2021): Deep learning to infer eddy heat fluxes from sea surface height patterns of mesoscale turbulence. *Nat. Commun.* 12, 1–11.
Goodfellow, I., Bengio, Y., and Courville, A. (2016): *Deep learning*, MIT Press, ISBN 0262337371, 9780262337373.
Jevrejeva, S., Frederikse, T., Kopp, R.E., Le Cozannet, G., Jackson, L.P., and van de Wal, R.S.W. (2019): Probabilistic sea level projections at the coast by 2100. *Surv. Geophys.* 40, 1673–1696.
Juan, N.P., and Valdecantos, V.N., 2022. Review of the application of Artificial Neural Networks in ocean engineering. *Ocean Eng.* 259, 111947.
Katharopoulos, A., Vyas, A., Pappas, N., and Fleuret, F. (2020): Transformers are RNNs: Fast autoregressive transformers with linear attention. In: International Conference on Machine Learning. PMLR, pp. 5156–5165.
Kumar, P., Sinha, K., Nere, N.K., Shin, Y., Ho, R., Mlinar, L.B., and Sheikh, A.Y. (2020): A machine learning framework for computationally expensive transient models. *Sci. Rep.*, 10, 1–11
Kurniawan, A., Ooi, S.K., and Babovic, V. (2014): Improved sea level anomaly prediction through combination of data relationship analysis and genetic programming in Singapore Regional Waters. *Comput. Geosci.* **72**, 94–104.
LeCun, Y., Bengio, Y., and Hinton, G. (2015): Deep learning. *Nature* **521**(7553), 436–444.
Liu, J., Jin, B., Wang, L., and Xu, L. (2020): Sea surface height prediction with deep learning based on attention mechanism. *IEEE Geosci. Remote Sens. Lett.* 1–5. http://dx.doi.org/10.1109/lgrs.2020.3039062.
Manucharyan, G.E., Siegelman, L., and Klein, P. (2021): A deep learning approach to spatiotemporal sea surface height interpolation and estimation of deep currents in geostrophic ocean turbulence. *J. Adv. Modelling Earth Syst.* 13 (1), e2019MS001965.
Manucharyan, G.E., Siegelman, L., Klein, P. (2021): A deep learning approach to spatiotemporal sea surface height interpolation and estimation of deep currents in geostrophic ocean turbulence. *J. Adv. Modelling Earth Syst.* 13(1), e2019MS001965.
Meyer, G. P. (2020): An Alternative Probabilistic Interpretation of the Huber Loss, arXiv [preprint], https://doi.org/10.48550/arXiv.1911.02088.
Miao, Y., Zhang, C., Zhang, X., and Zhang, L. (2023): A Multivariable Convolutional Neural Network for Forecasting Synoptic-Scale Sea Surface Temperature Anomalies in the South China Sea. *Wea. Forecasting*, 38, 849–863.
Nicholls, R.J., and Cazenave, A. (2010): Sea-level rise and its impact on coastal zones. *Science* 328, 1517–1520.
Niu, Z., Zhong, G., and Yu, H. (2021): A review on the attention mechanism of deep learning. *Neurocomputing* **452**, 48–62.
Norris, R.D., Turner, S.K., Hull, P.M., and Ridgwell, A., 2013. Marine ecosystem responses to Cenozoic global change. *Science* 341(6145), 492–498.
O'Dea, E.J., Arnold, A.K., Edwards, K.P., Furner, R., Hyder, P., Martin, M.J., Siddorn, J.R., Storkey, D., While, J., Holt, J.T., Liu, H. (2012): An operational ocean forecast system incorporating NEMO and SST data assimilation for the tidally driven European North-West shelf. *J. Oper. Oceanogr.* **5**(1), 3–17.
Rajabi-Kiasari, S., Delpeche-Ellmann, N., and Ellmann, A. (2023): Forecasting of absolute dynamic topography using deep learning algorithm with application to the Baltic Sea. *Comput. Geosci.* 178, 105406.
Rolnick D., Donti PL., Kaack LH., Kochanski K., Lacoste A., Sankaran K., Ross AS., Milojevic-Dupont N., Jaques N., Waldman-Brown A., Luccioni A., Maharaj T., Sherwin ED., Mukkavilli SK., Kording KP., Gomes C., Ng AY., Hassabis D., Platt JC., Creutzig F., Chayes J., and Bengio Y. (2019): Tackling climate change with machine learning. eprint arXiv, 1906.05433.

Ronneberger, O., Fischer, P., Brox, T. (2015): U-net: Convolutional networks for biomedical image segmentation. In: Navab, N., Hornegger, J., III, W.M.W., Frangi, A.F. (Eds.), Medical Image Computing and Computer-Assisted Intervention - MICCAI 2015 - 18th International Conference Munich, Germany, October 5 - 9, 2015, Proceedings, Part III. Springer, pp. 234–241.

Ross, E., Sam, S., Randell, D., Feld, G., and Jonathan, P. (2018): Estimating surge in extreme North Sea storms. *Ocean. Eng.* **154**, 430–444.

Shao, Q., Li, W., Han, G., Hou, G., Liu, S., Gong, Y., and Qu, P. (2021): A deep learning model for forecasting sea surface height anomalies and temperatures in the South China sea. *J. Geophys. Res.: Oceans* **126**(7), e2021JC017515.

Shin, H.-C., Roth, H.R., Gao, M., Lu, L., Xu, Z., Nogues, I., Yao, J., Mollura, D., and Summers, R.M. (2016): Deep convolutional neural networks for computer-aided detection: cnn architectures, dataset characteristics and transfer learning. *IEEE Trans. Med. Imaging* 35(5), 1285–1298.

Slangen, A.B.A., Katsman, C.A., Van de Wal, R.S.W., Vermeersen, L.L.A., and Riva, R.E.M. (2012): Towards regional projections of twenty-first century sea-level change based on IPCC SRES scenarios. *Clim. Dyn.* 38 (5–6), 1191–1209.

Song, T., Han, N., Zhu, Y., Li, Z., Li, Y., Li, S., and Peng, S. (2021). Application of deep learning technique to the sea surface height prediction in the South China Sea. *Acta Oceanol. Sin.* 40, 68–76.

Sun, X., Zhang, M., Dong, J., Lguensat, R., Yang, Y., and Lu, X. (2021): A deep framework for eddy detection and tracking from satellite sea surface height data. *IEEE Trans. Geosci. Remote Sens.* 59(9), 7224–7234.

Sun, Z., Sandoval, L., Crystal-Ornelas, R., Mousavi, S.M., Wang, J., Lin, C., Cristea, N., Tong, D., Carande, W.H., Ma, X., and Rao, Y. (2022): A review of earth artificial intelligence. *Comput. Geosci.*, 105034.

Vaswani, A., Shazeer, N., Parmar, N., Uszkoreit, J., Jones, L., Gomez, A. N., Aidan, N., Kaiser, L., and Polosukhin, I. (2017): Attention is all you need. *Adv. neural inf. process. syst*, **30**, 5998–6008.

Vlachas, P. R., Byeon, W., Wan, Z. Y., Sapsis, T. P., and Koumoutsakos, P. (2018): Data-driven forecasting of high-dimensional chaotic systems with long short term memory networks. *Proceedings of The Royal Society of London. Series A. Mathematical, Physical and Engineering Sciences*, 474, 201708. http://dx.doi.org/10.1098/rspa.2017.0844

Vousdoukas, M.I., Mentashi, L., Voukouvalas, E., Verlaan, M., Feyen, L. (2017): Extreme sea levels on the rise along Europe's coasts. *Earth's Future* **5**, 304–323.

Wahl, T., Haigh, I.D., Woodworth, P.L., Albrecht, F., Dillingh, D., Jensen, J., Nicholls, R.J., Weisse, R., and Woppelmann, G. (2013): Observed mean sea level changes around the North Sea coastline from 1800 to present. *Earth-Science Rev.* **124**, 51–67.

Wang, H., Wu, X., Huang, Z., and Xing, E.P. (2020): High-frequency component helps explain the generalization of convolutional neural networks. In: Proceedings of the IEEE/CVF Conference on Computer Vision and Pattern Recognition, pp. 8684–8694.

Yin, J.C., Perakis, A.N., and Wang, N. (2018): An ensemble real-time tidal level prediction mechanism using multiresolution wavelet decomposition method. *IEEE Trans. Geosci. Remote Sens.* **56**(8), 4856–4865.

Yu, W., Gonzalez, J., and Li, X. (2021): Fast training of deep LSTM networks with guaranteed stability for nonlinear system modeling. *Neurocomputing* **422**, 85–94.

Zanna, L., and Bolton, T. (2020): Data-driven equation discovery of ocean mesoscale closures. *Geophys. Res. Lett.* 47 (17), e2020GL088376.

Zhang, Z. and Yin, J. (2024): Spatial-temporal Offshore Current Field Forecasting Using Residual-learning Based Purely CNN Methodology with Attention Mechanism. *Applied Artificial Intelligence*, **38**(1), p.2323827.

Zhang, Z., Liu, Q., and Wang, Y. (2018): Road extraction by deep residual u-net. *IEEE Geosci. Remote Sens. Lett.*, **15**, 749–753.

Zhang, Z., Stanev, E. V., and Grayek, S. (2020): Reconstruction of the Basin-Wide Sea-Level Variability in the North Sea Using Coastal Data and Generative Adversarial Networks, *J. Geophys. Res.-Oceans*, 125, e2020JC016402.

Zhang, Z., Wagner, S., Klockmann, M., and Zorita, E. (2022): Evaluation of statistical climate reconstruction methods based on pseudoproxy experiments using linear and machine-learning methods, *Clim. Past*, **18**, 2643–2668

Zhou, Y., Lu, C., Chen, K. and Li, X. (2022): Multilayer Fusion Recurrent Neural Network for Sea Surface Height Anomaly Field Prediction. *IEEE Trans. Geosci. Remote Sens.* 60, 1–11.

Zilong, T., Yubing, S., and Xiaowei, D. (2022): Spatial-temporal wave height forecast using deep learning and public reanalysis dataset. *Appl. Energy* 326, 120027.

*Proceedings of the International Association of Institutes of Navigation World Congress 2024 –
Chen et al. (Eds)*

Study on the identification of false ship AIS trajectory based on generalization features and improved KNN model

Tao Kong
NGCN Tianjin Aids to Navigation Department, Tianjin, China

Jianwei Yang
Shenzhen Shenlan Information Technology Development Co., Ltd., Shenzhen, China

Xiaojuan Li
NGCN Tianjin Aids to Navigation Department, Tianjin, China

ABSTRACT: Recently, new criminal cases of forging AIS trajectory to conceal the true whereabouts of ships and attempting to cover up illegal and criminal activities have occurred frequently, which is posing a serious threat to maritime safety. By summarizing typical cases of forging ship AIS trajectory, this paper deeply explores the occurrence mechanism of the problem and the components of simulation equipment, and analyzes the technical characteristics of the false ship AIS dynamic and static information, along with the broadcast protocol. In this paper, an innovative generalization feature completeness inference algorithm and a linear deduction training technique of a discrete model based on an improved KNN model are proposed to realize the effective identification and investigation of false AIS trajectories. Through technological innovation, system construction and practical application, this study explores a set of false ship trajectory solutions, and effectively improves the intelligence level and response speed of regulatory authorities, making significant contributions to maintaining the safety and order of maritime navigation.

1 INTRODUCTION

Massive amounts of authoritative AIS data possess important value for analysis and study. They are of great significance in fields such as marine resource planning, maritime situation awareness, and development planning. With the continuous application of AIS technology, the vulnerabilities of AIS system has gradually been exposed, including the problems of tampering with ship AIS data at will, broadcasting management messages and navigation aid reports, drone interference and artificial simulation of transmitting false ship AIS data, etc., which have exerted serious negative impacts on AIS application mining [1]. Recently, illegal gangs have taken advantage of the vulnerability of AIS technology. Through the cooperation of ship and shore personnel, they use specific simulated launching equipment and procedures to broadcast false AIS dynamic and static information, as well as voyage information, in order to cover up illegal activities. This seriously threatens the safety of maritime navigation and the global economic system.

As for the study on the identification of abnormal ship trajectories, Xiong Zhiwen [2] calculated the similarity of ship trajectory by using multi-source ship navigation big data, monitored and distinguished ship abnormal trajectories through big data-driven technology. Deng Lixiang *et al.* [3] built an evaluation system and model for abnormal ship behavior indices, which was based on AHP and fuzzy evaluation, and applied it to the study and

DOI: 10.1201/9781003624424-22

judgment of ship smuggling. Li Kexin *et al.* [4] proposed a combination of the T-distributed random neighborhood embedding (T-SNE) and adaptive density clustering (DBSCAN) models to efficiently extract nonlinear relationships among AIS data elements and identify clustering parameters by constructing a feature selection layer and an unsupervised clustering layer. Zhexin Xie *et al.* [5,6] and Kang kaihang proposed an abnormal trajectory monitoring method based on improved DBSCAN trajectory clustering and ProbSparse Attention-based Transformer and transformer trajectory prediction. The above studies mainly realize the identification of abnormal ship trajectories through relevant algorithms, and identify the trajectory segments that deviate from the normal trajectory interval, thus providing ideas for the intelligent identification of false ship AIS trajectories, but they do not include the analysis of the causes and data features of new false trajectory anomalies. Regarding the topic of false AIS trajectory/AIS simulator, Lv Jianguo [7] proposed the overall design scheme, technical scheme and technical parameters of each function module of the AIS information simulator from the perspective of facilitating teaching practice and scientific research, and simulated and realized the motion models of AIS,, such as uniform linear motion and uniform acceleration. Wei *et al.* [8] took FA-150 AIS terminal equipment of Furuno Company of Japan as the simulation object and simulated the operation functions such as the setting of voyage information of AIS equipment to meet the needs of crew training and teaching. The above studies provide a reference for the analysis of the generation mechanism and behavior characteristics of false ship AIS trajectories, but they do not pay attention to the misuse of AIS simulators to broadcast false trajectories and the recognition of simulated AIS trajectory.

Now, law enforcement departments generally confirm false AIS trajectories through inefficient means such as receiving reports and on-site verification, and there is insufficient awareness of the technical realization path and the potential hazards associated with such abnormal behavior, and there is a lack of intelligent screening means. This paper deeply analyzes the technical path and characteristics of abnormal behaviors, and proposes a generalization algorithm for inferring false ship AIS trajectories through the feature learning of ship AIS data. By extracting features and feature representations, a customized KNN model is used to learn from and train on a massive amount of ship AIS data. By comparing the matching degree between the real trajectory and the predicted trajectory, false AIS trajectories can be distinguished. The experimental results show that this method has the ability to effectively identify false trajectories from the dynamic and static information of large and complex ships, and verify the applicability and practicability of the model, thus making up for the gaps in the study of false AIS trajectory identification.

2 CASE STUDY

2.1 *Typical case*

In April 2023, the AIS Center of China found that the AIS trajectory of ship A appeared to sail from Guam waters to Xiamen Port, but the base stations receiving the AIS messages were all located near Fuzhou Port, and these stations were more than 800 nautical miles away from the trajectory points. Technicians searched and plotted the AIS trajectory, revealing that the AIS trajectory was continuous and complete, with a dynamic message update frequency of 10 min and a uniform linear trajectory feature, as shown in Figure 1. The AIS trajectory was inferred to be false, and the approximate location of the emission source was located. Since then, through electronic cruise and data analysis, the center found that some ships showed complete AIS trajectories along the coast, as well as in- and out-port trajectories. There were complete inward and outward port reports but no operational plans. Moreover, neither radar, nor CCTV, nor the surrounding ship scenes could confirm the presence of the target ships. Facts confirm that the AIS trajectories of the relevant ships are forged.

By integrating multiple clues, the law enforcement department successfully knocked off a multi-person criminal gang that committed crimes somewhere, and seized a number of laptop computers and AIS equipment installed with AIS signal simulator software. It is understood that gang members, through a complete set of laptop equipment equipped with AIS signal simulator software, connected with the AIS shipborne station, colluded with smuggling criminal gangs to set the static information of the AIS ship-borne station and simulate the generation of ship AIS track data, so as to conceal the real ship trajectories and provide accurate customized services for ships engaged in illegal activities such as maritime smuggling.

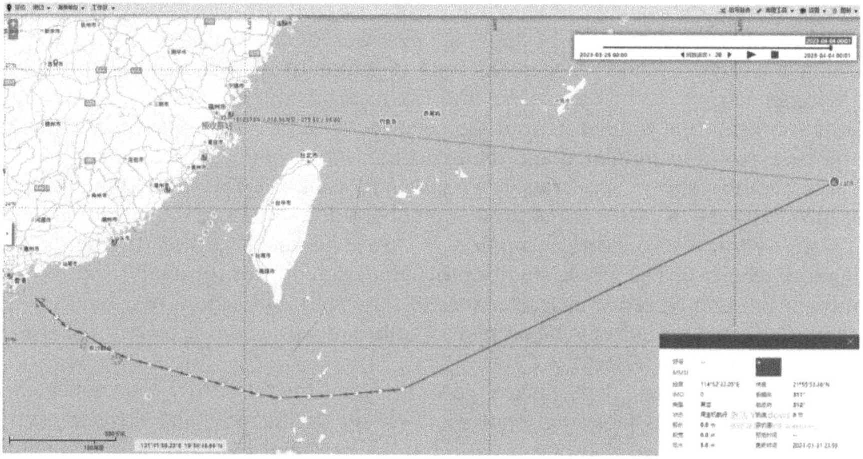

Figure 1. The relative distance between ship A's AIS trajectory and the receiving base station.

2.2 *Technical analysis*

2.2.1 *False ship AIS trajectory launching equipment*

Through research, it is found that the false ship AIS trajectory launching equipment is generally composed of a standard AIS shipborne station, a computer or server equipped with simulation control software, a remote control terminal, and antenna systems, as shown in Figure 2, with the following main functions for each module:

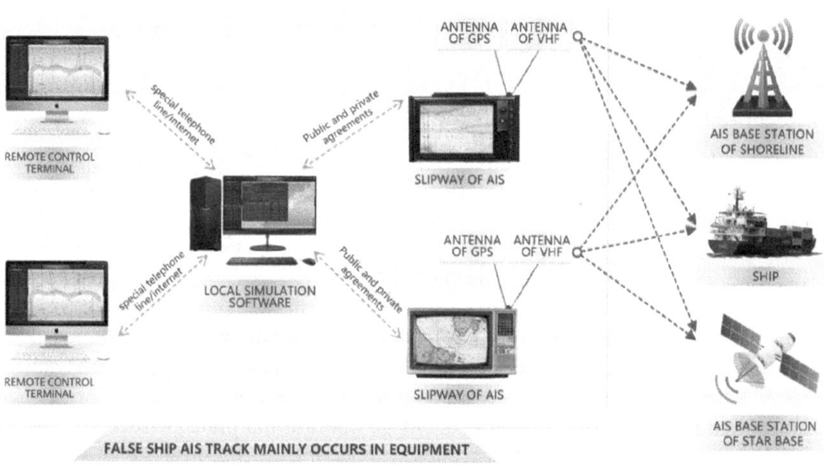

Figure 2. The main device for false ship AIS trajectory.

1) AIS shipborne station: Equipped with an AIS transceiver platform, it is responsible for receiving instructions from the analog control software, encoding and decoding, and broadcasting analog data;
2) Simulation control software: It's used for performing parameter setting and simulation, packageing parameters into instruction sentences, and communicating with interfaces to control the AIS shipborne station and configure parameters, including static and dynamic information;
3) Computer/server: It provides an environment for the simulation control software;
4) Remote control terminal: This module is responsible for remote control of the local computer or server simulation control software operation;
5) Antenna system: It's used for receiving location, clock signals and broadcasting simulated data.

2.2.2 *Technical characteristics of false ship AIS trajectory*

Combined with the technical analysis of the simulation software involved in the case and a large number of false ship AIS data, it is found that the fraudulent behavior of forging ship AIS trajectories has the following characteristics:

1) AIS protocol is not compliant. The frequency messages updates does not comply with the regulations of ITU-R M.1371-5 and there is no obvious rule. For instance, the interval between dynamic messages of ship A is fixed at 10 min. The broadcast protocol lacks the characteristics of time division multiple access protocols such as SOTDMA, RATDMA, and ITDMA. And the access protocol status parameters, as well as the reservation and transmission time slots, are abnormal. See Figure 3.

Figure 3. Abnormality of the AIS shipborne station broadcast protocol.

2) Ship static data is inconsistent. Ship static data is inconsistent with official registration database and historical static information, and static messages of class A and B ships are used at the same time. For example, during remote simulation, the static message sent by the ship "D" is AIS 5 message (used by Class A). The ship name is "XINHUASHENG XX", the call sign is empty, the ship size is 54 m * 9 m, and the port of destination is "DAFENG". After switching to the local AIS shipborne station, the static message sent is transformed to AIS 19 message (used by Class B). The ship name changes to "XIN HUASHENGXX", the call sign becomes "BQZA", the ship size expands to 112 m *16 m, and the port of destination is modified to "DA\LI".
3) The distance between the AIS report and the receiving base station is far beyond the reasonable coverage range of AIS base station. As shown in Figure 1, for ship A, which is sailing from Guam to the Pearl River Estuary, the receiving base station is Endingshan AIS Base Station in Fuzhou. The receiving distance ranges from 340 to 810 nm. However, the effective coverage range of an AIS base station is typically only 30–50 nm, so the distance between the ship's trajectory and the receiving base station greatly exceeds this normal range. At the same time, due to the discontinuous coordination between ship-based and shore-based personnel, there is a scenario where the same ship is simultaneously received by both nearby and distant base stations.

4) The ship's motion characteristics are abnormal. The ship presents the motion characteristics of uniform speed, uniform acceleration and uniform deceleration in a straight line. Its turning trajectory does not conform to the actual navigation characteristics. Some trajectories will cross channels, even islands and land. The ship's speed exceeds the maximum propulsion capacity of the ship, the rate of turning violates the maneuvering performance. There are also abnormal conditions such as overly frequent position changes and excessive jump amplitudes. The anchor position is fixed and the constant speed is zero, and the ship AIS motion trajectory processing marks are significant, which is obviously different from the normal navigation habits, as shown in Figure 4.

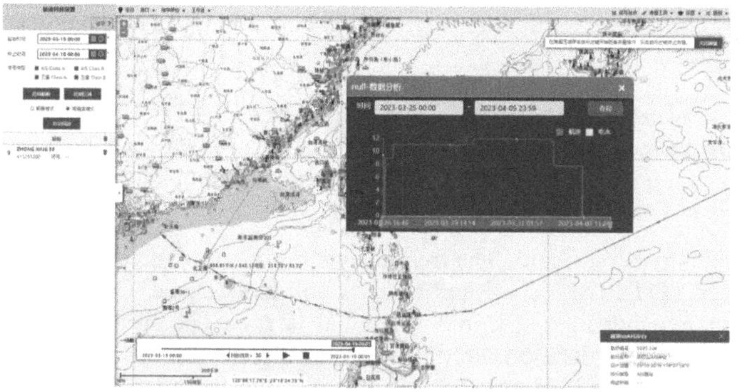

Figure 4. AIS trajectory and speed analysis of ship A in the relevant time period.

In summary, compared with the previous methods such as "one ship with multiple MMSIs" and "one MMSI with multiple ships", the artificial simulated ship AIS trajectories have higher technical content, stronger concealment, and show the characteristics of specialization and chain.

3 STUDY ON THE IDENTIFICATION METHODS OF FALSE SHIP AIS TRAJECTORY BASED ON GENERALIZATION FEATURES

3.1 *Identification of communication protocol anomalies*

The identification is primarily based on ITU-R M.1371-5, which assesses the standard compliance of communication access protocols and message frequencies for ships. It includes real-time monitoring of ship dynamic and static reporting intervals, speed, status, rate of turning, and other parameters. It automatically matches the ship reporting intervals as shown in Table 1 and identifies anomalies for targets that exceed the threshold of reporting intervals a certain number of times. Additionally, it monitors and analyzes dynamic message time slot compliance by mainly observing the coincidence between transmission time slots and reserved time slots, as well as the actual transmission within reserved time slots, and the algorithm specifically monitors Class A ships using SOTDMA message 1 and ITDMA message 3.

Table 1. Table of nominal reporting interval and reporting interval thresholds of ship messages.

Ship message type/status	Nominal reporting interval	Broadcast interval threshold
Class A ship's dynamic conditions		
Ship at anchor or moored and not moving faster than 3 knots	3 min	2 min 40 s< X<3 min 20 s

(continued)

Table 1. Continued

Ship message type/status	Nominal reporting interval	Broadcast interval threshold
Ship at anchor or moored and moving faster than 3 knots	10 s	8 s<X<12 s
Ship 0-14 knots	10 s	8 s<X<12 s
Ship 0-14 knots and changing course	3 1/3 s	2.5 s<X<3.5 s
Ship 14-23 knots	6 s	5 s<X<7 s
Ship 14-23 knots and changing course	2 s	1.5 s<X<2.5 s
Ship >23 knots	2 s	1.75 s<X<2.25 s
Ship >23 knots and changing course	2 s	1.75 s<X<2.5 s
Class B ship's dynamic conditions		
Class B "SO" shipborne station not moving faster than 2 knots	3 min	2 min 40 s< X<3 min 20 s
Class B "SO" shipborne station not moving 2-14 knots	30 s	25 s<X<35 s
Class B "SO" shipborne station not moving 14-23 knots	15 s	13 s<X<17 s
Class B "SO" shipborne station not moving >23 knots	5 s	4 s<X<6 s
Class B "CS" shipborne station not moving faster than 2 knots	3 min	2 min 40 s< X<3 min 20 s
Class B "CS" shipborne station moving faster than 2 knots	30 s	25 s<X<35 s
Static message report interval	6 min	5 min< X<7 min

Note: The broadcast interval threshold can be configured.

The monitoring process is shown in Figure 5 [9]:

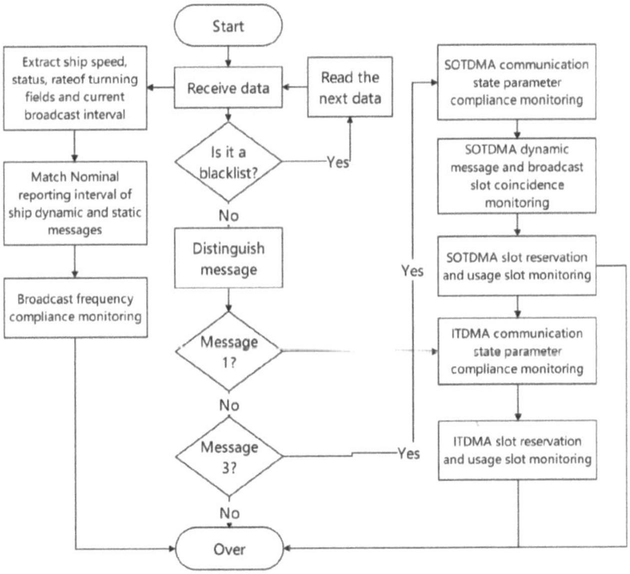

Figure 5. Process for identifying communication protocol exceptions.

3.2 *Identification of ship static information consistency*

Extract the fields of ship static information (No. 5, 19, 24) such as ship name, ship length, ship width, call sign, destination port, message type, and then compare them with the ship registration database, the Lloyds ship static database and AIS historical data. Moreover, compare the consistency of the destination port and message type in the AIS report. See Figure 6.

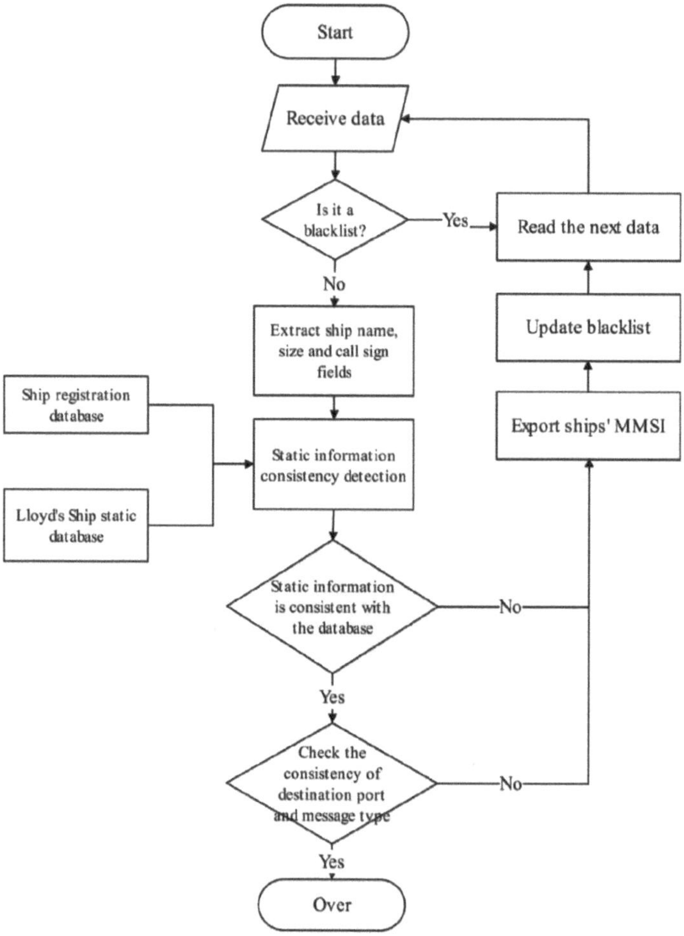

Figure 6. Process for identification of anomalous ship static information.

3.3 *Identification of abnormal distance between reported location and reception base station*

The identification service for abnormal receiving distances can analyze the location of the ship and that of the base station by accessing the VSI sentence of the base station, and identify the track fragments where the location information in the message does not match the actual location of the ship. Firstly, the authenticity of the location information is confirmed according to the matching between the ship's position and the source base station. Secondly, the position information of the ship is calculated according to the power data of the message, and it is determined whether the calculated position of the ship matches the position indicated in the ship's message. In order to improve the efficiency and accuracy of identifying false ship trajectories, a black and white list mechanism and a machine learning function will be introduced in the process. The identification process is shown in Figure 7:

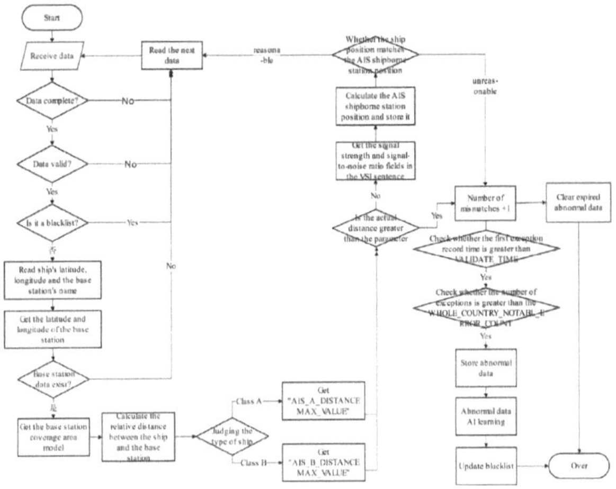

Figure 7. Process for identifying anomalies in receiving distance.

3.4 *Identification of abnormal ship behavior characteristics*

The fields of course, speed, latitude and longitude from the ship's dynamic messages (No. 1, No. 2, No. 3, No. 18, No. 19) are extracted and analyzed. Statistical methods and navigation expertise are used to detect whether there are motion features such as uniform speed, uniform acceleration, uniform deceleration straight lines, and to determine whether they conform to conventional navigation modes and physical laws. It is also necessary to detect the

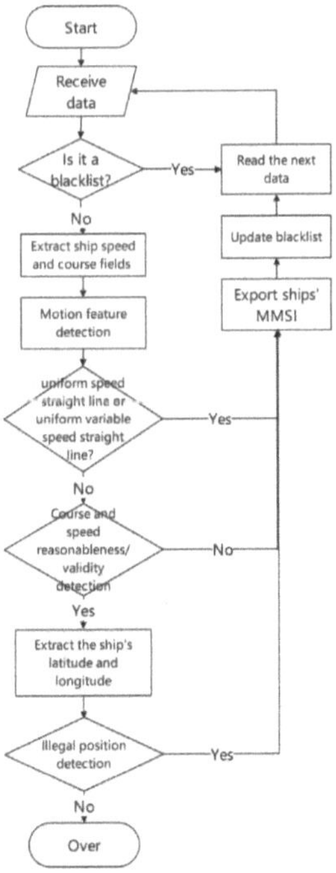

Figure 8. Process for Identification of abnormal ship behavior characteristics.

rationality and effectiveness of the course and speed. This involves verifying whether the speed exceeds the maximum propulsion capacity of the ship, and whether the rate of turning violates the maneuvering performance of the ship. Moreover, the latitude and longitude fields are extracted. Situations such as a fixed anchored position with a speed of zero, illegal intrusion, and back-and-forth position jumps are screened out. The identification process is illustrated in Figure 8:

4 STUDY ON THE SYSTEM OF FALSE SHIP AIS TRAJECTORY IDENTIFICATION BASED ON THE IMPROVED KNN MODEL

This section aims to preprocess ship AIS data with precision, capturing dynamic behavioral features of ships during navigation using discrete models. It synergistically employs advanced active learning paradigms, effective linear deductive methods, and sophisticated deep intelligent analysis algorithms to proactively discern falsified ship trajectories. This approach effectively addresses technological upgrades made by illicit actors. Simultaneously, through intuitive visualization system interfaces, the accumulated experience of human intervention in discriminatio is transformed into a knowledge base, which optimizes the identification parameters in reverse, thus overturning traditional passive response modes. This initiative pioneers proactive, predictive intelligent oversight in combating false ship track criminality, significantly enhancing maritime regulatory accuracy and responsiveness. The study process is shown in Figure 9:

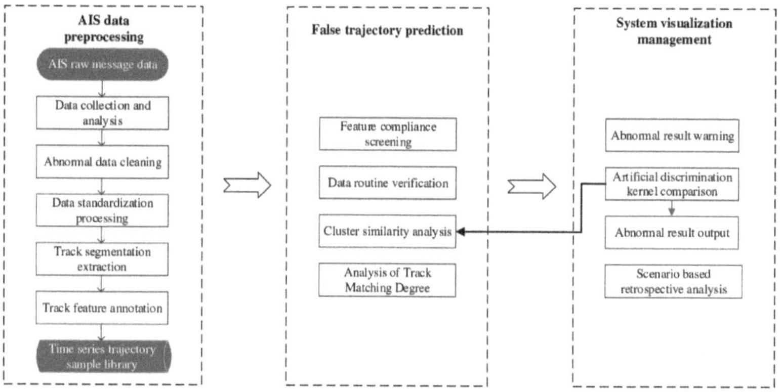

Figure 9. Study process of false ship AIS trajectory identification based on KNN.

4.1 *Technology selection and algorithm design*

4.1.1 *Technical selection direction*

In the field of ship trajectory prediction, current study focuses primarily on employing Long Short-Term Memory (LSTM) networks, Transformers (self-attention models), as well as their enhanced or combined variants. The core idea is to leverage their capabilities in handling time-series data, thereby achieving highly accurate ship trajectory prediction. In terms of anomaly detection, the research emphasis lies in effectively identifying noise and providing early warning for navigational anomalies. This study adopts the KNN (K-Nearest Neighbors) model with ship distance as a key feature for ship trajectory prediction. This model aligns well with the repetitive and trend-based characteristics of ship navigation. It exhibits smaller average distance iteration errors, particularly suitable for scenarios where future travel trends are predicted by searching for historically similar trajectories.

Furthermore, this study focuses on the analysis of generalized characteristics of false ship AIS trajectories and predicts future technological advancements in this area. By understanding the underlying mechanisms of false AIS signal generation, it delves into analyzing the patterns, motives, and technical means in generating false trajectories. The research identifies and records the conventional generalized abnormal ship trajectories, utilizing a dual verification mechanism from points to lines, to accurately identify current false trajectories. It also anticipates potential new fraudulent methods, thereby providing regulatory agencies with targeted decision-making support.

4.1.2 *Improved algorithm design*

Due to the fundamental principle of the KNN model, which searches for the nearest neighbors in the feature space, the time and resources required to compute distances between a new sample and all training samples significantly increase with the scale of the dataset and the dimensionality of the features. This is primarily attributed to the fact that the KNN algorithm uses a KD-tree structure. In scenarios like ship trajectory spatial-temporal clustering retrieval, the efficiency advantage of KD-trees diminishes as the scale grows. To address this issue and prepare for the application of full-scale data, we adopt the i-Quadtree, an improved quadtree structure, to achieve more efficient data management and faster query response times.

In i-Quadtree, each node can be divided into different numbers of child nodes based on the density of points within its region. The division criterion typically compares the number of points in a node with a threshold. If the number of points exceeds this threshold, the node is further divided; otherwise, it remains a leaf node.

The formula for determining whether to split a node *N* in an i-Quadtree can be expressed as:

$$\text{if}|\mathrm{N}| > T\ \text{then Split}(N) \tag{1}$$

Where | N | represents the number of points stored in node *N*; *T* is the predefined threshold value, above which a node should be split.

The process of splitting a node *N* can be represented as:

$$Split(N) = \begin{cases} \textit{Create } 4 \textit{ child nodes if uniform split} \\ \textit{Create variable number of child nodes based on point density if optimized split} \end{cases} \tag{2}$$

For an optimized split, the decision on the number of children to create can be made using various strategies, such as:

$$\text{Number of Children} = \min\left(\max\left(1, \lceil\sqrt{|N|}\rceil\right), 4\right) \tag{3}$$

This formula suggests that at least one child should be created (to avoid infinite recursion) and at most four children (to maintain the quadtree structure). The square root function helps to distribute the children more evenly according to the point density in the node.

Once the number of sub-nodes has been determined, the next step is to calculate the coordinates of each sub-node. Assuming that the center coordinates of node *N* are (x_N, y_N) and its width is *W*, the center ordinates (x_S, y_S) of sub-node *Si* (i = 1,2,...,n) can be calculated as follows:

$$(x_{Si}, y_{Si}) = (x_N + \Delta xi,\ y_N + \Delta yi) \tag{4}$$

Here, Δxi and Δyi represent the offsets along the x-axis and y-axis respectively, from the center of the parent node to that of sub-node *Si*. These offsets are determined based on the layout and size of the sub-nodes.

Through dynamically adjusting the splitting strategy of nodes based on the distribution of data points, the i-Quadtree achieves more efficient storage and query operations for spatial data. The aforementioned formulas illustrate how the splitting strategy is determined based on the distribution of data points, thereby optimizing storage space usage while ensuring query efficiency.

4.2 *AIS data preprocessing*

AIS data preprocessing plays a crucial role in multiple aspects, including data collection, reception, deduplication, transformation, translation, standardization, and as well as the extracting, processing, and packaging of features from raw data. It ensures the stability of data quality, the uniformity data formats, the consistentency of data scales, and the interpretability of data content.

The AIS sample data for this study originates from the AIS Data Center of China. The scope covers the area within 500 nautical miles along the busy Chinese coastal region where trajectory receptions are dense. The time span ranges from January 1 to March 31 of 2024, encompassing a total of 726,527 ships and 10,441,291,752 ship records. This authoritative data resource provides a solid foundation for the study's exploration in the fields of artificial intelligence and big data learning.

4.2.1 *Data collection and parsing*

Data collection and parsing are prerequisites for data analysis and model training. They primarily involve the precise integration of data interfaces to collect raw AIS ship dynamic and static message information. Subsequently, following AIS binary encoding requirements, the raw message data is converted into readable, formatted data. The transformed data is stored in high-performance databases, ensuring reliability for subsequent data querying, analysis, and application.

4.2.2 *Anomalous data cleansing*

Anomalous data cleansing focuses on eliminating outlier, duplicate, interference, and noise data to enhance data quality. This process ensures data accuracy and consistency, reducing analysis biases caused by data quality issues.

Outlier data includes numerical errors and data points exceeding logical ranges, such as longitudes greater than $\pm 180°$ or speeds exceeding 30 m/s. Duplicate data refers to instances where vessel trajectory points overlap within a short timeframe. Interference data pertains to prolonged low-speed values observed when ships are anchored. Noise data occurs during ship navigation when positioning device interference or errors occasionally cause significant position shifts, followed by a return to normal conditions. The effectiveness of data cleaning is shown in Figure 10.

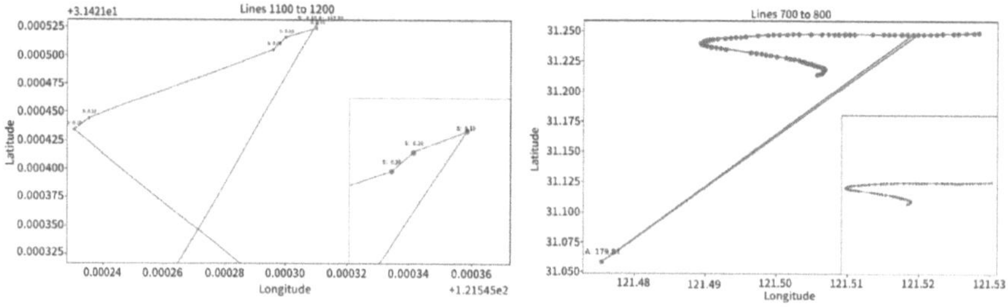

Figure 10. Anomaly data cleaning effect chart.

4.2.3 *Data standardization*

Data standardization involves interpolating missing trajectory data and resampling based on time intervals, providing a solid foundation for subsequent data analysis and machine learning model processing.

In cases of minor missing segments in ship trajectories, excessively sparse data lacks temporal continuity. Handling this with cubic spline interpolation can better restore smooth changes in the ship's trajectory during navigation.

If the time interval for data resampling is too large, it may lead to less accurate prediction results. Conversely, if it is too small, it could result in overly complex models. Typically, a time interval of around 10 seconds is condsidered representative for normal ship trajectory data. Considering a balance between prediction accuracy and model complexity, a resampling interval of 30 seconds is generally suitable, see Figure 11.

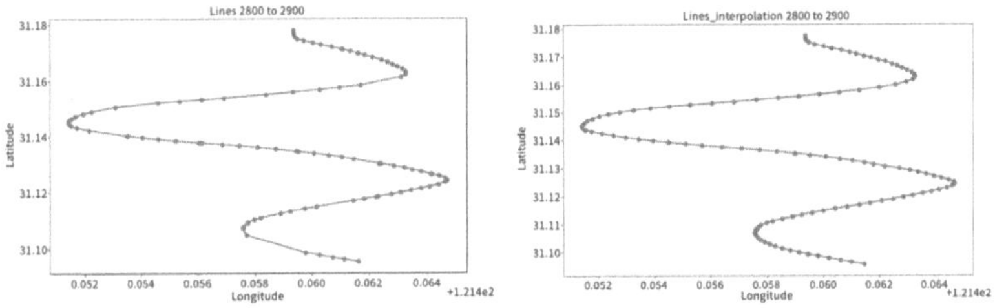

Figure 11. Data standardization processing effect chart.

4.2.4 *Trajectory segmentation*

Due to constraints such as communication interference, signal obstruction, and network congestion, vessel trajectories may experience prolonged periods of data loss. Interpolation alone may not accurately restore the navigation trends, and including such incomplete trajectories directly in the analysis could severely impact the accuracy of the results. To ensure the precision and reliability of the analysis, these trajectories are divided into two segments for separate processing.

Based on maritime experience, if the time interval between two consecutive trajectory points of a single vessel exceeds 10 min, the trajectory is segmented into two distinct parts. The segmented extraction chart of the trajectory is shown in Figure 12.

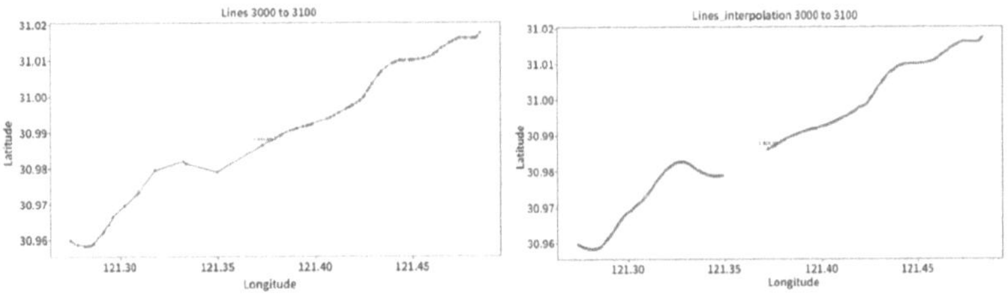

Figure 12. Trajectory segment extraction effect chart.

4.2.5 *Trajectory feature annotation*

Trajectory feature annotation involves processing dynamic ship data such as time, position, speed, heading, etc., using differential and standard deviation models to extract key features. These features include the stability of navigation speed, the continuity of heading, and the trend analysis of flow vectors. Attributes like time, position (latitude and longitude), speed, heading, ship type, and draft form the basic outline of ship dynamic behaviors. By annotating features such as departure/arrival, laden status, and adding markers for sharp turns, sudden accelerations/decelerations, and anchoring at critical trajectory points, the trajectory's trend, interpretability, and analytical value are enhanced.

4.3 *False trajectory prediction method study*

4.3.1 *Cluster similarity analysis*

Cluster similarity analysis utilizes the KNN (K-Nearest Neighbors) model to accurately extract the K nearest trajectory features from a dense trajectory network surrounding the current trajectory object's previous trajectory point target. Based on their feature distributions, it constructs a precise depiction of the feature set describing normal movement trajectories. Subsequently, the One-Class SVM (Support Vector Machine) algorithm is used. This algorithm builds a decision boundary that can distinguish between normal and abnormal behaviors by learning the feature distribution of normal trajectories. When the feature-processed result of a new trajectory point significantly deviates from the decision boundary of the normal mode, it is flagged as abnormal.

4.3.2 *Trajectory matching analysis*

4.3.2.1 *Linear deductive trajectory prediction*

When constructing the KNN-based ship trajectory prediction model, we have ingeniously designed an innovative labeling strategy to achieve more precise prediction results. Specifically, we use the last 10 latitude and longitude points within the preceding 5-min as input features to capture real-time ship navigation patterns. The subsequent 20 points constitute the prediction target, which depicts the ship's trajectory for the next 10 min. To meet the demands of the KNN classification model, we meticulously crafted a label-generation mechanism: using the endpoint timestamp of the input-eature sequence (the last 10 points) as a single label. This strategy simplifies the process of retrieving similar trajectories, ensuring the model swiftly locates the most closely matching historical data segment from the preceding 5 min. Subsequently, by leveraging this similar segment's navigational trends for the following 10 min, we are able to infer the future direction of the ship under prediction.

Furthermore, to enhance the model's accuracy and address challenges posed by sparse data, we introduced flexibility in trajectory direction by treating the reverse sequence of a trajectory segment as an independent data sample. By marking the timestamp of the starting point of the reverse trajectory as the label for this segment, we not only double the available training sample size but also enrich the model's learning perspective, thereby enhancing its predictive capabilities comprehensively. The partial code of the trajectory prediction function is shown in Figure 13.

```
n_in = 10
n_features = len(col_names)
n_out = 20
# 构造一个n_in->n_out的监督学习型数据，可以使用未标准化前的数据
reframed = data_processing_tools.series_to_supervised_KNN(data=values,
                                                          tag=tag,
                                                          n_in=n_in,
                                                          n_out=n_out,
                                                          column_names=col_names)
print(f'reframed_len {len(reframed)}')
print(reframed)
# 划分成训练数据集和测试数据集
values = reframed.values

# 取出tag_reverse标签
tag_reverse_index = reframed.columns.get_loc(f'tag(t+{n_out - n_in})')
# 新的列tag_reverse的数据
tag_reverse = values[:, tag_reverse_index] + '_reverse'
# 移除新增列的数据
values = np.delete(values, tag_reverse_index,  axis: 1)

# 取出tag标签，tag与reframed不是一一对应的，因为reframed去掉了带nan的行
tag_index = reframed.columns.get_loc(f'tag(t-1)')
```

Figure 13. Trajectory prediction code.

During the process of ship trajectory prediction, ensuring a high degree of consistency between the predicted and actual trajectories is crucial. To achieve this, we employ two key steps: trajectory shifting and scaling.

Firstly, we perform trajectory shifting to address potential discrepancies in the starting positions of identified similar trajectory segments from historical data. Even when a segment closely resembles the current trajectory, its starting point may still deviate. To resolve this, we simply adjust the position of the last 10 min of the identified similar trajectory segment to align with the endpoint of the current trajectory. This effectively eliminates position deviations and lays the groundwork for more precise adjustments.

Secondly, trajectory scaling is conducted to correct speed variations. Despite spatial alignment achieved through shifting, minor time deviations may still occur due to speed mismatches. Specifically, if the average speed of the similar trajectory over the first 5 min significantly differs from that of the current trajectory, the predicted trajectory for the subsequent 10 min may exhibit either accelerated or decelerated behavior. To address this issue, we introduce speed trends as an additional feature. By computing the ratio of average speeds (or equivalent distances) between the current trajectory and the similar trajectory over the first 5 min, we scale the shifted trajectory accordingly. This scaling ensures consistency in speed between predicted and actual trajectories, thereby further reducing prediction errors and enhancing accuracy. The comparison between the predicted trajectory and the actual trajectory is shown in Figure 14.

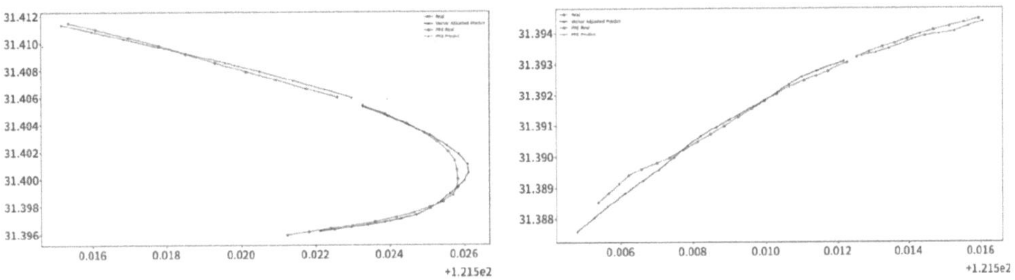

Figure 14. The comparison between the predicted trajectory and the actual trajectory.

4.3.2.2 *Identification of false trajectories*

In the detection of false trajectory, DTW is used to calculate the similarity between the predicted trajectory and historical real trajectories. DTW can flexibly adapt to differences in the length and shape of time series, identifying the optimal alignment between two sequences through non-linear mapping. This allows it to accurately measure the similarity even when there are minor differences in speed or starting points. By calculating the DTW distance between the predicted trajectory and multiple historical real trajectories, we can quantify the reasonableness and consistency of the predicted trajectory. If the DTW distance between a predicted trajectory and most historical trajectories is significantly larger than a predefined threshold, it will trigger the system's alert mechanism.

As shown in Figure 15, when there are abnormal labels among the ship labels, the trajectory of the abnormal point is tracked continuously for 15 min. By matching the predicted trajectory with the actual trajectory, the combination of trajectory points are expressed in a matrix form. The shortest Euclidean distance matrix between the points of the two trajectory sequences is calculated. If the matching degree is low, the trajectory will be considered a suspected false trajectory and will be flagged for further investigation.

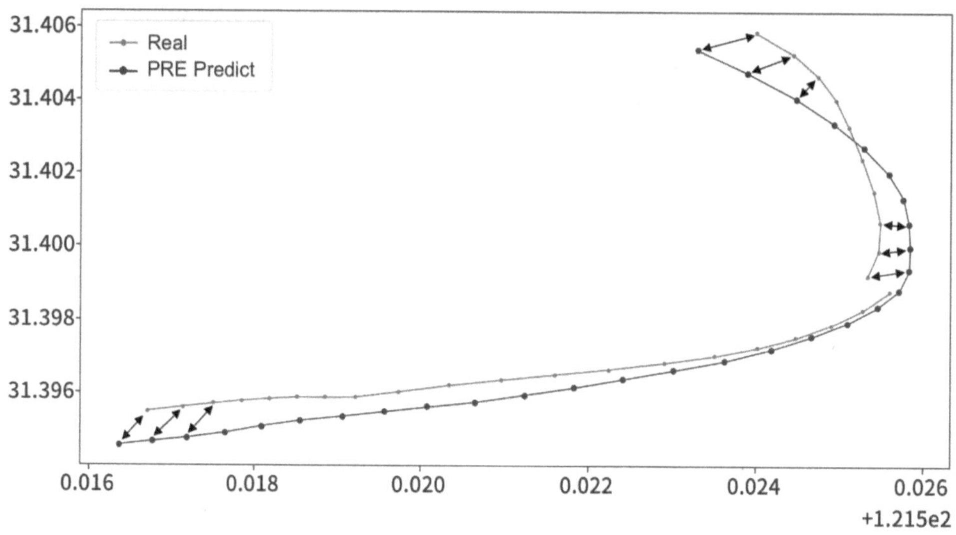

Figure 15. Trajectory similarity comparison Chart.

4.4 *Customization of visualization system functions*

To accurately address and manage abnormal situations related to false trajectories, we have developed a customized visual system that integrates four core functions: intelligent alerting, deep artificial discrimination tools, customized abnormal result outputs, and scenario-based retrospective analysis. Through innovative methods such as dynamic alert thresholds, multi-channel real-time notifications, expert knowledge assistance, interactive report generation, situational reconstruction technology, and event grid chart interactive analysis, the system not only swiftly identifies and alerts potential issues but also assists users in gaining deep insights into the complex causes of anomalies, refining generalized feature scenes, enhancing the continuous learning capabilities of the system, achieving efficient human-machine collaboration, ensuring data accuracy, and enabling reliable decision-making.

5 EXPERIMENTAL APPLICATION AND RESULT ANALYSIS

The study achieves the efficient parsing of a vast amount of ship AIS messages nationwide by integrating full-sample data into the system. This ensures that the position information of each ship was accurately matched with the reporting base station. It also enables the efficient identification of ship reports corresponding to VSI messages, and the precise identification of false ships trajectories.

In February, 2024, the system detected that one ship trajectory appeared near Lianyungang. The data was received by base stations near Zhoushan, including Taohuadao, Huangyantou, and Juexi, where the positions reported by the ship were over 330 nautical miles away from the receiving base stations, which was significantly beyond the normal reception range. There was a continuous large-scale broadcast, suggesting the possibility of falsified ship trajectories. Later, the Maritime Traffic Management Center conducted subsequent tracking and confirmed that the ship's AIS signal was indeeded falsified, as detailed in Figure 16.

Figure 16. Generalization features analysis results.

In order to verify the feasibility of identification method based on the improved KNN model, we have prospectively designed and simulated a set of sample data, as shown in Table 2:

Table 2. Simulated false trajectory sample data.

Time	Actual longitude	Actual latitude	Predicted longitude	Predicted latitude	Distance	Standard deviation
2024/01/02 19:30:00	121.2563031	30.9634138	121.2562871	30.96316174	28.07	0.43
2024/01/02 19:30:30	121.25549	30.96323803	121.2555069	30.96300417	26.05	0.39
2024/01/02 19:31:00	121.2546697	30.96305561	121.2547117	30.96284977	23.24	0.35
2024/01/02 19:31:30	121.2538362	30.96287036	121.2539012	30.96270488	19.42	0.29
2024/01/02 19:32:00	121.2529915	30.96268958	121.2530751	30.9625758	14.95	0.23
2024/01/02 19:32:30	121.2521673	30.96253408	121.252233	30.96246889	9.58	0.15
2024/01/02 19:33:00	121.2513516	30.96240922	121.2513748	30.96239034	3.05	0.05
2024/01/02 19:33:30	121.2505139	30.96231297	121.2505018	30.96234225	3.46	0.05
2024/01/02 19:34:00	121.2496471	30.96224699	121.2496177	30.96232135	8.73	0.13
2024/01/02 19:34:30	121.2487591	30.96221523	121.2487264	30.96232398	12.49	0.19
2024/01/02 19:35:00	121.2478306	30.96233042	121.2478315	30.9623465	1.79	0.03
2024/01/02 19:35:30	121.2469301	30.96237889	121.246937	30.96238529	0.97	0.01
2024/01/02 19:36:00	121.2460331	30.96247337	121.2460466	30.96243669	4.28	0.06
2024/01/02 19:36:30	121.2451455	30.96261381	121.2451642	30.96249708	13.10	0.20
2024/01/02 19:37:00	121.2442673	30.96278931	121.2442934	30.9625628	25.31	0.38
2024/01/02 19:37:30	121.2433974	30.96298715	121.2434382	30.96263022	39.88	0.60
2024/01/02 19:38:00	121.2425349	30.96319463	121.2426024	30.9626957	55.85	0.85

According to the design of each process node, the system has successfully pushed the early warning information of suspected false trajectory into the system visual interface, as shown in Figure 17, where red is the predicted trajectory and blue is the real one.

Figure 17. Results of artificial intelligence analysis.

According to the above experimental verification, this study has made some progress in identifying false ship AIS trajectories, and has detected some non-compliant or suspicious sailing trajectory. However, some false alarms were also found due to the lack of standardization of the AIS shipborne station and the inaccurate prediction results of the sparse part of the track, which may be caused by the limitations of the training dataset, or the deficiencies in the model structure and parameter settings. Therefore, it is necessary to make targeted improvements to the follow-up supervision mechanism and trajectory prediction algorithm. However, it is undeniable that the whole research process fully reflects the great potential of modern information technology in complex environmental monitoring and risk management, and has accumulated valuable experience for the next stage of iterative development and formal use. It is believed that with the improvement of accuracy and maturity, it can provide timely decision support to relevant departments and effective responce to such illegal incidents.

6 CONCLUSION

This study summarizes the various manifestations and generation mechanisms of false trajectories. It abstracts the commonalities and variations in the generation of false trajectories, constructs an identification system based on artificial intelligence deep learning algorithms, and develops a comprehensive system that integrates data collection, real-time monitoring, intelligent analysis, and visual display. Not only does it achieve innovation at the technical level, but also demonstrates disruptive transformative power in methodology, thinking modes, and industry applications, making it a paradigm of technological application innovation. In the future, we will further optimize the extraction of generalized features, explore machine learning algorithm models that combine multiple feature dimensions, and enhance the accuracy and reliability of identifying false ship AIS trajectories. Additionally, we plan to integrate multiple sources of maritime signals such as radar and ship stations for aggregation and comparison, providing strong technical support for the precise prediction of false ship AIS trajectories. To eliminate the generation of false ship AIS trajectories, it is essential to start from laws and regulations, improve international conventions, strengthen legal frameworks, enhance publicity and education, promote industry self-discipline, and build a multi-level defense system combined with system applications. This will effectively curb the generation of false ship AIS trajectories, maintain maritime navigation safety and order, and contribute to building a safer, more stable, and efficient global maritime system.

REFERENCES

[1] Kong tao, Yao gaole, Chen guangtong *et al.* (2022). The monitoring system design and implementation of AIS ship station dynamic information conformity based on AIS VDL[J]. *China Maritime Safety*, 1, 49–52+56.

[2] Xiong zhiwen. (2023). Research on abnormal detection of ship navigation path based on big data drive [J]. *Ship Science and Technology*, 45(05):152–155.

[3] Deng lixiang, Xu yiqun and Wan longjun. (2021). The Research of Ship Abnormal Behavior Evaluation Based on AHP and Fuzzy Evaluation[J]. *Journal of Guangzhou Maritime University*, 29(01):7–12.

[4] Li kexin, Guo jian, Wang yujun *et al.* (2023). Trajectory anomaly identification method of vessels based on dimensional-density reduction clustering[J]. *Packaging Engineering*, 44(11):284–292.

[5] Xie Z, Bai X, Xu X, *et al.* (2024). An anomaly detection method based on ship behavior trajectory[J]. *Ocean Engineering*, 293:116640.

[6] Kang kaihang. (2023). *Ship Trajectory Clustering Analysis and Anomaly Detection Based on Machine Learning[D]. Dalian Maritime University.*

[7] Lv jianguo. (2017). *Design of AIS information simulator[D]. Dalian Maritime University.*

[8] Wei wei, Liu jiazhao, Wang yong *et al.* (2013). Research and development of AIS simulator[J]. *Journal of Qingdao Ocean Shipping Mariners College*, 34(03):1–4.

[9] Recommendation ITU-R M.1371-5: Technical Characteristics for an Automatic Identification System Using Time Division Multiple Access in the VHF Maritime Mobile Frequency Band[S]. 2014.

Proceedings of the International Association of Institutes of Navigation World Congress 2024 – Chen et al. (Eds)

Fluid dynamics simulation and performance optimization research of aids to navigation system based on ANSYS LS-DYNA

Chuanying Zhu, Yu Guo, Chen Wang & Haiming Li
Tianjin Aids to Navigation Department, Binhai New Area, Tianjin, China

ABSTRACT: This study focuses on the fluid dynamics simulation and performance optimization of the aids to navigation system. A comprehensive fluid dynamics simulation model was established, including vessel and aids to navigation models. This paper conducts in-depth analysis of the fluid dynamics behavior of the aids to navigation system in water, including statics analysis, dynamics in water, and the process of a vessel colliding with an aids to navigation. By setting appropriate simulation parameters and boundary conditions, the behavioral characteristics of the aids to navigation system in water flow, as well as the speed and displacement characteristics of vessel colliding with the aids to navigation, are revealed, providing important basis for improving system performance.

Based on ANSYS LS-DYNA nonlinear finite element simulation technology, this study conducts simulated analysis of the collision process between vessels and navigation marks. It investigates the damage and collision resistance of navigation marks under collision conditions, providing a theoretical basis for the design and optimization of navigation marks.

Through in-depth research on fluid dynamics simulation and performance optimization of aids to navigation systems, this study provides theoretical support and practical guidance for improving the stability and overall performance of the system. Future research will continue to optimize the design of aids to navigation systems, enhance their performance in complex water flow environments, and effectively solve challenges under various navigation conditions.

1 INTRODUCTION

Aids to navigation is navigational facilities set up for the safety and convenience of vessel navigation or for the protection of specific facilities at sea. The integrity and availability of aids to navigation are directly related to the safety of maritime traffic. However, inevitably, the placement of aids to navigation itself increases the possibility of collisions with vessels. In the actual management of aids to navigation, it has been found that vessel collisions with aids to navigation do occur. Various types of vessels such as fishing vessels, passenger vessels, oil tankers, and engineering vessels can collide with aids to navigation. In the event of a collision with an aids to navigation, it can lead to damage to both the vessel and the aids to navigation, causing economic losses, as well as significantly affecting the navigational effectiveness of the mark, disrupting the normal maritime traffic environment, and potentially leading to maritime accidents [1]. Therefore, studying collisions between vessels and aids to navigation, and evaluating the performance during such collisions, helps understand the characteristics of aids to navigation, vessel structures, force situations, motion characteristics, and holds important theoretical and practical significance.

Currently, there is limited research on collisions between vessels and buoys both domestically and internationally. Conducting real-life collision studies could not only result in damage to vessels and aids to navigation but also incur significant costs in terms of resources

 DOI: 10.1201/9781003624424-23

[2,3]. Moreover, it would require a large number of sensing devices for monitoring, with actual results possibly not being very accurate. With the rapid development of information technology, simulation technology, artificial intelligence, and virtual reality technology have become effective research tools. With the rapid advancement of nonlinear finite element technology, numerical simulation applications in vessel collision analysis are becoming increasingly widespread. This paper utilizes the ANSYS LS-DYNA software, focusing on the "Ding Hai Shen Zhen X" model aids to navigation as the primary research object, to simulate vessel collisions with aids to navigation using finite element simulation technology and analyze the mechanical response and aids to navigation movement during the collision process.

2 SIMULATION MODEL BUILDING

2.1 *Vessel model*

This article takes the Bohai Bay as an example, where vessel collisions with aids to navigation are mainly caused by small construction vessels and fishing boats. Based on the actual size and structure of the vessels, a simplified vessel geometric model is created in SOLIDWORKS. The bow of the vessel, which is the main collision part, is in the aft of the vessel away from the impact area [4–6]. A rigid body simplification simulation is used to provide the influence of stiffness and mass. The model has a vessel length of 44.5 meters, a width of 12 meters, and vessel type depth 7.5 meters. The vessel model is shown in Figure 1.

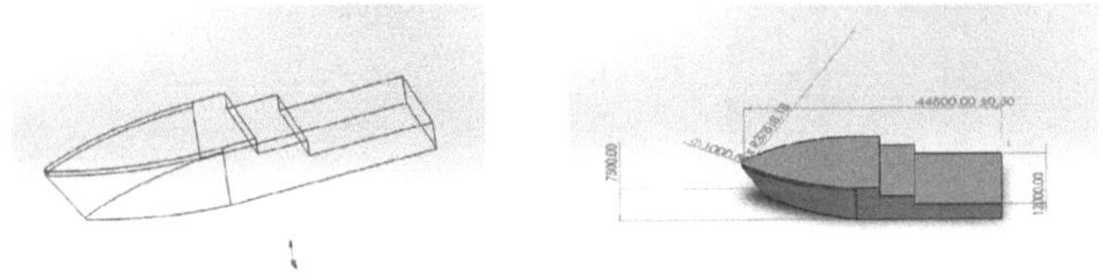

Figure 1. Vessel model and size.

The vessel model is made of alloy steel, linear elastic isotropic. The basic parameters are shown in Table 1 below:

Table 1. Basic parameters of vessel model materials.

attribute	numeric value	unit
modulus of elasticity	19e+1	N /m^2
In Poisson than	0.26	not applicable
Medium shear modulus	7.8e+10	N /m^2
mass density	7300	kg/m^3
Tensile strength	448082500	N /m^2
compressive strength		N /m^2
yield strength	241275200	N /m^2
coefficient of expansion due to heat	1.5e-05	/K
heat conductivity	38	W/(m·K)

2.2 *Aids to navigation model*

The aids to navigation model is simplified based on the "Ding Hai Shen Zhen X" model's actual size and shape, using solid elements for discretization. The aids to navigation material selected is 316L stainless steel. Firstly, detailed design drawings and actual dimensions of the "Ding Hai Shen Zhen X" aids to navigation are thoroughly studied to identify key structural features such as the main structure, support components, identification sections, and possible connection details. Subsequently, based on these features, necessary simplifications are made, removing minor details like lifting rings that have little impact on the overall mechanical performance to reduce model complexity and increase computational efficiency.

In the discretization stage, solid elements are chosen as the modeling foundation in this study. Solid elements accurately simulate volume and shape in three-dimensional space and are suitable for structural analysis under complex stress states. Through a reasonable mesh partitioning strategy, such as adaptive mesh refinement and boundary layer mesh refinement, the mesh density at crucial locations (such as stress concentration areas) is ensured to capture subtle stress changes while maintaining overall mesh economy [7].

2.2.1 *Body model*

For the body model of the "Ding Hai Shen Zhen X" aids to navigation made of 316L stainless steel, relevant material manuals and experimental data are meticulously consulted in this study to accurately set key parameters such as the material's elastic modulus, Poisson's ratio, yield strength, tensile strength, and density. 316L stainless steel, known for its excellent corrosion resistance and good mechanical properties, performs well in marine environments, making it an ideal material for aids to navigation manufacturing. By accurately setting these material properties, the model's behavior in simulation analysis aligns with the actual dimensions of the aids to navigation. The actual size and physical appearance of the aids to navigation are shown in Figure 2.

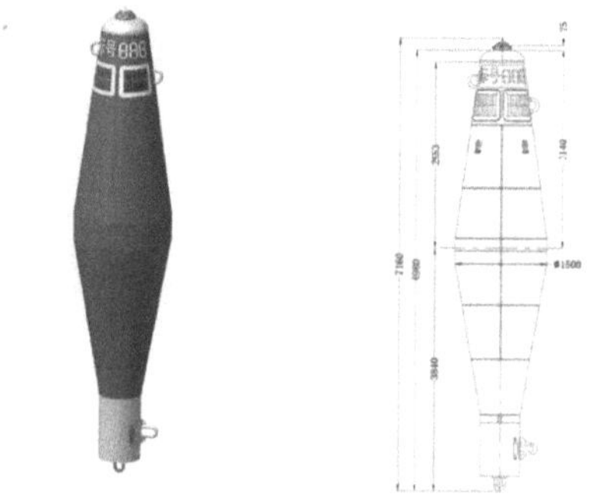

Figure 2. Actual dimensions and physical objects of navigation AIDS.

The structural dimensions of the aids to navigation model are based on a 1:1 scale model, with the material selected being 316L stainless steel. The model and material parameters are shown in Figure 3 and Table 2.

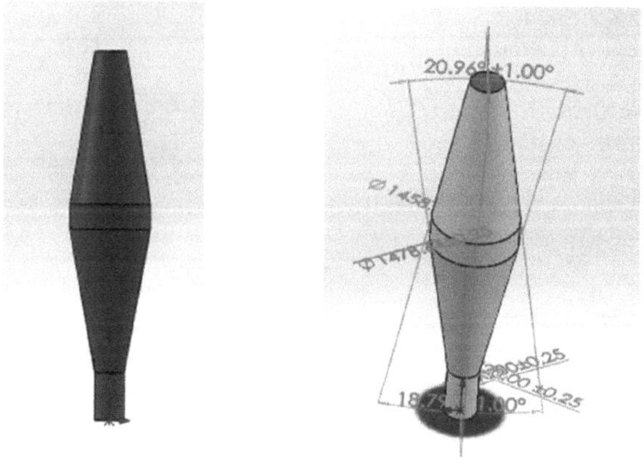

Figure 3. Structure and dimension of the model of the aids to navigation.

Table 2. Aids to navigation model material parameters.

attribute	numeric value	unit
modulus of elasticity	2e+11	N /m^2
In Poisson than	0.265	not applicable
Medium shear modulus	8.2e+10	N /m^2
mass density	8027	kg/m^3
Tensile strength	485000000	N /m^2
yield strength	170000000	N /m^2
coefficient of expansion due to heat	1.65e-05	/K
heat conductivity	14.6	W/(m.k)

2.2.2 *Basic structure of the anchor chain*

In accordance with the size and structural requirements of the anchor chain outlined in the "Buoy Anchor Chain," this study opts for the commonly used 38mm diameter cast iron anchor chain in actual applications by aids to navigation management units. For simplification in calculations, regular links are used instead of shackles, swivels, and enlarged links. The basic structure of the anchor chain is depicted in Figure 4.

Figure 4. Basic structure of the anchor chain.

2.2.3 *Basic structure of sunk stones*

The structure of the sunk stones is shown in Figure 5.

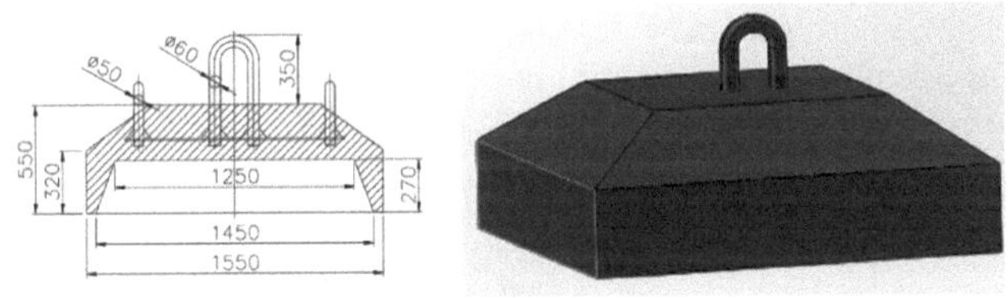

Figure 5. Basic structure of the sunk stones model.

The material of the sunk stone is made of grey cast iron, and the parameters are shown in Table 3.

Table 3. Material parameters of the sunk stones.

attribute	numeric value	unit
modulus of elasticity	6.61781e+10	N /m^2
In Poisson than	0.27	not applicable
Medium shear modulus	5e+10	N /
mass density	7200	KG/m^3
Tensile strength	151658000	N /m^2
compressive strength	572165000	N /m^2
yield strength		N /m^2
coefficient of expansion due to heat	1.2e-05	/K
heat conductivity	45	W/(m.K)

The model of the assembled aids to navigation is shown in Figure 6.

Figure 6. The complete aids to navigation.

3 THE COLLISION SIMULATION SETTING

Using the contact algorithms in ANSYS LS-DYNA software, the contact surfaces between the vessel and the aids to navigation are defined. Automatic single-side contact (ASSC) and automatic surface-to-surface contact (ASTS) methods are employed, with static and dynamic friction coefficients set at 0.1. Suitable collision velocities and angles are defined to simulate collision processes under different conditions [8–11].

3.1 *Hydrostatic pressure*

Taking the Bohai Bay as an example, the seawater density ranges from 1020 to 1070 kg/m^3. In this study, a seawater density of 1050 kg/m^3 is selected. The free surface is set at the middle section of the aids to navigation, which is also the nominal design draft position. The static hydrostatic acceleration is set at 9.8 m/s^2.

3.2 *Initial velocity*

Since aids to navigation are generally deployed in coastal and nearshore areas where vessel speeds mostly range around 10 knots per hour or lower, a velocity of 5 m/s is set in this study for analyzing the collision effects of vessels.

3.3 *Mesh division*

To simplify the calculation of the vessel model, the cell size is 2m, the growth rate is 1.5, the minimum curvature size is 1m, the normal Angle of curvature is 72°, and the minimum edge length is 0.6.

In order to accurately the cell size is 0.1m, the growth rate is 1.5, the minimum curvature size is 0.05m, the normal angle of curvature is 72°. The mesh of the partition is shown in Figure 6.

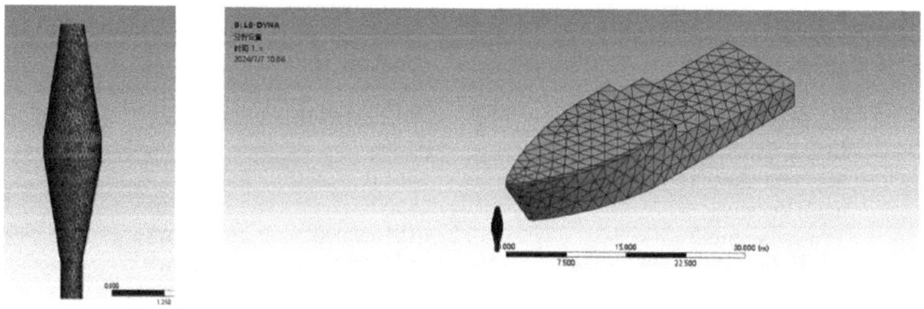

Figure 7. Mesh division of aids to navigation and vessel model.

4 SIMULATION RESULTS AND ANALYSIS

4.1 *Static mechanical analysis of AIDS*

Through buoyancy simulation, the stress on the aids to navigation in water primarily concentrates at the middle-lower end where the diameter of the aids to navigation is larger. Under the combined forces of buoyancy, gravity, and the anchor chain, reinforcement ribs should be installed in this section and the body of the mark should be thickened. The static simulation results are shown in Figure 7.

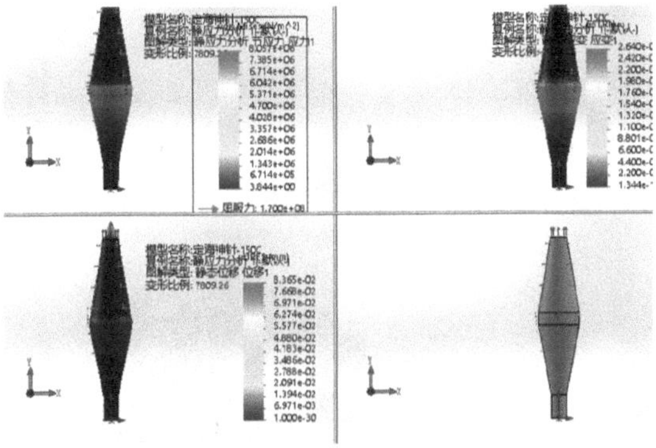

Figure 8. Results of the statics simulation.

4.2 *The sinking process simulation*

Currently, the general process for deploying aids to navigation at sea by setting units involves utilizing vessel-based positioning systems or mobile positioning systems to accurately locate the design position, then releasing the anchor chain to allow the sunk stones to freely descend to the seabed. Simulation results indicate that during the sinking process, the sunk stones is not only affected by gravity but also by the buoyancy of the water. Due to the sunk stones being designed with a flat-shaped structure on the bottom surface, it enters the water in a swaying, slipping manner. This diving posture causes the final grounding position to deviate from the position determined by the positioning system. The dynamic simulation of the sunk stones entering the water is illustrated in Figure 8.

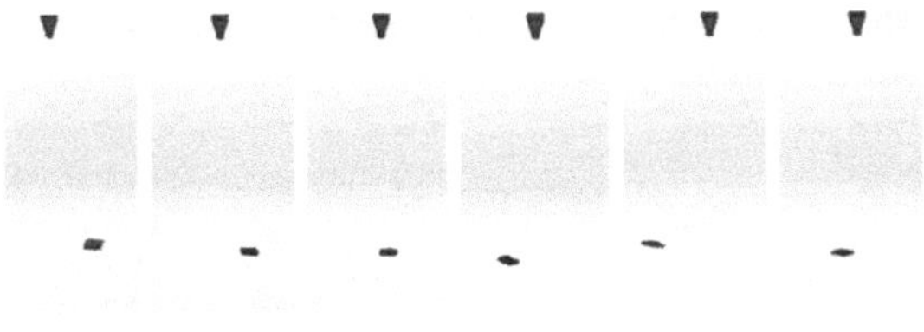

Figure 9. The sinking process simulation.

4.3 *Analysis of collision stress*

Due to the buoyant characteristics of the aids to navigation, simulation results show that during the moment of impact between a vessel and the aids to navigation, it often goes through a continuous process of "collision-bounce-collision-bounce." The collision process only ends when the vessel departs from the aids to navigation's turning range or when the mark moves away from the vessel's side after the initial collision. The maximum principal stress reaches its peak at the moment of vessel contact, as shown in Figure 9, and the stress cloud map is illustrated in Figure 10.

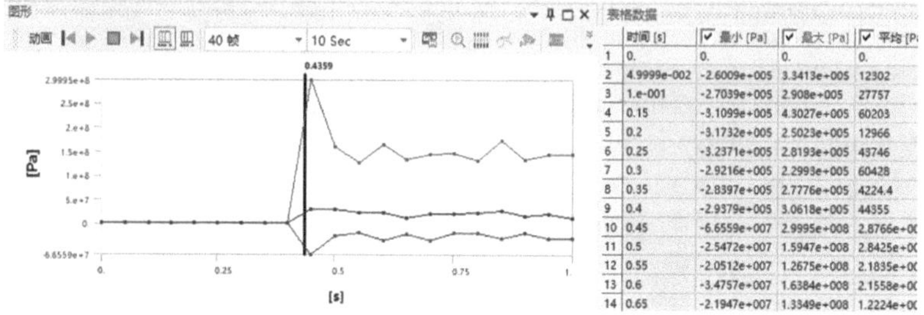

	时间 [s]	最小 [Pa]	最大 [Pa]	平均 [P
1	0.	0.	0.	0.
2	4.9999e-002	-2.6009e+005	3.3413e+005	12302
3	1.e-001	-2.7039e+005	2.908e+005	27757
4	0.15	-3.1099e+005	4.3027e+005	60203
5	0.2	-3.1732e+005	2.5023e+005	12966
6	0.25	-3.2371e+005	2.8193e+005	43746
7	0.3	-2.9216e+005	2.2993e+005	60428
8	0.35	-2.8397e+005	2.7776e+005	4224.4
9	0.4	-2.9379e+005	3.0618e+005	44355
10	0.45	-6.6559e+007	2.9995e+008	2.8766e+0(
11	0.5	-2.5472e+007	1.5947e+008	2.8425e+0(
12	0.55	-2.0512e+007	1.2675e+008	2.1835e+0(
13	0.6	-3.4757e+007	1.6384e+008	2.1558e+0(
14	0.65	-2.1947e+007	1.3349e+008	1.2224e+0(

Figure 10. Maximum principal stress.

Figure 11. Stress cloud diagram [ALT].

4.4 *Aids to navigation displacement analysis*

The simulation results indicate that during the collision process, the aids to navigation undergoes a strong impact, leading to overall displacement. The direction of displacement is determined by the angle of impact, as well as hydrological and meteorological conditions such as wind and waves. The magnitude of displacement is related to the vessel's collision speed, and this type of impact generally does not cause a change in the position of the sinker. Analysis confirms that a complete change in the aids to navigation's position only occurs under two circumstances: when the anchor chain is severed, causing the mark to drift; or when the vessel drags the mark, leading to the mark dragging the anchor chain and the anchor chain moving the sinker, resulting in a deviation from the designed position. Simulation results show that as the impact speed increases, the rate of movement of the aids to navigation also continues to rise. The movement of the aids to navigation with changes in vessel speed is depicted in Figure 11.

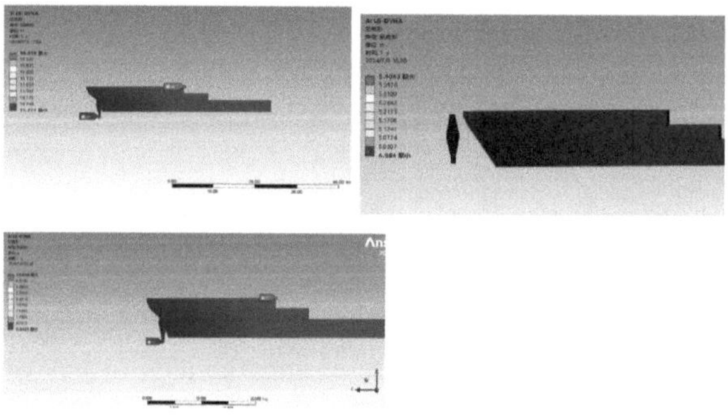

Figure 12. Schematic diagram of the aids to navigation displacement.

According to the simulation results, the changes of aids to navigation displacement with vessel speed are shown in Table 4.

Table 4. Aids displacement changes.

velocity (m/s)	20	10	5
displacement (m)	15.771	6.343	4.984

The basic formula for the displacement with the collision velocity can be summarized:

$$S = \mathrm{k}\bullet V^2 + \mathrm{b} \tag{1}$$

Where S is the aids to navigation displacement, V vessel velocity, k is the displacement coefficient, and b is constant.

The induction formula of k is:

$$k = \frac{\sum_{i=1}^{m}\left(\frac{S_i - b}{V_i^2}\right)}{m} \tag{2}$$

The induction formula of b is:

$$b = \frac{\sum_{i=1}^{m}(S_i - \mathrm{k}V_i^2)}{\mathrm{m}} \tag{3}$$

4.5 *Speed analysis*

When a vessel collides with an aids to navigation at a certain speed and angle, it disrupts the mark's original state of equilibrium. This sudden change causes the mark to undergo an acceleration, the magnitude of which depends on the speed and direction of the vessel's impact, which generally does not align with the direction of the vessel's movement. Different collision speeds lead to varying instant accelerations of the aids to navigation, with the maximum instantaneous acceleration reaching hundreds of gravitational accelerations, which can be quite shocking to perceive. Such a high instantaneous acceleration directly results in deformation or even damage to the body of the aids to navigation.

Simulation results show that as a vessel moves at a certain speed, it exerts a certain amount of pressure on the surrounding water, leading to the formation of a surge wave. This wave makes the aids to navigation in front of the vessel gradually move forward at an increasing speed in an approximate straight line before actually contacting the hull. Once the vessel makes contact with the hull, there is a sudden change in the speed of the aids to navigation. According to the principle of conservation of momentum, the vessel transfers

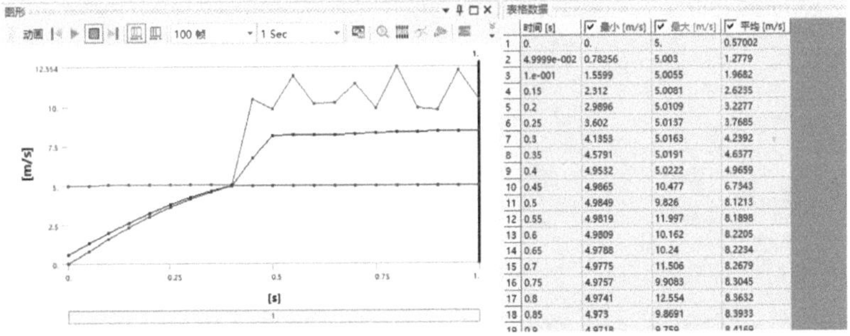

Figure 13. Speed change of aids to navigation.

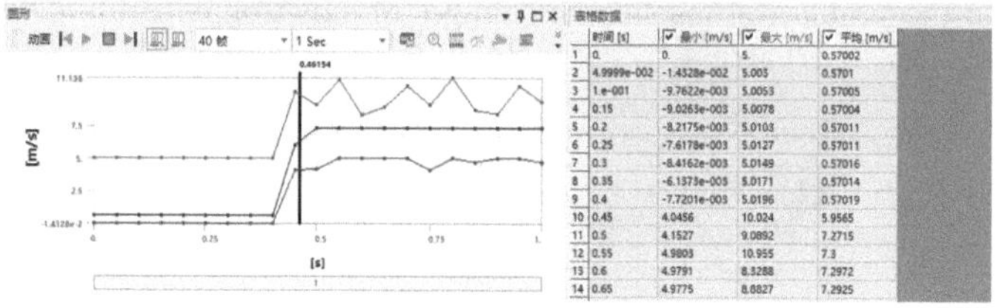

Figure 14. The directional velocity change of the aids to navigation along the Z-axis.

momentum to the aids to navigation. Due to the small volume and light weight of the mark, it gains a speed far surpassing that of the vessel in an instant. As a result of the vessel's impact, the aids to navigation rapidly moves away from the vessel. Influenced by multiple collisions with the vessel and turbulent flow effects, the speed of the aids to navigation undergoes various fluctuations in magnitude. The changes in the speed of the aids to navigation are illustrated in Figures 12 and 13.

4.6 *Acceleration analysis*

Simulation results show that the instantaneous acceleration of the aids to navigation when struck by a vessel reaches its maximum value. This is caused by the high-speed impact of the vessel, which disrupts the equilibrium of the mark, resulting in the unique instantaneous acceleration of the mark. The acceleration graph from the simulation indicates that after reaching the maximum acceleration, the aids to navigation experiences several significant increases in acceleration.

The simulation images reveal that due to the inclination angle of the vessel's hull and the conical shape of the aids to navigation, the mark experiences a downward force after being struck by the vessel. Under the influence of this downward force, the mark rapidly descends, storing a considerable amount of buoyancy. When the downward force is not sufficient to

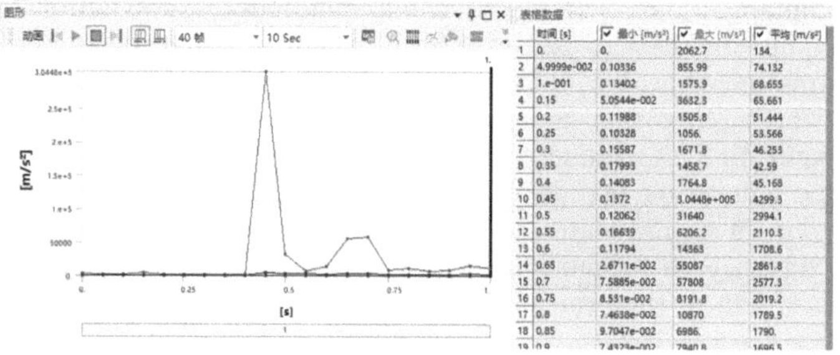

Figure 15. Change of aids to navigation acceleration.

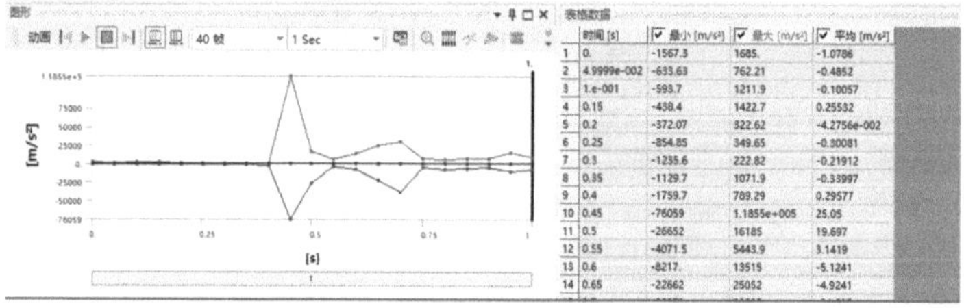

Figure 16. Directional acceleration of aids to navigation.

keep the mark sinking, buoyant force acts on it, causing the mark to gain an upward acceleration. This fluctuation in acceleration of the aids to navigation continues until it reaches a new dynamic equilibrium, as shown in Figures 14 and 15.

5 CONCLUSIONS AND OUTLOOK

This paper utilized ANSYS LS-DYNA simulation technology to conduct a simulated analysis of the collision process between vessels and aids to navigation. The study presents the displacement, stress, and acceleration conditions of a vessel colliding with an aids to navigation. It also analyzes the state of a sunken stone entering the water. The research results propose improvements for the operational and maintenance units responsible for setting up and maintaining aids to navigation. Additionally, they offer reference points for manufacturers to enhance and upgrade aids to navigation equipment.

In the future, it is recommended to further consider the influence of fluid media, enhance collision simulation models, increase the number of simulation calculations, and improve the accuracy of simulation results. Moreover, conducting more research on collision prevention devices for aids to navigation could enhance their resistance to collisions.

REFERENCES

[1] Chunguang Li, Hu Guo, high-rate group. (2024), Simulation analysis of vessel collision [J]. *China Water Transport (second half of the month)*, 24 (02): 32–34.

[2] Manjun. Liao, (2023). Numerical simulation study of vessel and buoy collision [J]. *Chinese Water Transport*, (07):152–155.DOI:10.13646/j.cnki.42-1395/u.2023.07.052.

[3] Shuyou Zhang. (2023). Study on the dynamic response of the collision between vessel and jack-up platform [D]. *Harbin Engineering University*, DOI:10.27060/d.cnki.ghbcu.2023.000208.

[4] Liang Yan, Min Wan, Shuguang Zhang, *et al.* (2022), Simulation simulation of single pile collision facility of offshore fan based on LS-DYNA [J]. *Jiangsu vessel*, 39(02):4–6+29+60.DOI:10.19646/j.cnki.32-1230.2022.02.002.

[5] Wenwen Chen, Feng Lu, Lian Chen, *et al.*, (2023). Simulation study on the hydrodynamic performance of vessel cycloid thruster [J]. *Vessel Science and Technology*, 45 (12): 14–19.

[6] Guangxia Chen, (2024). Comparison of the modeling methods between SolidWorks and UGNX based on shaft-like parts [J]. *Southern Agricultural Machinery*, 55 (13): 140–143 + 167.

[7] Guilong Li. (2024), Parametric method of mechanical products based on SolidWorks [J]. *Manufacturing and Upgrading Today*, (04): 88–90.

[8] Chunrong Wang, Erdong Xia, Long Wu, *et al.* (2024), Design and analysis of steel mesh shear device based on SolidWorks [J]. *Equipment Engineering of China*, (06): 164–165.

[9] Zhikang Zhang, Xing Gao, Xinghong Ye, *et al.* (2023), Standardization and procedure development of vessels [J]. *Vessel Sea Engineering*, 52 (06): 44–48.

[10] Xiaofang Zhao, (2023). Simulation of vesselbuilding structure design based on finite element analysis [J]. *Vessel Science and Technology*, 45 (24): 57–60.

[11] Ven Hu, (2023). Direct calculation method of local hull structural strength based on ANSYS [J]. *Chinese Water Transport*, (04):80–81.DOI:10.13646/j.cnki.42-1395/u.2023.04.028.

Proceedings of the International Association of Institutes of Navigation World Congress 2024 – Chen et al. (Eds)

The analysis of the impact of bridge lighting on the recognizability of aids to navigation on the bridge

Xiaowei Gao
Tianjin AtoN Division, The Navigation Guarantee Center of North China Sea, Tianjin, China

ABSTRACT: This article starts with multiple ship collision bridge accidents in recent years, introduces the functions and design principles of aids to navigation on the bridge, and analyzes the adverse effects that bridge lighting may have on the recognizability of aids to navigation on the bridge. This impact is mainly reflected in the following three aspects: unreasonable brightness setting, improper color selection, and excessive glare phenomenon. Unreasonable brightness settings and glare phenomena can interfere with the visual perception of ship drivers, and improper color selection can mix the background light and the light color of aids to navigation on the bridge, producing new mixed color lights. All three of these effects can lead to a decrease in the recognition of aids to navigation on the bridge, thereby affecting the safety of ship navigation at night.

Based on the above analysis, the article further proposes a series of targeted optimization measures. For example, precise control of brightness can prevent strong or weak background light from affecting the recognition of aids to navigation on the bridge, ensuring that the aids to navigation on the bridge are clear and visible; using reasonable color matching can coordinate the colors of bridge lighting and aids to navigation on the bridge, improving the overall visual effect, and not confusing the signals of aids to navigation on the bridge; by introducing anti glare technology, unnecessary visual interference is reduced to ensure that aids to navigation on the bridge can still efficiently perform their navigation functions in complex background light environments. In addition, measures such as introducing laser technology, constructing AIS aids to navigation on the bridge, and strengthening supervision by relevant departments can also improve the adverse effects of bridge lighting on the recognizability of aids to navigation on the bridge to a certain extent.

Keywords: bridge lighting, aids to navigation on the bridge, marine light pollution, brightness control, color coordination

1 INTRODUCTION

According to statistics, there were a total of 21 “ship collision bridge” accidents in China from 2014 to 2022, resulting in direct economic losses of over 100 million yuan. After analyzing the causes of these accidents and interviewing authoritative experts, it was found that several of them were caused by the misjudgment of bridge height by ship drivers who failed to clearly identify aids to navigation on the bridge when crossing the bridge at night [1]. Taking the 2019 Foshan “April 20th” “Huijinyi 21” ship collision with the Heshun Bridge as an example, during the ship’s nighttime navigation, it was affected by the excessively bright and overly mixed colors of the lighting used for bridge illumination. This disrupted the ship driver’s ability to recognize aids to navigation on the bridge. As a result, the ship took the risk of crossing the bridge without leaving any extra water depth, ultimately leading to the occurrence of the collision accident.

DOI: 10.1201/9781003624424-24

In the current environment where bridge lighting projects are widely implemented, the adverse effects of the background light generated on the recognizability of aids to navigation on the bridge cannot be ignored. For example, unreasonable brightness settings can lead to a decrease in the recognizability of aids to navigation on the bridge, and may also cause glare. Glare can make it difficult for ship drivers to clearly see the specific indication information of aids to navigation on the bridge, thereby interfering with their judgment of water surface conditions. Improper color selection can also lead to a decrease in the recognizability of aids to navigation on the bridge, and may cause mixed light phenomena. The light color of aids to navigation on the bridge mixes with the background light to form a new mixed color light, changing the properties of the original aids to navigation on the bridge, resulting in visual confusion, making it difficult for ship drivers to quickly and accurately distinguish the information conveyed by the aids to navigation on the bridge. Based on the above reasons, it is particularly crucial to deeply analyze the adverse effects of bridge lighting on the recognizability of aids to navigation on the bridge, and explore measures for their coexistence with aids to navigation on the bridge.

Although bridge lighting projects and aids to navigation on the bridge belong to different research categories, their interdisciplinary research has significant innovation and challenges. This study will integrate knowledge and methods from multiple disciplines such as management, optics, and transportation engineering. By analyzing the adverse effects of bridge lighting on the recognizability of aids to navigation on the bridge, it can provide new ideas for related academic research and promote communication and integration between disciplines. This study also aims to provide valuable decision support for engineering management, designers, and decision-makers, helping them plan bridge lighting projects more scientifically, improve the design quality of such projects, effectively reduce their adverse effects on aids to navigation on the bridge, ensure the normal operation of aids to navigation on the bridge, and further improve the navigation safety of bridge water bodies.

2 THE FUNCTIONS AND DESIGN PRINCIPLES OF AIDS TO NAVIGATION ON THE BRIDGE

Aids to navigation on the bridge are a type of inland navigation marker. They are located in the center of the bridge on the facing side of the navigation bridge opening. They mainly indicate the position and navigation scale of the ship's navigation bridge opening, thereby reducing the possibility of collision between ships and bridges [5].

2.1 *The role of aids to navigation on the bridge*

As an important component of bridge safety facilities, the role of aids to navigation on the bridge is mainly to ensure the safety of navigation vessels and the safety of cross sea bridges [7]. The setting of aids to navigation on the bridge aims to guide ship drivers to correctly identify the outline and range of the bridge through clear identification. Especially during nighttime or under adverse weather conditions, the role of aids to navigation on the bridge becomes particularly critical.

In addition, in case of accidents or emergencies in bridge waters, aids to navigation on the bridge can provide accurate location information for rescue personnel, helping them quickly reach the rescue scenes and carry out rescue operations. This can minimize accident losses and protect people's lives and property safety.

2.2 *Principles for setting aids to navigation on the bridge*

In the national standard "Aids to Navigation on the Bridge in Navigable Waters of China" (GB 24418-2020), there are clear provisions on the principles of setting aids to navigation on the bridge. Specifically, the setting of aids to navigation on the bridge must meet the principles of accurate positioning and conspicuous clarity. They should not be obstructed or confused by other objects to ensure that they can be clearly seen by ship drivers in various weather and light conditions. The information conveyed by aids to navigation on the bridge must be accurate and without ambiguity or misunderstanding; the position and height must be clear enough to ensure that the ship's pilot can identify the bridge position and waterway boundaries in advance and accurately; the color and shape should also comply with relevant regulations to reduce visual interference to ship drivers [9].

It should be noted that the setting of aids to navigation on the bridge needs to be dynamically adjusted according to the actual situation. The position or type of aids to navigation on the bridge need to be adjusted with the maintenance or reconstruction of the bridge; it is necessary to optimize or update aids to navigation on the bridge when traffic flow changes. The setting and management of aids to navigation on the bridge should ensure that they adapt to changes in the traffic environment and safety needs, while also establishing a sound system and mechanism.

2.3 *The degree of influence of bridge lighting on the recognizability of aids to navigation on the bridge*

In order to better understand the different types and characteristics of bridge lighting projects and their impact on the recognizability of aids to navigation on the bridge, this article divides them into two categories: landscape lighting and functional lighting.

Landscape lighting focuses on beautifying bridges and enhancing the visual appeal of urban night scenes, highlighting the artistic and ornamental value of lighting. This type of lighting projects often uses colorful lighting, creating a colorful light and shadow effect through dynamic transformation and color combination, which is highly visually impactful. So landscape lighting sometimes interferes with ship drivers' recognition of aids to navigation on the bridge to a certain extent.

Functional lighting is designed to ensure the safety of the bridge pavement and facilitate the passage of pedestrians and vehicles. This type of lighting projects focuses more on the practicality and safety of lighting, usually using white or yellow light sources to ensure sufficient brightness, and provide clear vision for pedestrians and drivers on the bridge, thus reducing the occurrence of traffic accidents. The impact of functional lighting on the recognizability of aids to navigation on the bridge is relatively small, but in practical operation, attention should be paid to selecting appropriate light colors to prevent the occurrence of mixed light and avoid affecting the visual perception of ship drivers.

3 ANALYSIS OF THE IMPACT OF BRIDGE LIGHTING ON AIDS TO NAVIGATION ON THE BRIDGE

From the discussion in the previous chapter, it was found that bridge lighting has certain side effects. No matter how the brightness and color of the lights are controlled during the design process, it is inevitable that it will affect the recognition of aids to navigation on the bridge. Therefore, this chapter will analyze in detail the impact of bridge lighting on aids to navigation on the bridge from three aspects: brightness, color, and glare.

3.1 *The influence of brightness contrast on the recognizability of aids to navigation on the bridge*

Unreasonable brightness settings can have a significant impact on the recognizability of aids to navigation on the bridge, which is a safety hazard that cannot be ignored. As an important symbol for guiding ships to safely pass through bridges, the recognizability of aids to navigation on the bridge directly affects navigation safety. According to visual ergonomics, when the brightness contrast is between -0.13 and 1.3, there will be a situation where the background light covers the lights of aids to navigation on the bridge, which will cause strong interference to the sight of ship drivers. They will temporarily become blind under strong light, making it difficult to accurately identify the indication information of aids to navigation on the bridge, including but not limited to the direction of the waterway, height restrictions, and necessary warning signals, greatly increasing the risk of ship bridge collision accidents during nighttime navigation.

In optics, in order to quantitatively describe the brightness difference between the object and the environment, the brightness contrast is defined as:

$$C = \frac{|L_0 - L_b|}{L_b} \tag{1}$$

In the formula: C is the brightness contrast; L_0 is the object brightness, cd/m^2; L_b is the background brightness, cd/m^2.

When the human eyes are just able to recognize an object, the difference in brightness between the object and the background is called the critical brightness difference, represented by $\triangle L_t$, then the critical contrast C_t is

$$C_t = \frac{\Delta L_t}{L_b} \tag{2}$$

The ratio of the actual brightness contrast C between the object and the background to the critical contrast C_t is called the visibility level VL, which is

$$VL = \frac{C}{C_t} = \frac{\Delta L}{\Delta L_t} \tag{3}$$

In the formula: VL is the visibility level, and the larger the value of the visibility level VL, the clearer the object can be seen; $\triangle L$ is the actual brightness difference between the object and the background, in cd/m^2; $\triangle L_t$ is the critical brightness difference between the object and the background when the human eyes are just able to recognize the object, in cd/m^2.

If equations (1) and (2) are substituted into equation (3) and simplified, then

$$VL = \frac{C}{C_t} = \frac{\Delta L}{\Delta L_t} = \frac{|L_0 - L_b|}{\Delta L_t} \tag{4}$$

When the aids to navigation on the bridge work normally at night, under certain weather conditions, the brightness L_0 of the aids to navigation on the bridge is a certain value. The closer the background brightness L_b of the bridge lighting is to the brightness L_0 of the aids to navigation on the bridge, the smaller the visibility level VL of the aids to navigation on the bridge. The lower the recognizability of the aids to navigation on the bridge, the more difficult they are to be recognized by ships [2].

3.2 *The influence of color contrast on the recognizability of aids to navigation on the bridge*

The national standard "Aids to Navigation on the Bridge in Navigable Waters of China" (GB 24418-2020) clearly specifies the light colors of aids to navigation on the bridge, including four colors: red, green, white, and yellow. When ships navigate at night in navigable waters of bridges, the light color of aids to navigation on the bridge is an important basis for ship drivers to take collision avoidance actions and ensure safe navigation of ships.

When there is a bridge lighting project in the navigable water area of the bridge, the mixing of background light color and the light color of aids to navigation on the bridge can lead to deviation in the judgment of the light color of aids to navigation on the bridge by ship drivers. For example, originally intended to be a warning sign for bridge piers, its light quality is a continuous flashing yellow light. However, under the background light of a large number of warm tones around it, its warning effect may be greatly weakened, making it difficult for ship drivers to quickly and accurately judge the condition of the waterway, and thus unable to take necessary avoidance measures in a timely manner.

When studying color contrast, there are two systems that can be used, namely the *RGB* system and the *CIE-XYZ* standard chromaticity system. Firstly, introduce the *RGB* system. According to Glass's color mixing law, the three stimulus values of mixed light *S* can be obtained as

$$\begin{cases} R = R_1 + R_2 \\ G = G_1 + G_2 \\ B = B_1 + B_2 \end{cases} \tag{5}$$

In the formula: R_1, G_1, B_1 and R_2, G_2, and B_2 are the tristimulus values of the two lights S_1 and S_2 that are being mixed.

When there is a significant difference in the color properties of the colored light S_1 and S_2, there is a certain difference between the color of the light observed visually and the original color of the light. In addition, color only depends on the proportion of *R*, *G*, and *B*, rather than their size. A color can be accurately expressed by two quantities in its corresponding chromaticity coordinates *r*, *g*, and *b*. The conversion formula between *R*, *G*, *B* and their corresponding chromaticity coordinates *r*, *g*, and *b is* as follows:

$$\begin{cases} r = R/(R+G+B) \\ g = G/(R+G+B) \\ b = B/(R+G+B) \end{cases} \tag{6}$$

Then introduce the *CIE-XYZ* standard chromaticity system, in which *CIE* (International Commission on Illumination) represents any color *C* as follows:

$$C = xX + yY + zZ \tag{7}$$

In the formula: *X*, *Y*, and *Z* represent the hypothetical colors corresponding to red, green, and blue, respectively; *x*, *y* and *z* are the reference color quantities for matching color *C*, and their standard formulas are as follows:

$$\begin{cases} x = X/(X+Y+Z) \\ y = Y/(X+Y+Z) \\ z = Z/(X+Y+Z) \end{cases} \tag{8}$$

In the formula: the parameters *x* and *y* are called chromaticity values, where $X+Y+Z=1$. Therefore, the chromaticity coordinates (x, y) can represent any two-dimensional color [3].

The conversion between *RGB* system and *XYZ* system can be achieved through the following equation:

$$\begin{cases} x = 0.490r + 0.310g + 0.200b/(0.667r + 1.132g + 1.200b) \\ y = 0.117r + 0.812g + 0.010b/(0.667r + 1.132g + 1.200b) \\ z = 0.000r + 0.010g + 0.990b/(0.667r + 1.132g + 1.200b) \end{cases} \tag{9}$$

In the *XYZ* color model, the *CIE* chromaticity map can be obtained by drawing the chromaticity coordinates (*x*, *y*) of the light color, as shown in Figure 1 [4].

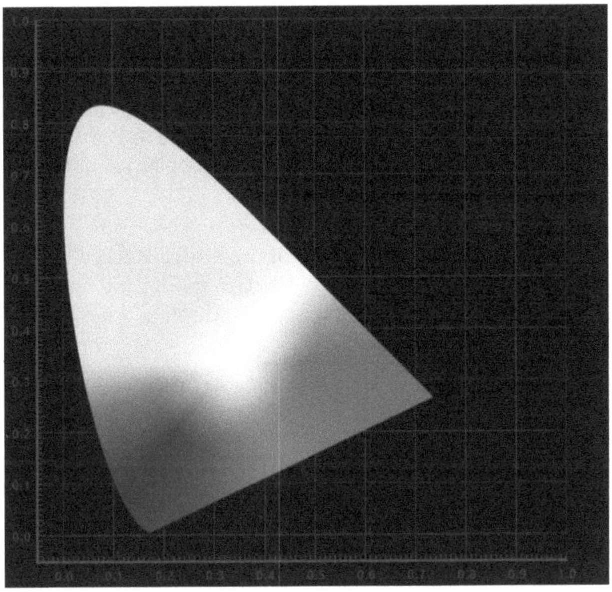

Figure 1. *CIE* chromaticity chart.

From Figure 1, it can be seen that the middle area of the chromaticity map is white, with lower color saturation closer to the middle white area and higher colors saturation closer to the outer contour of the boundary. When the colors of the aids to navigation on the bridge differs significantly from the background light colors of the bridge lighting project, the mixture of the two will produce new colors, resulting in a change in the colors of the aids to navigation on the bridge, known as the phenomenon of mixed light. This phenomenon requires ship drivers who navigate at night to spend energy and time searching and identifying aids to navigation on the bridge in these various lights, interfering with their acquisition of aids to navigation on the bridge indication information, which may lead to maritime traffic accidents.

3.3 *Impact of glare on the recognizability of aids to navigation on the bridge*

The impact of glare on aids to navigation on the bridge is mainly reflected in the following situations: when there is glare in the night navigation environment of ships, it can affect the ship drivers' acquisition of information from aids to navigation on the bridge. Glare is divided into direct glare, reflected glare, disabling glare, and unpleasant glare [5]. The glare discussed in navigable waters near bridges primarily concerns direct glare and reflected glare.

The lighting design standard for outdoor work sites in China (GB50582-2010) defines "glare" and "glare value" as follows: glare refers to a visual phenomenon that causes discomfort or reduces the ability to observe details or targets due to inappropriate brightness distribution or range in the field of view, or extreme contrast; glare value is a psychological parameter that measures the subjective response of outdoor lighting devices to discomfort in the human eyes [10].

At present, there are three commonly used glare evaluation models [6], and the applicable ranges of the three glare models are shown in Table 1:

Table 1. Scope of application of three glare models.

Model category	Applicable scope
Unified Glare Index (*UGR*)	Suitable for indoor glare evaluation
Glare value (*GR*)	Glare evaluation suitable for outdoor use
Threshold Increment (*TI*)	Glare evaluation applicable to road lighting

Due to the fact that the navigable water area of the bridge studied in this article belongs to the category of outdoor lighting, the glare index *GR* model recommended in the lighting design standards for outdoor work sites in China is adopted. The *GR* value should be calculated according to the following formula:

$$GR = 27 + 24\lg\frac{L_g}{L_b^{0.9}} \tag{10}$$

In the formula: L_g is the brightness of the glare source, in cd/m^2; L_b is the background brightness, in cd/m^2.

According to equation (10), the glare value (*GR*) can be calculated. The magnitude of the *GR* value reflects the intensity of the glare, and the smaller the value, the weaker the glare intensity. A light environment with a glare value less than 30 has almost no glare phenomenon, while a light environment with a value of 70 will interfere with human vision. The relationship between glare value (*GR*) and discomfort glare is shown in Table 2 [3].

Table 2. Relationship between GR value and uncomfortable glare intensity.

GR value (cd/m^2)	Discomfort level
<30	Almost no impact
30-40	There is a slight sensation
40-50	Slightly uncomfortable
50-60	uncomfortable
60-70	Feeling dizzy
70-80	Generating interference
>80	greatly interfere disturb

4 COUNTERMEASURES AND SUGGESTIONS

4.1 *Countermeasures*

Based on the analysis of the impact of brightness contrast, color contrast, and glare in the previous chapter, corresponding countermeasures can be proposed for these three aspects.

4.1.1 *Accurate brightness control*

By adopting an intelligent control system, the brightness of bridge lights is automatically adjusted according to time, weather, and seasonal changes, ensuring that there is no excessive direct or reflected light on the aids to navigation on the bridge at any time, causing light pollution or visual confusion. For example, when there are ships passing under the bridge at night, the unnecessary decorative bridge lights should be appropriately reduced to enhance the brightness contrast of the aids to navigation on the bridge, ensuring that the lights of aids to navigation on the bridge can provide the most eye-catching guidance for ships sailing at night.

4.1.2 *Scientific color matching*

Hire professional lighting designers to conduct in-depth research on color psychology and visual communication principles, ensuring that the colors of bridge lighting complement rather than conflict with the standard colors of light of aids to navigation on the bridge. For example, if red or green light is used for aids to navigation on the bridge, the surrounding bridge lights should use similar tones as much as possible to avoid mixing due to significant colors differences between the two, which can interfere with the judgment of ship drivers. In addition, the colors of bridge lighting can also be adjusted by using an intelligent control system. Even if the colors of bridge lighting differs significantly from that of aids to navigation on the bridge, when ships pass by, the lights around the aids to navigation on the bridge can be quickly adjusted to similar colors, reducing the impact of mixed light on the recognizability of aids to navigation on the bridge.

4.1.3 *Application of anti glare technology*

When designing and installing bridge lighting, advanced anti glare lamps and optical lenses are used, and through precise calculation and simulation, the projection angle and diffusion range of the light can be effectively controlled to focus on the structural parts that need to be illuminated on the bridge. At the same time, the light is limited to spread horizontally or vertically, ensuring that it directly shines on the bridge structure rather than spreading to the waterway, reducing light pollution to the navigable water area of the bridge, avoiding strong glare to the cockpit of passing ships, thereby reducing eye fatigue for drivers and improving the safety factor of night navigation.

4.2 *Suggestions*

In addition, after consulting materials from different fields and seeking inspiration from multiple disciplines, the following three suggestions can be proposed, aiming to take measures from different perspectives to alleviate the adverse effects of bridge lighting on the recognizability of aids to navigation on the bridge. At the same time, it can also promote the development of bridge lighting engineering towards a more intelligent, environmentally friendly, and aesthetically diversified direction.

4.2.1 *Application of laser technology*

Laser is widely used due to its good directionality, high brightness, and good monochromaticity. Its strong visual effect is currently lacking in aids to navigation on the bridge under complex nighttime backgrounds. However, due to the blinding effect of lasers on the human eyes, the output power of the laser used should not be too high, with the limit of not damaging the human eyes. Rotating laser lights can be installed on aids to navigation on the bridge with complex background lights to emit laser beams into the air or water surface. By adjusting the brightness and angle of the beam, the identification effect of aids to navigation on the bridge at night can be enhanced while protecting the eyes of ship drivers [8].

4.2.2 *Construction of AIS aids to navigation on the bridge*

If conditions permit, AIS aids to navigation on the bridge can also be constructed to ensure that when ships navigate in low visibility conditions such as rainy or foggy days, AIS at the ship's end can be used to timely detect the bridge piers ahead, avoid collisions, and obtain navigation data on the net height of the bridge openings through electronic screens, further increasing bridge identification. In addition, AIS aids to navigation on the bridge can also integrate warning functions, automatically sending warning signals to the ship's AIS when the ship approaches the preset safe distance, reminding crew members to take necessary avoidance or deceleration measures. The application of AIS aids to navigation on the bridge not only effectively reduces navigation accidents caused by severe weather, but also greatly improves the safety and efficiency of navigation by providing accurate navigation information and warning services.

4.2.3 *Intensive supervision*

Proper supervision can also reduce the impact of bridge lighting on the recognizability of aids to navigation on the bridge. The regulatory measures to strengthen supervision are as follows:

(1) Relevant departments should introduce relevant laws and regulations to define background light interference, which should include but are not limited to the definition of light pollution, monitoring standards, maximum allowable light intensity and color range, as well as punishment measures for violations, and include marine light pollution in marine environmental protection.
(2) Marine environmental protection agencies and maritime supervision departments should jointly supervise; the public and media play a supervisory role through public opinion and reporting, and improve the mechanism for managing marine light pollution. Build a comprehensive and multi-level marine light pollution management system with government leadership, departmental collaboration, and social participation to ensure that bridge lighting projects not only beautify the city but also reduce the impact on navigation safety in navigable water areas, achieving harmonious coexistence between economic development and environmental protection.

5 CONCLUSION

After the above analysis, the research conclusion can be drawn as follows:

(1) Unreasonable brightness settings can affect the recognizability of aids to navigation on the bridge, and when the brightness contrast is between -0.13 and 1.3, the background light will significantly mask the lighting of aids to navigation on the bridge, causing serious visual interference to ship drivers. In response to this impact, an intelligent control system can be used to adjust the background light in real time, improve the visibility of aids to navigation on the bridge, and enhance the contrast between the two.
(2) Improper color selection can lead to mixing of background light and the light of aids to navigation on the bridge, increasing the risk of night navigation for ship drivers. In response to this phenomenon, in lighting design, it is advisable to set up lights with similar colors near aids to navigation on the bridge as much as possible to reduce the occurrence of mixed light; intelligent control systems can also be used to dynamically adjust the colors of the background light.
(3) The glare phenomenon caused by brightness exceeding a certain threshold can also have a certain adverse effect on the recognition of aids to navigation on the bridge. When the glare value is greater than 70, it will interfere with human vision. By introducing anti glare technology and using advanced anti glare lamps and optical lenses to control the projection angle and diffusion range of light, glare can be effectively reduced, thereby reducing the degree of visual interference to ship drivers.

The work done in this article can provide some help in seeking a balance between bridge lighting and the recognizability of aids to navigation on the bridge, and promoting the harmonious coexistence of urban landscape beautification and water traffic safety. However, due to various limitations such as time and conditions, there are still some shortcomings. Further research can be conducted from the following aspects:

(1) Further arguments support the feasibility of applying laser technology to aids to navigation on the bridge, exploring reasonable installation angles, positions, and output power. The aim is to enhance the recognizability of these aids to navigation on the bridge at night without causing eye damage and protecting the eyesight of ship drivers.
(2) Further validation is needed to confirm the feasibility of establishing AIS aids to navigation on the bridge. At present, AIS responders are mainly used on coastal aids to navigation, and there are almost no equipped on aids to navigation on the bridge. Therefore, it is necessary to explore the reasons for the lack of equipment, which may involve multiple considerations at the technical, cost, demand, regulatory, and practical operational levels. Next, we will analyze these potential factors one by one and propose corresponding improvement strategies to promote the application of AIS in aids to navigation on the bridge.
(3) If the use of various technologies to reduce the impact of bridge lighting on the recognizability of aids to navigation on the bridge is a temporary measure, then defining background light interference and legislating and regulating marine light pollution is a fundamental measure. However, the procedures are relatively complex (especially involving international legislative issues), and it is necessary to conduct in-depth technical research and cooperate with multiple disciplines, departments (and even governments) to promote the early introduction of relevant legislation.

REFERENCES

[1] Chen Hui (2024): Why did ships "crash" into bridges after 21 accidents in 9 years? *China Emergency Management Report.*

[2] Li Dazhi, Li Chunan, Wan Xu, *et al.* (2021): The impact of background lighting on the recognizability of navigation beacon lights and its regulatory measures. *China Maritime*, 10, 58–60.

[3] Li Chunnan (2022): *Research on the Impact of Marine Light Pollution on the Recognizability of Navigation Beacon Lights. Dalian Maritime University.*

[4] Li Zhirong, Zhu Jinshan, Zhang Wenshuan, *et al.* (2020): Analysis of the impact of bridge lighting engineering on the safety of night navigation of ships in water. *Safety and Environmental Engineering*, 27 (02), 118–125.

[5] Wang Aimin (2018): Design points for bridge and culvert markers crossing inland waterways. *The Pearl River Water Transport*, (5): 90–91.

[6] Wang Yajing, Huang Haijing and Chen Gang (2020): Effects of lighting on ECG, visual performance and psychology of the elderly. *Optik*, 203 (C).

[7] Zhang Aiping (2009): Research on the Application of Donghai Bridge Culvert Markings. The Coastal Navigation Markings Group of the Navigation Markings Professional Committee of the Chinese Society of Navigation, the Radio Navigation Group of the Navigation Markings Professional Committee of the Chinese Society of Navigation, and the Inland Navigation Markings Group of the Navigation Markings Professional Committee of the Chinese Society of Navigation. *Proceedings of the Annual Meeting and Academic Exchange Conference of the Coastal Navigation Aids Group, Radio Navigation Group, and Inland Navigation Aids Group of the Navigation Aids Professional Committee of the Chinese Society of Navigation* [Publisher unknown].

[8] Zhang Chen, Li Xin, Wang Hui, *et al.* (2021): Research on Improving the Identification of Navigation Bridge Holes for Multi porous Navigation Bridges on the Yangtze River Main Line. *China Water Transport and Navigation Technology*, 03, 26–30.

[9] The State Administration for Market Regulation, the National Standardization Administration (2020): GB 24418-2020. *Navigation Aids for Bridges in Navigable Waters in China's Seas.*

[10] Ministry of Housing and Urban Rural Development of the People's Republic of China, General Administration of Quality Supervision Inspection and Quarantine of the People's Republic of China (2010): GB 50582-2010. *Lighting Design Standards for Outdoor Work Sites.*

Proceedings of the International Association of Institutes of Navigation World Congress 2024 – Chen et al. (Eds)

The applied research of state of health on the batteries of marine aids to navigation

Meng Ji & Jinhua Xu
Tianjin Aids to Navigation Division, The Navigation Guarantee Center of North China Sea, Tianjin, China

ABSTRACT: Recent of years, lithium-ion battery(LIBs) products are widely used, and LIBs have become the main choice of power supply for energy storage systems. With the development of marine aids to navigation(AtoN) technology, The Lead-acid batteries, which provide power for AtoN, are gradually being replaced by LIBs. At the meantime, some problems with the use of LIBs are also gradually emerging, after repeated charge and discharge of the battery, the active material of the internal electrode and the stock of lithium-ion continue to consume and lose, resulting in the decline of its capacity, charging and discharging power and other performance, the battery will gradually aging. Finally, it will affect the efficiency of the AtoN. This will have a significant impact on the safety of ship navigation in the channel and near shore at night. For the AtoN which work at sea, almost all power are provided by LIBs. So, it is very necessary to detect the health state of the battery. State of health（SOH）is the embodiment of the healthy life of the battery, It can display the battery's charge, energy, charging and discharging power and other states.SOH is the basis for estimating the performance time of AtoN. It is also an important parameter to assist the correction and improve the accuracy of state of charge estimation. At the same time, it is the basis of predictive maintenance. The internal resistance of the battery increases after aging, which will produce more heat and cause thermal runaway more easily. Therefore, monitoring the SOH can not only fully understand the current status of the battery, to provide data support for future research on improving the stability and life of energy storage systems, but also have greater help to the planned maintenance of the AtoN in the jurisdiction. It can reduce thefrequency of maintenance at sea, avoid the waste of manpower and material resources, save unnecessary expenses and so on. This article analyzes and discusses the feasibility and specific practices applies SOH on the batteries of AtoN.

Keywords: marine aids to navigation(AtoN), efficiency, lithium-ion batteries(LIBs), state of health(SOH), maintenance cost

1 INTRODUCTION

As an efficient, lightweight and environmentally friendly battery technology, Lithium-ion battery(LIBs) has a wide range of applications in modern society. From portable electronic devices to clean energy storage systems, LIBs are changing the way we live and industrial production, and they are widely used in electric vehicles, drones, mobile devices, aerospace and other fields.As technology develops and costs decrease, application areas will continue to expand [5]. So far, the marine aids to navigation(AtoN) equipped with lithium battery has been widely used, and the application of lithium battery in AtoN has become a trend. Although lithium battery technology has been relatively mature, it's application in AtoN still needs to be actively tried and explored, so that the application technology of lithium

 DOI: 10.1201/9781003624424-25

batteries in this field of AtoN tends to be perfect.At present, the latest AtoN lithium battery technology has realized the battery overcharge and overdischarge protection, high and low temperature charge protection functions,and it's supporting battery monitoring system has been able to real-time monitor the battery's working current, working voltage, charging current, charging voltage, and remaining power and other data. As is shown in Figure 1,

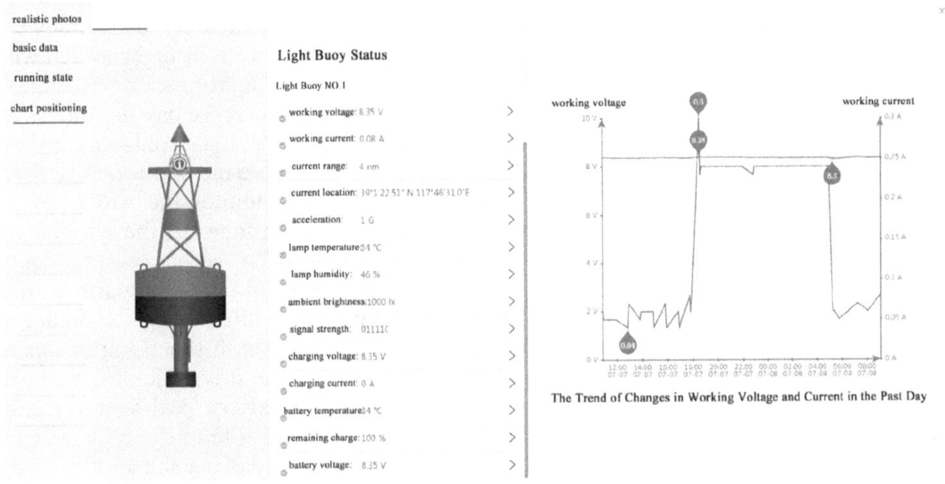

Figure 1. State of telemetry and remote control of AtoN.

However, the existing battery monitoring system does not monitor the battery State of health (SOH), which has been equipped on the AtoN. Therefore, it is impossible to accurately judge the aging of lithium-ion batteries, can not provide a reference for the replacement of lithium-ion batteries,.This has a serious impact on the of the efficiency of AtoN. To solve this problem, we conducted an in-depth study of the battery SOH,and developed a monitoring system that can monitor it from time to time.

2 THE EXISTING PROBLEMS IN THE APPLICATION OF LITHIUM BATTERIES FOR ATON

The existing technology has realized the practical application of intelligent AtoN. With the marine aids to navigation as the carrier, it realizes the "intelligence" of AtoN through sensor, automatic control, microprocessor and other technologies [3]. Intelligent AtoN are equipped with GPS or Beidou modules, the AIS data, the lit and unlit status of the AtoN, the charge and discharge status of the battery, the hydrometeorological information and other information are returned to the shore base through the satellite network. And it can be displayed on the telemetry and remote control system to achieve real-time monitoring of the AtoN. However, for the development of new intelligent AtoN and the monitoring and management requirement, the existing functions are still insufficient. It's more practical and convenient functions still need to be explored. In particular, the corresponding functions of the battery on AtoN need to be improved.

2.1 *The current telemetry remote control system has no battery health status data*

The current telemetry and remote control system only provides the data monitoring of the remaining battery power, charging voltage and charging current,as shown in Figure 1. The

SOH of the battery is not monitored, that way we don't know if the battery is currently healthy or not. The battery health could be 100%, or it could be 80%. These data are very important for the daily maintenance and management job andit also has an important impact on ensuring the normal performance of AtoN. Especially when the battery health is close to the critical value, the battery capacity after a long time of charging and discharging, accompanied by the aging of the battery, even if the battery is fully charged during the day time through the solar panel, the power is showed 100% of the current battery capacity, but in fact, the 100% power at this time is only 80% of the new battery capacity, or even lower. The stored electricity is not enough to support the AtoN continuous lit at night, which will have a serious impact on the navigation safety of passing ships at night time.

In practice, it is possible that after the battery is fully charged during the day time, and the power can support the AtoN to work normally for most time of the night, only an hour or two before the sun comes out lose power supply. At this time, because the power outage time is too short, the telemetry and remote control system can not monitor the AtoN power failure and no signal received, so that it can not send an alarm in time;Or, the system can monitor the signal interruption caused by the power failure, but after the sun rises, the solar panel recovers the power supply, the AtoN sends a normal signal, and the alarm in the system will be lifted in a short time.No matter in which case, it is difficult for the management personnel to detect it in time and make a quick response;Even if the alarm of signal interruption can be found by AtoN management personnel in time, it is difficult to make a correct judgment because there is no corresponding data of battery health at present. Therefore, it has led to the failure of the AtoN to play a normal role in the hour or two before the sunrise, which seriously affects the navigation safety of passing ships.

Because there is no data on the battery's health status, AtoN management personnel do not know exactly when the battery is not able to meet the normal operation of the light buoy. The action can only be taken after the battery level gives a continuous alarm.According to the provisions of the relevant system documents, the inspection interval of the AtoN is generally not more than three months, and the inspection interval of the wreck AtoN outside the port is generally not more than six months.If the battery life reaches its limit within a short time after the last inspection, the AtoN management personnel have to do the maintenancejob again, this inevitably leads to the waste of manpower and material resources and the increase of maintenance costs.And because of the distance from the AtoN to the port and the weather and sea conditions, sometimes measures can not be taken in time, which will also affect the navigation safety of passing ships.

2.2 *There is currently no actual service life data of the AtoN lithium battery*

Because lithium batteries have only been applied to AtoN in recent years, there is no actual use data on the life of the batteries.With the growth of the service life of the AtoN battery, its service life will also reach the limit.For the renew and maintenance of the AtoN battery, it can only be judged by the signal loss alarm on the telemetry and remote control system and the AtoN maintenance personnel according to the design life of the lithium battery.In this way, not only the accuracy of the AtoN lithium battery life judgment can not be guaranteed, but also the timeliness of replacing the lithium battery can not be guaranteed, thus affecting the normal performance of the AtoN. The general service life of lithium batteries currently in use is four to six years.This range of service life is based on estimates under conventional use conditions and is twice the life of lithium batteries compared to traditional lead-acid batteries.The service life of lithium batteries is affected by a variety of factors, including but not limited to usage habits, environmental conditions, charge and discharge times, and battery maintenance.

- Charge and discharge times:Most lithium batteries charge and discharge around 500 times. AtoN solar power lithium battery can be calculated on the charging and discharging once a day, it can be used about 500 days, in fact, the AtoN battery can not achieve deep discharge at night, that is, all the battery power is used up. And affected by the weather, it is impossible to charge the battery on rainy days. In this way, the charge and discharge process of the lithium battery can be approximately regarded as shallow charge and shallow discharge.Their service life will be extended, and some high-quality batteries can withstand more charging and discharging times.

- Environment of use:Temperature and humidity are two important factors affecting the life of lithium batteries.High temperature, high humidity environment will accelerate the aging process of the battery.In the high humidity environment at sea, if the seal protection of the AtoN battery is not in place, or the seal is damaged due to collision, it will seriously affect the service life of the battery.
- Frequency of use:If the lithium battery is used frequently, the battery life will be affected. For example, if the battery is used ten times a day, it may only last half as long as originally expected.

The above can be used as an important factor to judge the service life of the AtoN lithium battery, but in reality, actual data is still needed to provide support for the maintenance of the battery.Therefore, it is particularly important to monitor the SOH of the AtoN lithium battery.

3 APPLICATION INDUSTRY AND TEST METHOD OF BATTERY STATES OF HEALTH(SOH)

SOH (State of health) is commonly used to indicate the remaining life of the battery.As the number of battery cycles increases, the battery SOH decreases, the remaining life decreases, and the available capacity gradually decreases. When the battery capacity decreases to 80% of the standard capacity, the battery is judged to be invalid and needs to be replaced [6].

3.1 *At present, the SOH state of lithium batteries is mainly applied in the following industries, and the technology has been relatively mature*

3.1.1 *Electric vehicle industry*

The battery of an electric vehicle is a key energy storage device, and its health has a direct impact on vehicle performance and endurance.By monitoring and evaluating the SOH of the battery, battery health problems can be detected in time, and early warning or maintenance can be carried out to ensure the safety and reliability of electric vehicles.

3.1.2 *Energy storage system industry*

Lithium batteries are widely used in energy storage systems to balance power grid loads and reserve backup power.Monitoring the SOHof the battery can help optimize the performance and operational management of the energy storage system, detect battery health problems in time, and improve the stability and life of the energy storage system.

3.1.3 *Mobile device industry*

Lithium batteries are widely used in mobile devices such as smartphones and tablets.By monitoring the SOHof the battery, users can be reminded of the health status of the battery, helping users better manage battery usage and extend the battery life of mobile devices.

3.1 *There are several main test methods to check the battery Status of Health (SOH)*

3.2.1 *Open circuit voltage method*

The health of the battery is inferred by measuring the open circuit voltage of the battery when it is not under load.In general, the open circuit voltage of the battery is related to its

State of Charge (SOC), but after a long time of use, the open circuit voltage of the battery may be reduced, reflecting the health of the battery.

3.2.2 *Charge and discharge test method*

The health of the battery is assessed by testing the battery's charge and discharge cycle to measure the capacity change of the battery under different SOCs.In general, battery capacity decreases with increasing use time, and this rate of reduction can reflect the health of the battery.

3.2.3 *Internal resistance test method*

The health of the battery is assessed by measuring its internal resistance.The internal resistance is a comprehensive reflection of the electrochemical reaction inside the battery and the performance of the material, and the greater the internal resistance, the worse the health of the battery.The internal resistance test can be performed by the AC impedance method or the DC impedance method [2].

4 THE APPLICATION OF SOH IN THE LITHIUM BATTERY OF ATON

4.1 *The concrete calculation method of SOH*

At present, the application of SOH in lithium battery has been relatively mature, and it can be applied to the monitoring of lithium battery of AtoN. Taking the charging and discharging test method as an example, the specific calculation method of SOH is as follows:

The enhanced Coulomb counting algorithm is used to estimate the SOC and SOH parameters of the lithium-ion battery.The initial SOC is obtained from the loaded voltage (charge and discharge) or the open circuit voltage. The loss is compensated by taking into account the charging and discharging efficiency.By dynamically recalibrating the maximum releasable capacity of the working battery, the SOH of the battery can be estimated at the same time.

Technical principle:

The creleasable capacity of a working battery is the capacity released when it is fully discharged.Accordingly, SOC is defined as the percentage of the releasable capacity relative to the battery rated capacity (Crated) provided by the manufacturer.

$$SOC = \frac{\text{Creleasable}}{\text{Crated}} \times 100\%$$

A fully charged battery has a maximum releasable capacity (Cmax), which may differ from the rated capacity. In general, the Cmax of the newly used battery will have a certain difference with the Crated, and will decay with the passage of time. It can be used to evaluate the SOH of the battery.

$$SOH = \frac{\text{Cmax}}{\text{Crated}} \times 100\%$$

When a battery is discharged, the depth of discharge (DOD) can be expressed as a percentage of the discharged capacity relative to Crated

$$DOD = \frac{\text{Creleased}}{\text{Crated}} \times 100\%$$

Where Creleased is the capacity of current discharge.

Using the measured charging and discharging currents (lb), the DOD difference in the working period (dt) can be calculated as follows:

$$\Delta DOD = \frac{-\int_{t_0}^{t_0+\tau} I_b(t)\mathrm{d}t}{Crated} \times 100\%$$

In particular, lb is positive when charging and negative when discharging. DOD adds up over time.

$$DOD(t) = DOD(t_0) + \Delta DOD$$

In order to improve the estimation accuracy, it is necessary to consider the work efficiency (expressed as ŋ), and the DOD expression becomes accordingly:

$$DOD(t) = DOD(t_0) + ŋ\Delta DOD$$

ŋ is equal to ŋc in the charging phase and it's equal to ŋd in the discharging phase. Without considering working efficiency and battery aging, SOC can be expressed as:

$$SOC(t) = 100\% - DOD(t)$$

Considering SOH, SOC is estimated as follows:

$$SOC(t) = SOH(t) - DOD(t)$$

Figure 2 is the flow chart of the enhanced Coulomb counting algorithm. At the beginning, the historical data of the used battery is retrieved from the relevant memory.With a new battery, there's no historical information available. Assuming its SOH is good and has a value of 100%, the SOC is initially estimated by testing the open circuit voltage or the load voltage (depending on the starting conditions).

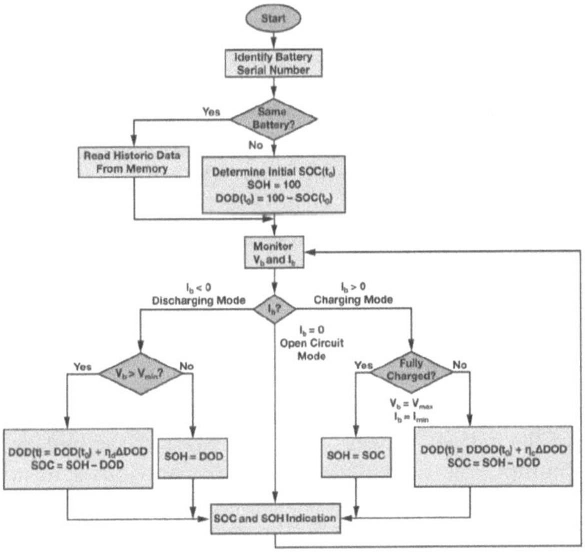

Figure 2. Flow chart of enhanced Coulomb counting algorithm.

The estimation process is based on the monitoring of the battery voltage (Vb) and Ib. The operating mode of the battery can be known by the magnitude and direction of the operating current.In discharge mode, DOD accumulates the charge consumed;In charge mode, DOD decreases as the accumulated charge enters the battery.After correcting for charge and discharge efficiency, more accurate estimates can be achieved.Then subtract the amount of DOD from the SOH to estimate the SOC.When the battery open circuit, the current is zero, SOC is directly derived from the relationship between OCV and SOC.

It should be noted that SOH can be re-evaluated when the battery is completely depleted or fully charged;The operating current and voltage of the battery are specified by the manufacturer.During discharge, when the loading voltage (Vb) is less than the lower limit (Vmin), the battery is exhausted.In this case, the battery can't use, shall be recharging.At the same time, the SOH value can be re-estimated by the cumulative DOD in the depleted state, thus recalibrating the SOH.On the other hand, during the charging period, if (Vb) reaches the upper limit (Vmax) and (lb) falls to the lower limit (lmin), the battery used is fully charged.The new SOC value can be calculated by accumulating the total charge of the battery to obtain a new SOH value. In practice, fully charged and depleted states occasionally occur.When the battery is frequently charged and thoroughly discharged, the accuracy of the SOH assessment is improved.

Because the calculation is simple and the hardware requirements are not complicated, in addition to commonly used electric vehicles, all devices containing lithium-ion batteries can easily implement the enhanced coulomb counting algorithm.In addition, the estimate error can be reduced to 1% in the working cycle immediately following the SOH reassessment. [1]

4.2 *Specific design of SOH application in AtoN battery*

At present, there is a certain technical and equipment foundation for the telemetry and remote control of AtoN. Considering the economic practicality, SOH monitoring of the AtoN battery can be integrated into the existing battery management system (BMS), and integrated into the AtoN telemetry and remote control system as a separate module for display. Figure 3 shows the page of the self-designed battery monitoring system, which monitors the important parameters of the AtoN battery and develops corresponding auxiliary functions around these parameters.

4.2.1 *Remaining power display and charge control*

The remaining power of the current battery (the state of charge of the battery) is used to characterize the available power of the battery.During the day, the charging current can be controlled through BMS, when the battery is charged to 80%, the current is reduced, and the battery is charged to 100% in trickle charging mode;After the battery is full, stop charging to avoid overcharging of the lithium battery, so as to protect the battery and extend the battery life.At the same time, the current charging situation is displayed, such as charging current, charging voltage, etc., so as to provide various data for AtoN management personnel to judge the current charging state of the battery.In addition, the system will also display the battery power curve, working voltage, working current curve, etc. in the past week, according to which can intuitively see the working state of the AtoN at that time.

4.2.2 *Monitoring of battery SOH and battery cycle times*

Displays the SOH of the current battery. When the SOH of the battery is 85% or less, the system will give a prompt, and the battery data in the previous state will be marked with yellow to remind the AtoN management personnel to pay special attention to the SOH of the battery.When the SOH of the battery is 80%, the system will give an alarm, and the current state of the battery data is marked in red, at this time, the AtoN management personnel should start to replace the battery as soon as possible.At the same time, this system monitors and displays data such as the number of battery cycles, battery design capacity, current

Figure 3. Battery health monitoring page.

actual capacity, production date, and battery running time, which can more intuitively reflect the health of the battery.

4.2.3 *Battery temperature display and control*

Lithium batteries are more sensitive to temperature, the suitable operating temperature range of lithium batteries is about 0 ° C ~45 ° C, in this temperature range, the performance of lithium batteries is the best, the service life will be longer, and it is also safer.When the ambient temperature is too high, too much heat is generated inside the battery, which may affect the battery life and even cause accidents.When the ambient temperature is too low, the chemical reaction speed inside the battery will slow down, reducing the output voltage and power of the battery, thus affecting the battery performance.Therefore, in this system, the temperature of the battery is monitored in real time, and when the temperature of the battery is too high or too low, the temperature is controlled by BMS.In actual use, when the battery temperature exceeds the safe range, the battery temperature can be controlled within the normal range by stopping charging and discharging, reducing the current and other measures to ensure the safety and life of the battery.

5 THE EXPECTED EFFECT AND THE ANALYSIS OF SOCIAL AND ECONOMIC BENEFITS OF SOH APPLICATION ON THE ATON LITHIUM BATTERY

5.1 *The expected effect of SOH application on the buoy lithium battery*

5.1.1 *Reflect the working state of the AtoN light*

As SOH can intuitively reflect whether the current battery is healthy or not, after mastering the SOH data of the lithium battery of the AtoN, the management personnel can understand the energy, charge and discharge power of the current battery according to it, and further judge the remaining service life of the battery and whether the AtoN light is working normally. In addition, according to the energy state of the battery, it can also be estimated whether the beacon can ensure the normal performance and duration of the whole night.This can not only reduce the unnecessary work of the AtoN management personnel, but also enable them to have a global grasp of the working status of the AtoN in the entire area and facilitate management.

5.1.2 *Provide data support for making maintenance plan*

Due to the lack of actual service life data of the AtoN lithium battery, after mastering the SOH of the battery, the management personnel can directly judge thebattery aging degree according to it.Whether its current energy state can support the normal performance of AtoN, whether it needs to be maintained, etc.If the battery really needs to be replaced, the management personnel can make a targeted maintenance plan based on the inspection schedule.

5.2 *Social and economic benefit analysis*

5.2.1 *Improve the normal rate of AtoN*

Most of the failures of AtoN are caused by the failure of energy system [4]. After monitoring the SOH of the AtoN battery, the status of the battery can be understood in real time.Once it is found that there is a problem, or it's energy may not be enough to support the normal work of the AtoN in the short term, it can immediately arrange personnel for maintenance and rapid response, so that the failure rate of the AtoN can be greatly reduced.At the same time, more targeted maintenance programs have also improved the normal rate of AtoN and ensures the navigation safety of passing ships.

5.2.2 *Reduce the maintenance cost of AtoN*

After monitoring the battery SOH, it avoids the embarrassment of having to replace the battery again in the short term due to battery aging after inspection.This not only reduces the number of maintenance, but also avoids the waste of manpower and material resources.For example, one voyage will take 8 hours for inspection AtoN at sea. The AtoN maintenance ship Fuel consumption in 5-6 tons, fuel prices fluctuate between 7000-8500RMB. Together with the food of the crew at sea, the salary of the relevant personnel, the depreciation fee of the ship, etc., one voyage can reach 50,000-60,00RMB. Under the background of the global economic downturn,it is also very necessary to reduce the cost of AtoN maintenance.

6 CONCLUSION

The application of lithium battery SOH can reduce the work intensity of AtoN management personnel. Help them have a comprehensive grasp of the operating status of AtoN in the jurisdiction, and the maintenance plan is more targeted,At the same time, it can further improve the normal rate of AtoN, ensure the safety of ship navigation, and save the AtoN maintenance cost. At present, battery SOH monitoring has been widely used in many fields, and the technology is relatively mature, which can be promoted and used on AtoN. On the basis of the existing telemetry and remote control system to expand the SOH monitoring function, the cost is also very low. In addition, digital AtoN is also an important direction for the future development, and the development and application of AtoN energy and related technologies can provide important support for digital AtoN.SOH as an important indicator of AtoN energy monitoring, it's practical role is also extremely important.It is believed that in the near future, AtoN energy and its related technologies can also get more development and innovation.

REFERENCES

[1] Adel Ghazel and Martin Murnane (2023): *In-depth study of battery state of charge (SOC) and state of health (SOH) estimation techniques [P]*, www.analog.com/cn/resources/technical-articles/a-closer-look-at-state-of-charge-and-state-health-estimation-tech.html.

[2] Dian Chetu (2024): *Voltage Characteristic Assessment for SOH Battery Health Assessment* [Z],www.new.qq.com/rain/a/20240703A09D5D00.

[3] Jia Shiyao (2017): *Study on management and technologyapplication for Marine Aids to Navigation basedon energy saving and environmental protection [D]*, Jimei University, 48.

[4] Wang Song (2023): Design and application of integrated green energy system for aids to navigation [J], *China Maritime Safety*, 4, 62.

[5] Xin Daneng (2024): *Three broad application areas of lithium-ion batteries* [Z], www.baijiahao.baidu.com/s?id=1794468834990963856&wfr=spider&for=pc.

[6] Yue Yeyuefeng (2023): *Description of power battery SOC, SOH, SOP, SOF* [Z], www.baijiahao.baidu.com/s?id=1785240435381325805&wfr=spider&for=pc.

Proceedings of the International Association of Institutes of Navigation World Congress 2024 –
Chen et al. (Eds)

Research on ship trajectory prediction model based on GRU-ABiLSTM

Zewei Li
School of Navigation, Wuhan University of Technology, Wuhan, China

Zhitao Yuan, Kezhong Liu & Mozi Chen
School of Navigation, Wuhan University of Technology, Wuhan, China
Hubei Key Laboratory of Inland Shipping Technology, Wuhan, China

Hang Yuan
School of Navigation, Wuhan University of Technology, Wuhan, China

ABSTRACT: In order to address the issue of low precision in ship trajectory prediction within complex navigable waters, a ship trajectory prediction model named GRU-ABiLSTM, which is based on the encoder-decoder architecture, is proposed. The prediction model employs gated recurrent units (GRU) in the encoder part of the model to capture the temporal features within the trajectory sequences, and utilizes a bi-directional long and short-term memory network (BiLSTM) in the decoder, and incorporates an attention mechanism to adjust the weights of the data features. The inputs of the model take the longitude, latitude, speed and heading of the ship at historical moment as basic features, while the density of ships in the waters, which has undergone a median filter smoothing process, is introduced as an additional feature. The AIS data collected from the core port area of Ningbo Zhoushan Port in March 2024 is selected for model training and validation, and compared quantitatively and qualitatively with GRU, LSTM, Seq2Seq-LSTM, and Attention-BiLSTM models, which provide better prediction results under different prediction duration and sailing scenarios.

Keywords: complex navigable waters, Ship trajectory prediction, GRU, BiLSTM, attention mechanism

1 INTRODUCTION

In complex navigable waters characterized by restricted fairways, heavy traffic, intersecting routes and more static obstacles (Zhixiang Fang *et al.* 2019), the ship trajectory prediction model helps to issue early warnings of potential danger by precisely predicting the future trajectories of ships, thereby safeguarding the safety of the ship, crew and cargo. Therefore, an efficient and accurate ship trajectory prediction model is indispensable for ensuring the safety of maritime transportation (Zhao Jiansen *et al.* 2023).

In the field of ship trajectory prediction research, the traditional research approach of mathematical modelling is widely adopted. This method involves simulating the behavioural characteristics of a ship by formulating specific physical equations. Houenou *et al.* (Houenou, A. *et al.* 2013, November) developed a curvilinear motion model and used an extended Kalman filter algorithm to predict the position, velocity and acceleration of ships. This process is computationally intensive and is not applicable to different scenarios.

The advancement of artificial intelligence has enabled machine learning methods to discern the relationship between historical ship motion patterns and their future trajectories. Rong

 DOI: 10.1201/9781003624424-26

et al. (Guo Yijie *et al.* 2024) decomposed the ship motion into longitudinal and latitudinal directions, and used Gaussian regression models to estimate the probability density function of longitudinal acceleration and latitudinal position, and predicted the uncertainty of the ship's trajectory. Zhang *et al.* (Zhang, X. *et al.* 2019, July) proposed a wavelet-based hidden Markov model (HMM) trajectory prediction algorithm, which addresses the drawbacks of Kalman filter and linear regression, such as low accuracy and large time lags.

In recent years, neural networks have attracted wide attention in the field of trajectory prediction due to their excellent performance in other fields. Among them, models such as Long Short-Term Memory Network (LSTM) and Gated Recurrent Unit (GRU) are often used to address the challenge of capturing temporal features, which have been difficult to handle in traditional methods. LIU (Liu *et al.* 2021) proposed an Atten-LSTM model combining the attention mechanism and LSTM based on the navigational environmental conditions of a complex inland waterway. This model is capable of efficiently and accurately predicting ship trajectories with guaranteed reliability. JU (Ju Cong 2023) proposed a hybrid ship trajectory prediction model named CNN-GRU based on CNN and GRU, which significantly improves the prediction accuracy and efficiency by improving data processing and model construction. Meanwhile, compared with the traditional unidirectional LSTM, bidirectional LSTM (BiLSTM) and its variants perform better in many trajectory sequence modelling tasks. GUO *et al.* (Guo Yijie *et al.* 2024) integrated the TCN network and the Bi-LSTM network to learn the before and after conditions of the trajectory sequences. This approach enables the extraction more information from the sequences. As a result, it allows for a more accurate prediction of the future trajectories of ships. HAN (Han Yuyin 2023) proposed a PSO-Bi-LSTM model, which combines the particle swarm algorithm, a hybrid PSO-BiLSTM prediction model with higher prediction accuracy than the PSO-LSTM model. This improvement is mainly due to BiLSTM's ability to take into account both past and future information of the sequence data, thus capturing the temporal characteristics of the data more comprehensively.

Currently, although deep learning methods have demonstrated greater effectiveness in addressing ship trajectory prediction issues, they still grapple with problems related to insufficient accuracy and reliability. In order to further improve the accuracy of ship trajectory prediction, a GRU-ABiLSTM model based on the Encoder-Decoder architecture is proposed. This model integrates GRU, BiLSTM, and the attention mechanism to capture the temporal features of the trajectory, and at the same time adding the ship density information extracted based on the AIS data as an additional feature.

2 BILSTM AND GRU MODELS

2.1 *Bidirectional Long Short Term Memory network (BiLSTM)*

Long Short-Term Memory (LSTM) (Greff Klaus *et al.* 2017) controls the flow of information through a gate mechanism to better capture and remember the dependencies within time series data. However, in the traditional LSTM structure, information can only be transmitted unidirectionally along the time series. and early information still gradually decays as it is transmitted over long sequences. In order to capture the contextual information in sequence data more comprehensively, in 2005, Alex Graves (Alex Graves and Jürgen Schmidhuber 2005) proposed Bidirectional Long Short-Term Memory (BiLSTM) by introducing two independent LSTM networks that process along the forward and reverse directions of the time series respectively. At each time step, BiLSTM takes into account the contextual information both before and after the current moment, thus capturing the bidirectional correlations in the sequence data more comprehensively. The structure of BiLSTM is shown in Figure 1:

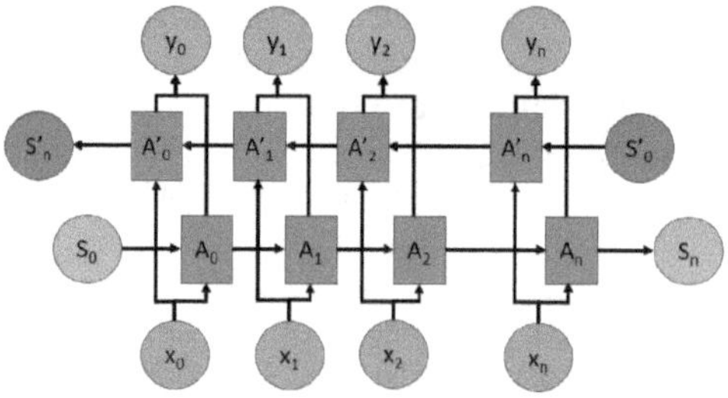

Figure 1. Schematic diagram of BiLSTM structure.

The working principle of BiLSTM is shown in Figure 1: at each time step n, the forward LSTM unit A_0, A_1, A_2 … A_n receives the input X_n of the current time step and the hidden state S_{n-1} of the previous time step, and calculates the forward hidden state of the current time step. Similarly, at each time step n, the backward LSTM unit A'_n, A'_{n-1}, A'_{n-2} … A_0 receives the input X_n of the current time step and the hidden state S'_{n-1} of the next time step, and calculates the backward hidden state of the current time step. Concatenate the forward and backward hidden states to form the current output vector, which serves as the final output of BiLSTM (Sun Junfeng *et al.* 2023).

For each independent LSTM network, its unit structure is shown in Figure 2.

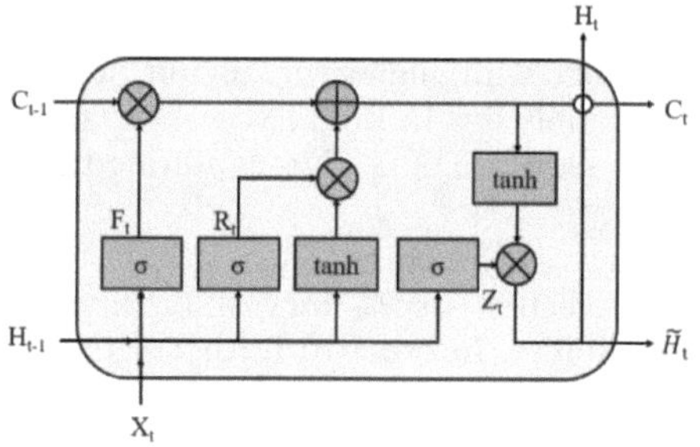

Figure 2. Schematic diagram of LSTM unit structure.

Among them, X_t is the input at the current time; C_t is the current cellular state at the moment; C_{t-1} is the cell state from the previous moment; H_t is the hidden state at the current time; H_{t-1} is the hidden state of the previous moment; $\tilde{H}_t$ is a candidate hidden state; F_t is the output gate output; R_t is the input gate output; Z_t is the output of the forget gate; σ(sigmoid) and tanh are two different activation functions; W is the coefficient matrix; b is the deviation term. The specific calculation process is shown in equations (1)–(3).

$$F_t = \sigma(W_R \cdot [H_{t-1}, X_t] + b_R) \tag{1}$$

$$R_t = \sigma(W_R \cdot [H_{t-1}, X_t] + b_R) \tag{2}$$

$$Z_t = \sigma(W_Z \cdot [H_{t-1}, X_t] + b_Z) \tag{3}$$

The updated cell state is represented as:

$$Q_t = \tanh(W_c \cdot [H_{t-1}, X_t] + b_c) \tag{4}$$

$$C_t = F_t \times C_{t-1} + R_t \times Q_t \tag{5}$$

The final output of the network is represented as:

$$H_t = Z_t \times \tanh(C_t) \tag{6}$$

2.2 *Gated Recurrent Unit (GRU)*

The Gated Recurrent Unit (GRU) not only mitigates the gradient vanishing problem in RNN models, but also outperforms Long Short Term Memory (LSTM) networks in some tasks (Junyoung Chung *et al.* 2014). The GRU unit structure is shown in Figure 3.

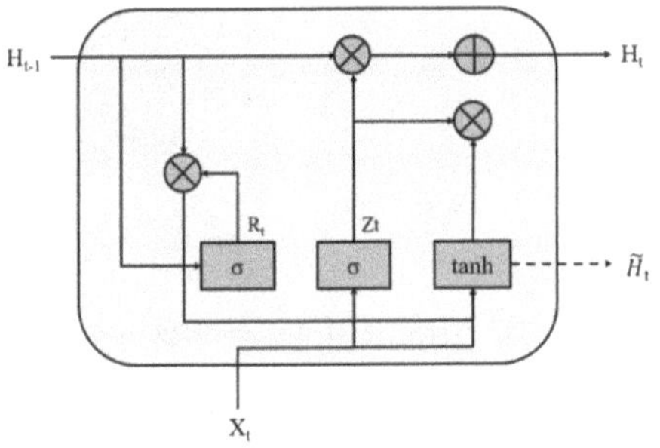

Figure 3. GRU structure diagram.

GRU consists of two gating units: an update gate and a reset gate. Among them, X_t is the input at time t; H_t is the output at the current time; H_{t-1} is the hidden state of the previous moment; $\widetilde{H}_t$ is a candidate hidden state; Z_t is for updating the door; R_t is for resetting the door; σ is the activation function sigmoid; tanh is another type of activation function. W is the coefficient matrix; b is the deviation term (Zhao Qi *et al.* 2023). The specific calculation process is shown in equations (7)–(10).

$$R_t = \sigma(W_R \cdot [H_{t-1}, X_t] + b_R) \tag{7}$$

$$Z_t = \sigma(W_Z \cdot [H_{t-1}, X_t] + b_Z) \tag{8}$$

$$\widetilde{H}_t = \tanh(W_R \cdot [R_t * H_{t-1}, X_t] + b_R) \tag{9}$$

$$H_t = (1 - Z_t)\widetilde{H}_t + Z_t H_{t-1} \tag{10}$$

From Figures 1 and 3, it can be seen that when compared with LSTM, GRU features fewer gate units and intermediate variables, and a simpler structure. Therefore, in some scenarios, the training speed is faster, which also helps to reduce overfitting during model training.

3 CONSTRUCTION OF GRU-ABILSTM MODEL BASED ON ENCODER DECODER

3.1 *Overall model framework*

The GRU-ABiLSTM ship trajectory prediction model is built upon an Encoder-Decoder architecture, where the encoder uses GRUs to extract temporal features from the trajectory sequence. Meanwhile, the decoder uses a bi-directional long- and short-term memory network (ABiLSTM), which incorporates an attention mechanism to adjust the feature weights. Specifically, five features – longitude (Lon), latitude (Lat), course-on-ground (COG), speed-on-ground (SOG), and the density of the ship in the area where the trajectory point is located at the current moment — are inputted into the encoder for encoding, and the trajectory of the ship at the future moment is finally outputted through the decoder module. Its overall structure is shown in Figure 4.

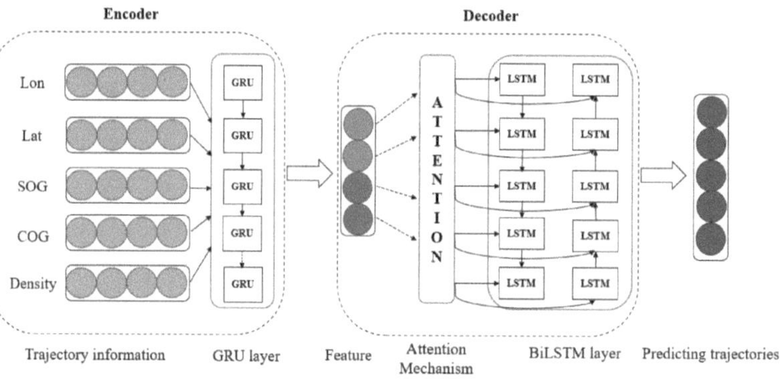

Figure 4. Schematic diagram of GRU-ABiLSTM model structure.

As shown in Figure 5, during the encoding phase, the model processes the historical trajectory sequences through feature extraction and information compression, and eventually provides a unified representation through intermediate semantic encoding vectors [15]. In the decoding phase, the model is tasked with retrieving the ship's potential motion pattern information and converting it into the predicted ship position.

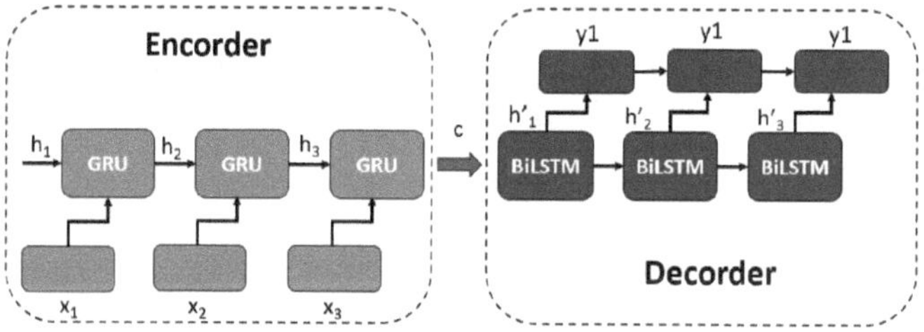

Figure 5. Schematic diagram of encoder decoder structure.

3.2 *GRU encoding process*

Each GRU unit in the GRU encoder section receives the hidden-layer state of the ship trajectory from a previous time step and the trajectory data input from the current time step. The hidden-layer state of the ship trajectory sequence at the current time step is generated by conducting a non-linear transformation on the input information.

$$h_t = f(x_t, h_{t-1}) \tag{11}$$

In the formula, h_{t-1} represents the state of the hidden layer at the previous moment; x_t denotes the input of time t; $f(\cdot)$ is the encoder. The encoder encodes the input sequence $X = (x_1, x_2, x_3 \ldots x_t)$ to obtain an intermediate semantic encoding vector c (Chen Xiaojian 2023).

$$c = z(x_1, x_2, x_3 \ldots x_t) \tag{12}$$

In the formula, $z(\cdot)$ is a nonlinear function.

3.3 *Attention module*

In the GRU-ABiLSTM modelling framework based on the Encoder-Decoder architecture, the commonly used intermediate semantic coding vectors are generated from the input sequence. However, the limited length of the context vector makes it difficult to comprehensively summarise the information of the entire ship trajectory sequence. Moreover, in real-world scenarios, the ship's sailing pattern changes over time. Relying solely on intermediate semantic coding vectors proves ineffective in efficiently capturing the temporal correlations between the output sequence and the hidden layer of the input sequence.

In the attention mechanism, each element within the input sequence interacts with others, and the corresponding weights between the elements are calculated, which can reduce the influence of irrelevant information and highlight key features through weight allocation.

Therefore, to address the issue mentioned above, the attention mechanism is introduced to take the encoder features at all moments as input. For each moment t = 1, 2, ..., t, the attention model is employed to compute the feature vector while taking into account the contextual information, thereby obtaining the personalised features of the decoder at that moment.

The operation of the attention mechanism consists of three steps: correlation calculation, weight normalization, and weighted summation. Firstly, calculate the correlation score s_{ti} between the h'_t of BiLSTM at time t and the hidden layer state h_i of GRU at time i (Tang Kun *et al.* 2023).

$$s_{ti} = v_a^T \tanh\left(W_a \cdot concat(h'_t, h_i)\right) \tag{13}$$

In the formula, W_a is the weight matrix; v_a is the transformation coefficient. Then normalize the correlation score and calculate the influence weight of GRU hidden layer state on bidirectional LSTM hidden layer state, as shown below.

$$a_t = softmax(s_{ti}) = \frac{\exp(s_{ti})}{\sum_j \exp(s_{ti})} \tag{14}$$

In the formula, a_t is the weight value calculated through the softmax function. Finally, by performing weighted summation using attention weights, the characteristic vector c_t of the bidirectional LSTM at the current time t is obtained as follows:

$$c_t = \sum a_t hi \tag{15}$$

3.4 *BiLSTM decoding process*

In the BiLSTM decoder part, the decoder also receives two inputs at each output moment: the feature vector computed by the attention mechanism and the output sequence that has been generated, which are nonlinearly transformed to output the ship's predicted trajectory points.

During the decoding phase, the decoder gradually decodes the encoded result to generate an output sequence $Y = (y_1, y_2, y_3 \ldots y_{t-1})$, where:

$$y_t = r(y_1, y_2, y_3 \ldots y_{t-1}, c) \tag{16}$$

Among them, $r(\cdot)$ is the activation function.

In the process of ship trajectory prediction, the bidirectional LSTM enables the system to efficiently learn the complex nonlinear relationship between the target output and the trajectory. This allows for generating the predictions of the ship's position at the future moment. The whole process constitutes an Encoder-Decoder-based GRU-ABiLSTM model for ship trajectory prediction.

4 EXPERIMENTAL SETUP

4.1 *Selection and processing of ship trajectory dataset*

Ningbo Zhoushan Port is one of the world's largest ports in terms of cargo throughput, with complex intersecting shipping routes and numerous islands and reefs, which are typical of complex navigable waters. The AIS data of the core port area of Ningbo Zhoushan Port in March 2024 is selected for model training and validation. The ship trajectory point visualisation results of this water area in March 2024 are shown in Figure 6.

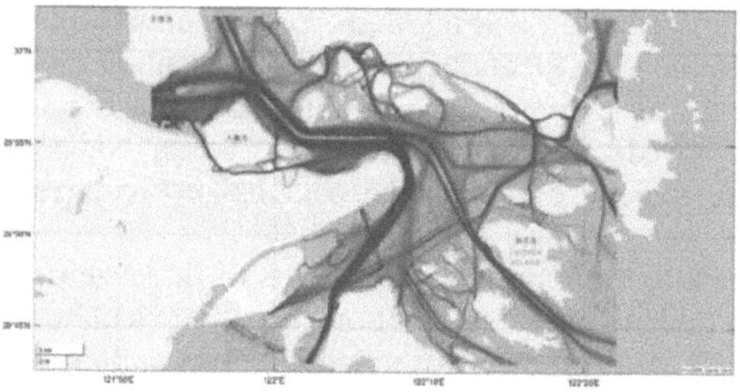

Figure 6. Visualization of ship trajectory data in the core port area of Ningbo Zhoushan Port in March 2024.

4.1.1 *AIS data preprocessing*

AIS data contains static information (e.g., MMSI, ship dimensions, ship type) and other static information, as well as dynamic information (e.g., latitude, longitude, heading, speed, rate of turn) and other dynamic information, which provides data support for trajectory prediction.

Since there is missing or even erroneous trajectory information in the original AIS data, it is necessary to preprocess the AIS data as follows:

(1) Delete abnormal data: Delete abnormal data whose MMSI is not 9-bit, whose latitude and longitude of track points are outside the range of research waters, and whose COG or SOG values are beyond the set range.
(2) Trajectory grouping: grouping the trajectories by MMSI and then placing each group of data in chronological order.
(3) Interpolation: Using the cubic spline interpolation method, the grouped trajectory data are interpolated to ensure the smoothness and continuity of the trajectory sequence.
(4) Trajectory sampling: each group of trajectories after interpolation is extracted at 1-minute intervals to reduce the amount of data while retaining the key information.

The original AIS data is pre-processed by the above steps to obtain the trajectories of 16,472 ships, containing 76,725 trajectory sequences with a total of 10,776,461 trajectory points, which are used for the training and validation of the model.

4.1.2 *Selection and processing of ship trajectory dataset*

The method of rasterization of trajectory data was used to extract ship traffic density features, aiming to convert continuous geospatial trajectory data into discrete and structured grid forms. A grid scale of 100×100m was chosen to divide the target waters into 364×496 grids, and then the number of track points on each grid was counted according to the ship track data in March 2024, and the statistical distribution of ship track points can reflect the density distribution of ships. The results of rasterisation of ship trajectory data are shown in Figure 7.

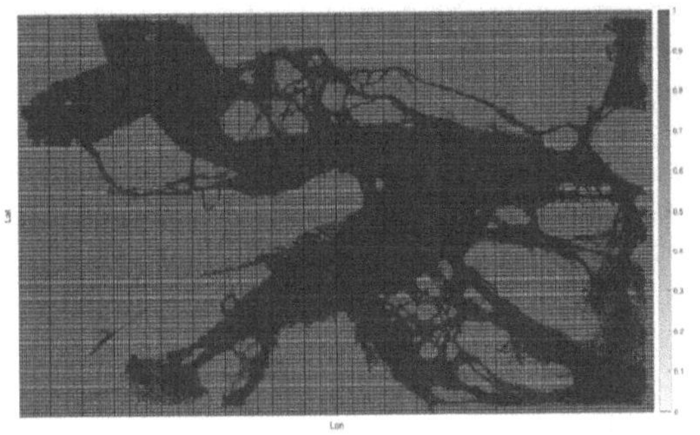

Figure 7. Rasterization results of ship trajectory data in experimental waters.

Due to the presence of noise and discontinuous pixel values in the original ship traffic density data obtained after rasterization of ship trajectory data, direct input to the model at this point will affect the accuracy of density feature construction. Therefore, further denoising and smoothing of the raster image is required. Median filtering is used to process the ship density raster image (Zhao 2023). As in equation (17), the median filter in the two-dimensional case can be defined as:

$$Y_{ij} = medianX_{in} = median\left[X_{i+m,j=n}; (m,n) \in W\right]\{X_{ij}; (i,j) \in Z^2\} \tag{17}$$

In the formula: W is the size of the plane window; m is the horizontal size of the window; n is the vertical size of the window; X_{ij} is a grid on the processed image plane, with coordinates (i, j); Y_{ij} is the median density of ships corresponding to all grids within window X_{ij} centered on W, which is the output value of median filtering processing; Z^2 is the sequence number of the two-dimensional data string, which is the coordinate set of the raster image.

The median filtering window has both horizontal and vertical dimensions of 3. As can be seen from Figure 8, some isolated pixel points and holes appear in the original image, while lacking smoothness. The outliers of the ship density are removed after the median filter denoising process, which results in a more reasonable distribution of the data and a clearer picture of the trend and pattern of the ship density changes.

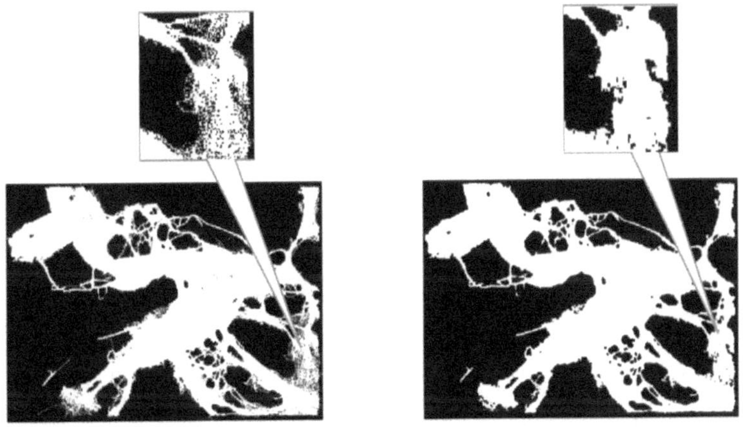

Figure 8. Comparison between the original image and the image processed by median filtering.

After median filter processing, traffic density features can be extracted based on the ship density raster image. Firstly, according to the ship's latitude, longitude and raster coordinates set at the current moment to search for the raster in which the ship is located, and calculate the ship density of this raster at the current moment, and then the density value and the ship's trajectory point at the moment of the four features of the longitude, latitude, the speed to the place, the direction of the ship's trajectory are spliced together, which together form the ship's trajectory point features at the current moment.

4.1.3 *Data normalization*

The ship trajectory data after preprocessing and adding ship traffic density information is divided into training set, validation set and test set in the ratio of 7:2:1. The training set is used to optimise the model parameters and learn the relationship between inputs and outputs; the validation set is used to evaluate the performance of the model on new data, prevent overfitting and select the best configuration. The test set is used to evaluate the generalisation ability and prediction accuracy of the model.

In order to speed up the model convergence as well as increase the stability of the data, the five feature dimensions of the ship trajectory data are normalised to the [0,1] interval. The normalisation formula is shown in equation (18).

$$Y = \frac{X - X_{\min}}{X_{\max} - X_{\min}} \tag{18}$$

In the formula, Y is the normalized data, X is the data of a certain feature dimension, and $X_{\min}$ and $X_{\max}$ are the minimum and maximum values of the data of that feature dimension, respectively.

4.2 *Model evaluation indicators*

The accuracy of the model prediction results is evaluated using the average displacement error (ADE) and the final displacement error (FDE). Where ADE denotes the average Euclidean distance of all corresponding trajectory points between the predicted trajectory and the true trajectory.FDE denotes the Euclidean distance between the last predicted trajectory point and the corresponding true trajectory point (Wang & He 2021). The specific calculations are shown in Eqs. (19) and (20).

$$ADE = \frac{\sum ||p_{true} - p_{pred}||}{N} \tag{19}$$

$$FDE = \frac{\sum ||p_{true(end)} - p_{pred(end)}||}{N} \tag{20}$$

Among them, p_{true} represents the true position, $p_{true(end)}$ represents the true endpoint position, p_{pred} represents the predicted position, $p_{true(end)}$ represents the predicted endpoint position, N represents the number of trajectory points, $||\cdot||$ represents the binary norm, used to calculate the Euclidean distance.

5 EXPERIMENT AND RESULT ANALYSIS

A 10-minute historical trajectory (i.e., 10 historical trajectory points) is used as model input, and multi-step prediction is used to directly output all predicted trajectory points. Where multi-step prediction means that the model outputs all predicted trajectory points simultaneously. The prediction results of the GRU-ABiLSTM model were compared with GRU, LSTM, Seq2Seq-LSTM, attention-BiLSTM and other models for quantitative and qualitative evaluation in order to analyse the difference in the performance of different models.

5.1 *Experiment and result analysis*

In the quantitative evaluation experiments, the historical trajectory duration of the input model was set to a fixed 10-minute period, and according to the characteristics of ship traffic in the waters of Ningbo Zhoushan Port and the needs of nautical practice, the trajectory prediction results of the different models without and with the addition of the ship density feature were outputted for the prediction durations of 10-minute, 15-minute and 20-minute periods, respectively, and the prediction accuracies were quantitatively evaluated by using the ADE and the FDE.

Table 1. AD(m)E/FDE (m) of different models at different prediction times (without incorporating ship density features).

	Time		
Model	10min	15min	20min
LSTM	362.1/581.2	634.9/809.8	846.1/1034.3
GRU	372.9/522.7	610.5/798.5	855.1/949.1
Seq2Seq-LSTM	312.4/535.5	398.1/772.7	773.2/890.1
Attention-BILSTM	295.3/489.2	431.4/699.1	685.1/884.3
GRU-ABiLSTM	231.1/423.4	302.5/506.8	415.9/569.7

Table 2. ADE/FDE (m) of different models at different prediction times (with incorporating ship density features).

	Time		
Model	10min	15min	20min
LSTM	349.2/537.9	565.3/753.9	796.2/901.2
GRU	356.3/544.2	542.23/735	801.9/877.6
Seq2Seq-LSTM	282.2/486.1	386.1/755.2	612.8/792.7
Attention-BILSTM	274.1/458.4	413.2/657.8	655.8/851.2
GRU-ABiLSTM	210.8/382.4	271.3/453.4	336.5/527.2

As can be seen from Tables 1 and 2, for different models, when the prediction duration and input features are the same, Seq2Seq-LSTM and GRU-ABiLSTM models are significantly better than LSTM and GRU models in ADE/FDE, which indicates that the introduction of Encoder-Decoder can effectively improve the prediction performance of the models. And the Attention-BILSTM model compares LSTM and GRU, because the introduction of bidirectional LSTM and attention mechanism, the prediction error is relatively small. However, the GRU-ABiLSTM model has the smallest ADE/FDE values and the highest prediction accuracy among all the compared models.

When the input features are the same, the ADE and FDE of all models increase with the increase of prediction duration, i.e., the prediction performance of all models decreases with the increase of prediction duration.

For GRU-AbiLSTM, the ADE/FDE growth rates between prediction durations are quantified as follows:

- Without ship density features: 30.9% (ADE) and 19.7% (FDE) from 10 to 15 minutes; 37.5% (ADE) and 12.4% (FDE) from 10 to 20 minutes.
- With ship density features: 28.7% (ADE) and 18.6% (FDE) from 10 to 15 minutes; 24.0% (ADE) and 16.3% (FDE) from 10 to 20 minutes.

These values indicate that GRU-ABiLSTM achieves the lowest error growth rates among all compared models, demonstrating its minimal sensitivity to extended prediction horizons and the most stable performance

For the same model and the same prediction time, the ADE/FDE of all models decreased when the regional ship density feature was added, which indicates that the addition of the regional ship density feature can effectively further improve the prediction effect of the model.

5.2 *Qualitative assessment*

In the qualitative evaluation experiments, the length of the historical trajectory and the length of the predicted trajectory are both 10 minutes, and the input trajectory point features include the regional ship density features. The prediction effect of different models is judged based on the visual comparison of the prediction results with the real trajectory (Ground Truth) in the three sailing scenarios of ship sailing in straight line, smaller steering sailing, and larger steering sailing, respectively, as shown in Figures 9, 10, and 11.

In the scenario of the ship sailing in a straight line, the predicted trajectory of GRU-ABiLSTM almost coincides with the real trajectory, and the ADE is less than 100 m. The predicted trajectories of the other comparative models deviate from the real trajectory to a larger extent, and the prediction effect is not as good as that of the GRU-ABiLSTM model.

In the scenario where the ship sails with smaller steering the GRU-ABiLSTM model and the Seq2Seq-LSTM model predict the steering amplitude of the ship closer to the real trajectory, but in terms of the average positional error, the GRU-ABiLSTM model is significantly better than the Seq2Seq-LSTM model, with the ADE of the former model being only 42% of that of the latter. The prediction errors of all other models are relatively large.

In the scenario where the ship sails with larger steering, the prediction results of the GRU-ABiLSTM model are still the closest to the real trajectory, while the deviation of the steering timing or steering magnitude of the trajectories predicted by the comparison models are larger.

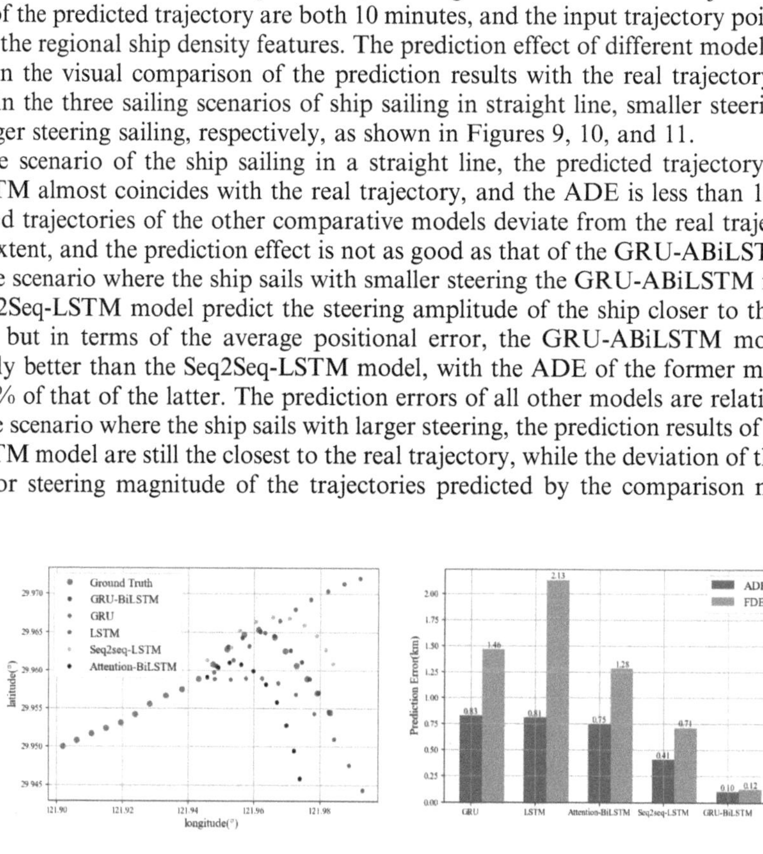

Figure 9. Comparison of straight-line navigation scene models.

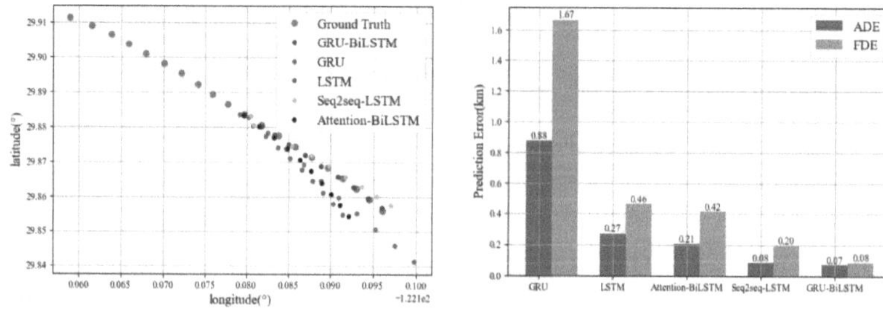

Figure 10. Comparison of small turn navigation scene models.

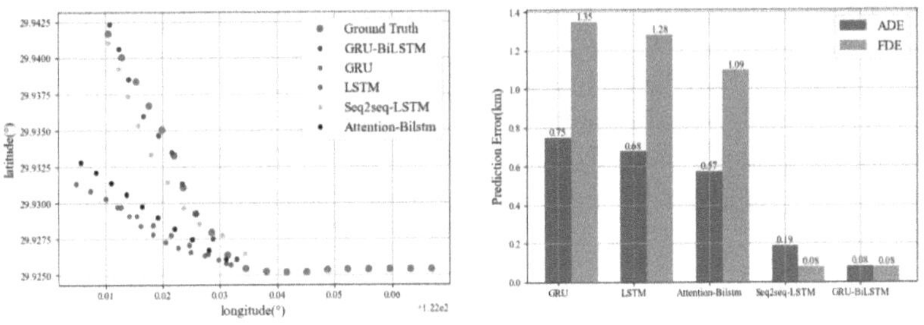

Figure 11. Comparison of models for large turning navigation scenarios.

The results of the quantitative and qualitative evaluation experiments show that the bidirectional LSTM and the attention mechanism improve the model's ability to capture the long-term dependencies of the AIS data, resulting in significantly better predictions than the other compared models.

5.3 *Ablation experiment*

To further investigate the role of each component in the GRU-ABiLSTM model, three variant models are constructed, the first variant model replaces the GRU module with LSTM and is named LSTM-ABiLSTM; the second variant model replaces the BiLSTM module with LSTM and is named GRU-ALSTM; and the third variant model removes the Attention module and named GRU-BiLSTM.The three variant models explore the important roles of GRU, BiLSTM and Attention modules in the GRU-ABiLSTM model, respectively. The historical trajectory duration and predicted trajectory duration of the ablation experiment are both 10 minutes, and the input trajectory point features contain regional ship density features. The prediction accuracy is characterised by calculating the ADE/FDE values for the three variants of the model at this time.

Table 3 shows the ADE/FDE values for GRU-ABiLSTM and its three variant models. From the data in the table, it can be seen that the GRU-ABiLSTM model reduces the ADE by 23.2% and the FDE by 14.1% compared to the LSTM-ABiLSTM model. This is due to the fact that GRU has a more concise structure compared to LSTM, which gives it better results in extracting complex features. Compared to the GRU-ALSTM model, the GRU-ABiLSTM model shows a reduction of 31.8% and 28.0% in the ADE and FDE metrics, respectively. This improvement is attributed to the bi-directional structure of BiLSTM, which is able to capture the dependencies and contextual information in the trajectory sequence more effectively. In addition, the GRU-ABiLSTM model reduces ADE by 31.3% and FDE by 18.6% compared to the GRU-BiLSTM model. This suggests that the attention mechanism can make the model more focused on key features, thus further improving the prediction accuracy.

The results of the ablation experiments illustrate that each module of the GRU-ABiLSTM model has a role in improving the ship trajectory prediction accuracy.

Table 3. ADE/FDE values of GRU-ABiLSTM and its three variant models.

模型	ADE(m)	FDE(m)
LSTM-ABiLSTM	260.5	436.8
GRU-ALSTM	278.2	489.1
GRU-BiLSTM	277.1	453.3
GRU-ABiLSTM	210.8	382.4

6 CONCLUSION

In order to further improve the accuracy of ship trajectory prediction in complex navigable waters, a GRU-ABiLSTM ship trajectory prediction model based on Encoder-Decoder is proposed. The typical complex navigable waters of Ningbo Zhoushan core harbour area are selected as the experimental waters, and ADE and FDE are used as the evaluation indexes. The experimental results of quantitative and qualitative evaluation show that GRU-ABiLSTM achieves higher prediction accuracy (ADE: 210.8 m, FDE: 382.4 m) compared with the comparison model under different prediction durations and navigation scenarios. Meanwhile, the introduction of the smoothed and denoised regional ship density as an additional feature input to the trajectory points further improves the prediction accuracy. Finally, the results of the ablation experiments show that all three modules, GRU, BiLSTM and Attention, are useful for improving the prediction performance of GRU-ABiLSTM.

REFERENCES

Alex Graves and Jürgen Schmidhuber. (2005). Framewise phoneme classification with bidirectional LSTM and other neural network architectures. *Neural Networks* (5/6), 602–610.

Chen Xiaojian (2023). Research on Gas Concentration Prediction Based on Encoder Decoder ILSTM Model. *Energy and Energy Conservation* (12), 102–105+176.

Greff Klaus, Srivastava Rupesh K, Koutnik Jan, Steunebrink Bas R and Schmidhuber Jurgen. (2017). LSTM: A Search Space Odyssey. *IEEE Transactions on Neural Networks and Learning Systems* (10), 2222–2232.

Guo Yijie, Zhang Junyi, and Wang Peng (2024). TCN BiLSTM ship trajectory prediction based on attention mechanism. *Computer Measurement and Control* (01), 30–36.

Han Yuyin (2023). *Research on Ship Trajectory Prediction Based on PSO BiLSTM* (Master's Thesis, Zhejiang Ocean University). Master's https://link.cnki.net/doi/10.27747/d.cnki.gzjhy.2023.000310.

Houenou, A., Bonnifait, P., Cherfaoui, V., and Yao, W. (2013, November). Vehicle trajectory prediction based on motion model and maneuver recognition. In *2013 IEEE/RSJ International Conference on Intelligent Robots and Systems* (pp. 4363–4369). IEEE.

Ilya Sutskever, Oriol Vinyals and Quoc V. Le. (2014). Sequence to Sequence Learning with Neural Networks. *CoRR*.

Ju Cong (2023). *Research on Ship Trajectory Prediction Based on CNN-GRU* (Master's Thesis, Dalian Maritime University). Master's https://link.cnki.net/doi/10.26989/d.cnki.gdlhu.2023.000415.

Junyoung Chung, Çaglar Gülçehre, KyungHyun Cho and Yoshua Bengio. (2014). Empirical evaluation of gated recurrent neural networks on sequence modeling. *CoRR*.

Liu Chengyong, Qiao Wenjie, Chen Shuzhe and Wan Yi (2021). *Research on Ship Trajectory Prediction Model Based on LSTM and Attention Mechanism. China Navigation* (04), 94-100+106.

Rong, H., Teixeira, A. P., and Soares, C. G. (2019). Ship trajectory uncertainty prediction based on a Gaussian Process model. *Ocean Engineering*, 182, 499–511.

Sun Junfeng, Li Chenghai and Cao Bo (2023). Network Security Situation Prediction Based on TCN BiLSTM. *Systems Engineering and Electronic Technology* (11), 3671–3679.

Tang Kun, Dai Yuqin, Xu Yongneng, Guo Tangyi and Shao Fei (2023). Online recognition method for abnormal driving behavior based on Encoder Decoder attention network. *Journal of Weapon Equipment Engineering* (08), 63–71.

Wang Senjie and He Zhengwei (2021). Ship trajectory prediction model based on generative adversarial networks. *China Navigation* (02), 72–77.

Zhang, X., Liu, G., Hu, C., and Ma, X. (2019, July). Wavelet analysis based hidden Markov model for large ship trajectory prediction. *In 2019 Chinese Control Conference (CCC)* (pp. 2913–2918). IEEE.

Zhao Fuqun (2023). Image denoising algorithm based on improved median filtering. *Computer and Digital Engineering* (02), 292–295+444.

Zhao Jiansen, Yan Zhongwei, Zhou ZhenZhen, Chen Xinqiang, Wu Bing and Wang Shengzheng. (2023). A ship trajectory prediction method based on GAT and LSTM. *Ocean Engineering*(P1).

Zhao Qi, Xu Zhiyuan and Ge Jiawei (2023). Ship trajectory prediction based on GRU DConvLSTM. *Ship Engineering* (06), 124–129+139.

Zhixiang Fang, Hongchu Yu, Ranxuan Ke, Shih-Lung Shaw and Guojun Peng. (2019). Automatic Identification System-Based Approach for Assessing the Near-Miss Collision Risk Dynamics of Ships in Ports. *IEEE Transactions on Intelligent Transportation Systems* (2), 534–543.

Author index

FSC
www.fsc.org
MIX
Papier | Fördert gute Waldnutzung
FSC® C083411

Kolibri 360 GmbH
Ferdinand-Jühlke-Straße 7
99095 Erfurt, Deutschland
produktsicherheit@kolibri360.de